ELEMENTARY

FOOD SCIENCE

SECOND EDITION

ELEMENTARY FOOD SCIENCE
SECOND EDITION

ELEMENTARY FOOD SCIENCE
SECOND EDITION

John T.R. Nickerson, Ph.D.

Professor Emeritus
Massachusetts Institute of Technology
Cambridge

Louis J. Ronsivalli, M.S.

Director
Northeast Utilization Research Center
Gloucester, Mass.

avi

AVI PUBLISHING COMPANY, INC.
Westport, Connecticut

© Copyright 1980 by
THE AVI PUBLISHING COMPANY, INC.
Westport, Connecticut

Second Printing 1982

Library of Congress Cataloging in Publication Data

Nickerson, John T
   Elementary food science.

   Bibliography: p.
   Includes index.
   1. Food handling. 2. Food industry and trade.
I. Ronsivalli, Louis J., joint author. II. Title.
TX537.N48 1980   641.3   79–22939
ISBN 0–87055–318–6

Printed in the United States of America by Eastern Graphics, Inc.

Preface to the Second Edition

We are pleased with the reception this book has received in many colleges and universities, not only in the United States, but in other countries as well.

In the Second Edition of *Elementary Food Science,* the following changes have been made: All measurements which were formerly listed as units only of the English system, have been herein recorded in both the English and the metric systems. The chapter on dairy products has been rewritten and brought up to date, especially from the standpoint of processing as applied to the various items included in this group of foods. A new section on whey has also been added. The chapter on fish and shellfish has been rewritten in such a manner as to exclude some of the data and detail concerning the biology of various marine species listed in the First Edition. In its place, a more detailed and up-to-date description of the processing of the different edible marine fish and shellfish has been included. In the chapter on food processing methods, a new section on glass containers has been added. In the chapter on food additives, the section on nonnutritive sweeteners has been expanded to include nutritive sweeteners. Minor changes have been made in all chapters in an attempt to make the general text easier to understand, and the index has been made more comprehensive than that of the First Edition.

JOHN T.R. NICKERSON
LOUIS J. RONSIVALLI

April 1980

Preface to the First Edition

This book was written with the hope that it will help to fill several needs. Food science, while taught in as many as 40 universities throughout the United States, is not widely recognized as a profession. Yet the subject is among the most relevant and important of endeavors. Food science is in a somewhat awkward classification, for it is not a discipline in the same sense that is either chemistry or mathematics. It is more nearly related to the class of professions that require a wide scientific background before an exacting study of the particular subject can begin, for instance, medicine.

Food science can be practiced at various levels of sophistication. As an illustration, the work of the analysts who try to interpret protein denaturation systems or to determine the effect of high energy such as radiation on the chemistry of foods is complex, while that of the individual in charge of quality control in a small food processing plant may, though not necessarily, be relatively simple. Medicine is practiced only at high levels of sophistication. Nurses, paramedics and hospital aides are those who deal with less exacting needs of the medical profession. The parallel between food science and medicine is only a convenient one, but it is interesting to note that these professions are much more related than is generally recognized.

Many food scientists categorize those concerned with the less difficult food applications as food technologists, perhaps for good reasons. This distinction has been avoided in this book.

It is the opinion of the authors that the handling of foods should be permitted only by knowledgeable people. Recently a newspaper reported that 200 people were victimized by food poisoning in an airliner. There is need for all people to have some understanding of the proper handling of foods. Yet it is probable that over 95% of all of those who work in this field have little knowledge of the subject. It is the opinion of the authors that the kind of information contained in Sections 1 and 2 of this text should be a required part of the curriculum of the last year of high school. It would prepare the student for two year courses in food science (Associate Degree) or even for further education in the regular four year food science college programs.

While this book is directed at those students in the last year of high school and students aspiring to obtain the two year Associate Degree, it should be definitely helpful to freshmen at the university level. Sections 1 and 2 would also prepare adults who never go to college and who are eventually employed in food handling establishments, restaurants, hotels, airlines, food processing plants, etc., to better qualify them for their work.

As with any introductory text, it was not possible to cover subjects thoroughly. For example, the chapter on fish and shellfish, while relatively long, is probably the least adequate considering the vast information which has been omitted. It is recognized that separate and complete texts could be written, and in some cases have been written, on a number of the subjects covered in this book, and appropriate references have been made. Also, some topics such as water, nuts and soft drinks have been omitted. This was not done because these subjects are not important, but because they could best be dealt with in separate texts.

JOHN T.R. NICKERSON
LOUIS J. RONSIVALLI

November 1975

Contents

Section I

Interrelated Food Science Topics

1

Why Food Science?

The scientific study of food is one of man's most important endeavors, mainly because food is his most important need. It is necessary for his survival, his growth, his physical ability and his good health. Food processing and handling is the largest of all of man's industries. Many factors require that those scientists who choose to study foods be prepared to absorb as many of the physical and life sciences and as much engineering as possible. Among these are the chemical complexity of foods, their vulnerability to spoilage, their role as a disease vector, and the varied sources of foods. The availability, nutritional adequacy and the wholesomeness of foods are also quite varied.

Whether we know enough of the facts to trace the development of food science from the beginning is questionable. History reports that the Romans realized, more than the Greeks, Egyptians, or any of the prior civilizations, that agriculture was a prime concern of the government. The Romans, as the Egyptians and the Greeks before them, were able to preserve a variety of foods by holding them in vinegar (with or without brine), in honey or in pitch. Some foods were dried, either by the sun or over a fire. These civilizations also produced cheeses and wines. Yet it is generally believed that until the latter part of the 18th century, the preservation of foods had evolved as an art handed down from generation to generation. Its development was slow, depending on accidental discovery, observation, trial and error, and attempts to reproduce and put into practice the newly found techniques. Drying, freezing, smoking, fermenting, cooking and baking had been practiced for centuries—even by illiterates. Foods frozen accidentally in cold climates and foods dried accidentally in dry climates were observed to have a longer "shelf-life" than foods which were neither frozen nor dried. Foods that might have

3

been put over a fire to hasten drying could easily have led to the smoking process. Thus, chance occurrences led to preservation methods that permitted man to conserve foods during times of glut so that he might survive the leaner spells. It can be said that those who made the observations and realized their impact, then put their interpretations to the test until the new practice was proven, were the first food scientists. Spallanzani (1765) and Appert (1795) were among the first to apply the quasi-scientific methods for preserving foods, and in 1809, Appert won a prize from the French government for developing a thermal processing technique for foods to be used by the military. Appert is credited with developing the canning process. Because of the scarcity of scientific information, Appert had to employ trial and error tactics, but his records attest to the accuracy of his observations and conclusions and show that he applied the scientific approach to gain his outstanding achievement even though he did not know why his method worked.

It was not until the discoveries of Pasteur in 1850 and the work of other microbiologists, such as Prescott and Underwood in 1895, that man learned that bacteria spoiled food and why thermal processing prevented food spoilage.

By 1875, man had learned to preserve foods by artificial refrigeration using first, natural ice, and later, manufactured ice, to preserve fish and meats. He also learned that brine could be made colder than 32°F (0°C) and this enabled him to freeze foods. By 1890, mechanical refrigeration had come into wide use, opening the way to the frozen storage of foods. Quick freezing was first used in 1924 to preserve fish. During the period 1932–1934, Clarence Birdseye, with laboratories in Gloucester, Mass., developed over 100 different frozen food items, and this achievement won for him the reputation and the credit for the beginning of the quick-frozen food industry. One of the most important ensuing technological developments was the invention of the fish blocks by Birdseye technologists. This is considered by many to have revolutionized the fish processing industry.

In 1898, it was noted that bacteria were destroyed by exposure to radioactive salts of radium and uranium. By 1930, the use of ionizing radiations to preserve food was patented by O. Wust. However, the irradiation preservation of foods was not actively investigated until the team of Proctor, Van de Graaf and Fram from the Massachusetts Institute of Technology undertook the project in 1943.

Modern technology has made possible such controlled, automated drying processes and sophisticated modifications as freeze-drying, drum drying, spray drying, and fluidized-bed drying. Controlled, automated versions of thermal and refrigeration processes have also been developed.

Radiation processing (by electron-, X-, and gamma-rays), microwave processing and aseptic canning have also been introduced.

Though many food processes alter foods so that the finished product is more palatable or otherwise more acceptable (to some at least) than the original raw material (sauerkraut, tuna, wine, roquefort cheese, etc.), in many cases it is desirable that preservation processes do not alter the food (fish fillets, beef steak, pork chops, etc.). Only refrigeration can preserve most foods without altering them substantially.

FOOD—MAN'S MOST IMPORTANT NEED

It is universally accepted that man's basic needs are food, clothing and shelter. Of course, such a list ignores man's need for oxygen and water, two critical requirements, but this is understandable since we take for granted the presence of adequate amounts of oxygen in the air we breathe, and we are ever aware of the copious supplies of water in the many rivers, lakes and wells in most parts of the world. It is only lately that a concern for water supplies is evident. It should be quite clear that food is listed before clothing and shelter because it is the most important of the three. In fact, like oxygen and water, food is a critical need without which man cannot survive. Clothing and shelter, on the other hand, are not critical to his survival, although their availability makes life more convenient for man and permits him to live in areas where the climate would be intolerable without them. Clothing may last for relatively long periods (months or years) and shelters may last for decades or even a lifetime; therefore, man has had to spend relatively little of his time in the procurement of these needs. His need for food, however, is relentless, and he is reminded to eat by the hunger sensations he feels at least three times each and every day of his life. Little wonder that primitive man, the hunter, spent a large part of his time foraging for food! Technological advances have made it possible for inhabitants of developed countries to spend considerably less time than ever before to earn enough to buy the food they need.

The availability of an abundant supply of food does not necessarily guarantee survival unless the food is nutritionally complete and contains no deleterious substances. Unfortunately, serious and sometimes fatal illnesses result from diets that lack sufficient proteins, vitamins or other nutritional components. Serious adverse consequences may also result from the consumption of foods containing such harmful substances as infectious microbes, microbial toxins, viable parasites, allergenic agents and a large number of chemical toxins. Thus, throughout his evolution man has had to concentrate on many factors affecting foods. He has had to increase the efficiency of food procurement to ensure a sufficient availability; to learn ways to preserve foods to carry him through times of scarcity or crop failures; and to learn specific processing methods, such as baking, pickling and fermenting to increase the variety and desirability of his food. Also, he has had to learn the rudiments of the nutritional and medical aspects of food diets to maintain his health, and to learn how to minimize food-borne illnesses. However, he has a long way to go, and the little that he already knows about foods only serves to make him aware of their complexity and of the ponderous work that needs to be done in food science.

Military leaders, throughout history, have been cognizant of the role of food in a military operation. An abundance of food has always been and will always be necessary to maintain the morale of the soldiers and to sustain invasion tactics; but the nature of the foods is also important, since the mobility of an army is affected by the mass of material it must carry. Thus, dried, compact foods enhance mobility. One of the outstanding facets of the military successes of Genghis Khan was the mobility of his army of mounted soldiers. With only a very scant food supply he was able to engage in swift cavalry attacks over long periods, which often caught his enemies off guard, too bewildered to rally an effective defense. Marco Polo is credited with reporting the Khan's solution to his food supply problem. Apparently, each of the Khan's horsemen carried two leather bags—one larger than the other. In the large one, he carried dried milk—produced by drying fluid milk in the sun during periods of rest. When sufficient dried milk was produced, the horsemen were prepared to start an offensive. During each morning of the offensive, some dried milk and water were put into the small bag wherein the dried milk was rehydrated, helped, in some measure, by the agitation resulting from the motion of the horse. The rehydrated milk was consumed at some time during the day. With a supply of dried milk, the lightly equipped army of the Khan could cover long distances in weeks, and when the supply of milk was exhausted, the men were able to continue, when necessary, by employing one more innovative technological tactic. They bled their horses once each day, taking about one pint of the animal's blood which they drank for nourishment. It is reported that the army was able to continue for at least one additional week by this scheme.

The full impact of a concerted technological effort in the food logistics for the military was first evident during World War II, when American troops were equipped with light, compact, nutritionally balanced food packets that could sustain them during a military action. The proven value of the application of food science for military purposes has resulted in a continuing effort by food scientists at the U.S. Army Natick Development Center, Natick, Mass., whose efforts are augmented by those of industrial and academic scientists.

FOOD SCIENCE AND HEALTH

The optimum physical and mental functioning of the body is dependent on the nutritional quality of the foods it receives. Man has observed this from the beginning of time, and certain diets have evolved as a result of these observations. The analysis and planning of diets were not possible until food science became established to a degree and produced the basic information that made these activities possible. From the knowledge acquired through the development of food science emerged conclusions that resulted in the classification of foods into nutritional groups, representatives of which are considered to be necessary in all diets to ensure the intake of a recommended minimum of protein, carbohydrates, vitamins, minerals, etc. Evidence of the links between diets and certain symptoms of ill health became easier to obtain as food science developed, and the potential of specific diets in corrective and preventive medicine has been gradually recognized and is now effectively practiced.

FOOD—THE LARGEST OF ALL INDUSTRIES

The food industry is the largest of all industries in the United States. It employs about 14,000,000 people. Its activities include farming, fishing, processing, transportation, wholesaling, retailing, warehousing and containerizing. These activities require many others. Included among them are those that supply work clothes, farm equipment, processing equipment, trucks, railroads, air transports, ships, and communication facilities. Also involved are industries that supply forklifts and other plant equipment; public utilities; recreation facilities; building materials and tradesmen to build the plants and install water, heat, refrigeration, computers, fishing gear, electrical and detection equipment, and other materials. It does not take much imagination to realize that the ramifications of the food industry reach into nearly all other industries. Although one might question which industry is ancillary to which, resolution of the question depends on which industry is most important to man. Obviously, it is the food industry.

The amount spent for food in the United States is about $100 billion annually, about $1/5$ of all spending. The largest of the many conglomerates that produce food products have annual sales in the hundreds of millions to billions of dollars and annual earnings in the tens to hundreds of millions of dollars. For example, the 25 largest food companies have annual sales totaling nearly $34 billion and annual earnings totaling over $1.2 billion.

Because the food industry is so large and involves such vast amounts of money, it has attracted large investments by giant diversified international corporations. The competition among these business giants has resulted in greater varieties of products and product forms, and intensified growth in prepared foods development, automatic vending, fast-order franchises, and other innovative strategies aimed at that $100 billion spent by American consumers for their food. Expansion of the food industry has accelerated under the impetus of growth by the new conglomerates, and considerable expansion has spread into other countries.

The advent of the large food store or supermarket, together with the growth of the automotive industry, has led to the large shopping centers —a concept that fosters the growth of the supermarket and the supermarket chains and the discontinuance of the small food stores. Supermarkets offer many attractions to the food shopper. Three major factors concerned with the success of supermarkets are: (1) large variety of foods, (2) large variety of brands, and most important, (3) lower prices due to buying in large quantities.

FOOD SCIENCE FOR SOCIETY'S SAKE

We are in an era where food and what is done to it are oftentimes topics for newspapers and other public communication media. Some of the publicity is authoritative and authentic, but some of it is less than authentic, even misleading and deceptive. In addition, the public is confused by the vacillation of authorities concerning the potential hazard of cyclamates, DDT and other compounds that are intentionally or unintentionally added to foods. The public is alarmed by the incessant warnings by authorities and pseudo-authorities against a variety of food additives, and it is bilked mercilessly by misleading assertions on the purity of organically grown foods and on the properties of certain foods that cause a decrease in obesity. Even in well-meaning drives to substitute oleomargarine for butter to lower the risk of heart disease, the public is misled to a degree. There is an awareness that public education on food science is sorely lacking, and the Institute of Food Technologists (the national society for food scientists) is intensifying a program initiated to remedy this. Professors of food science at a number of universities and scientists in private industry have also set up public education programs. As a result of the efforts of these dedicated people, a number of informative publications have been issued and are available at no cost to the general public. The Institute of Food Technologists has published short articles on the following subjects:

Botulism	Phthalates in Foods
Nitrites, Nitrates, and Nitrosamines	Nutrition Labeling
	Shelf-life of Foods
Carrageenin	The Effects of Food Processing
Mercury in Food	on Nutritional Values
Organic Foods	

Other relevant articles are scheduled for publication in the future. For those interested in food additives, an informative publication of about 64 pages was published by the Manufacturing Chemists Association, and it is reportedly available to the general public from that organization at no cost.

There is another basic reason to encourage the spread of knowledge of food science in society. Although we have an excellent network of regulatory agencies in municipal, state, and federal governments, we are still faced with a considerable incidence of food-borne illness. Botulism has been caused by mushrooms packed in a plant which did not have a food scientist on its staff. Countless cases of perfringens poisoning, salmonellosis, and staphylococcal poisoning due to foods eaten in restaurants, institutions, and even on airliners could be avoided were suitable expertise in food science exercised in the handling of the foods involved. It is recognized that small food businesses and small restaurants might not be financially capable of employing a food scientist. However, it would make a significant difference in the operations of these facilities if they obtained the services of a food consultant from time to time (e.g., to set up a process, to make periodic evaluations of time operations, etc.). It would also be helpful to subscribe to a food trade journal.

While it is not new to state that the increase of the world's population is more rapid than the increase in food supplies required to feed that population, it is only because of food science that the situation is not far more serious. Basic information on the growing of crops and on genetics has permitted significant increases in agricultural productivity. However, the widening discrepancy between supply and demand will ultimately be resolved by political or social action, and food science will serve only to delay or to cushion the effects of the inevitable if effective measures are not taken. In the United States, population growth is under control and there is presently no concern that the domestic food demand cannot be met. The worsening worldwide situation that resulted in recent large exports of grains has resulted in a dramatic decrease in the food reserves of this country. Thus, it is not difficult to see how the effects of food shortages worldwide will affect the United States. Food scientists, more than ever, must continue to raise the efficiency of food production, and they must turn to the conversion of waste and other low-value materials to synthesize foods and food analogs. They must obtain the highest productivity from the land, including deserts, and exploit marine and freshwater food sources heretofore unused or underutilized. They must explore hydroponics and closed systems of aquaculture. In this respect, food scientists should become involved in population control movements by initiating and/or supporting them, since unchecked population growth will eventually overshadow any and all gains made by food science to prevent further widening of the gap between the world's food supply and demand.

FOOD SCIENCE AS A PROFESSION

Food science may well be among the most important, the most timely and the most relevant professions of our time. It is most important because it is the device by which we can control the availability, the nutrition and the wholesomeness of food—the one commodity that is critical to the survival of man. It is timely because the gap between the accelerating food demand and inadequate food supply, on a worldwide basis, continues to widen and because the future looks even more bleak. Its relevance is apparent in some of the universities where special, practical courses in food science are available as electives to nonscience majors. This idea, that seemed worthy only because the subject matter appeared both relevant and timely to a few professors, has resulted in a positive student response that has surpassed all expectations. At the University of Massachusetts, the University of Florida, the University of Minnesota and Rutgers University, the specially tailored courses have attracted hundreds to thousands of students. Exposure to food science via these courses has resulted in the transfer of a few of the students from other fields to food science. At this writing, there are about 40 universities in the United States that teach food science, and most of them offer all three degrees (B.S., M.S. and Ph.D.) with only a few offering only one or only two.

Food science is not a discipline such as chemistry, mathematics, etc. It is rather a mixture of disciplines, with emphasis placed on the food-related aspects of the disciplines. Thus the food science student concentrates on the microbiology of foods, the biochemistry of foods, the rheology of foods, the applications of engineering principles to food processing and food preservation, etc., but he also studies some disciplines and some courses without emphasis on food-related aspects. The latter include mathematics, inorganic chemistry and basic physics. In some schools, laboratory courses include considerable training in the use of food processing equipment, such as a variety of heat exchangers, driers, homogenizers, comminuters, sealing machines, etc. Thus, it should be obvious that an engineering ability is useful for aspiring food scientists. The reasons for the broad academic preparation required by food scientists are as follows: (1) Foods, originating from animals and plants, are complex biochemical systems that continue to undergo change (mainly deteriorative) at rates that depend on such environmental conditions as temperature, humidity and presence of oxygen. (2) Foods are generally contaminated with a variety of microorganisms that subsist on the food components, creating changes in proteins, fats and carbohydrates that result in the formation of sometimes offensive, sometimes toxic, sometimes desirable by-products. (3) The food scientist may want to employ certain food additives to rectify nutritional deficiencies, to prevent the development of certain microbial toxins (e.g., botulinum toxin), to prevent spoilage, to improve texture, etc. (4) The food scientist needs to be proficient in plant operations and have a knowledge of process equipment and processes. (5) The food scientist must be concerned with pesticide and fertilizer residuals, as well as compounds of mercury and other elements; this is the case even when the presence of the contaminating material is as low as parts per million and less. (6) He needs to know something about parasites, insects and other foreign materials. In the literature search, the chemist goes to the Chemical Abstracts, the biologist goes to the Biological Abstracts, etc., but the food scientist very often covers all or most of the abstracts.

The professional society for food scientists is the Institute of Food Technologists (IFT). The 1975 Directory of that society lists a membership of about 10,000. Affiliated with the IFT are about 2500 food companies, educational institutions, research and development companies, etc.

2

Nutrition

Nutrition may be defined as a series of processes by which an organism takes in and assimilates food to promote growth, to expend energy, to replace worn or injured tissue, and to prevent some diseases. However, nutrition encompasses many processes, and thus, it may be given many definitions. Mendel, among others, has been quoted as defining nutrition as "The Chemistry of Life." Mendel's definition may be most appropriate from the scientist's point of view, because the processes by which food components are assimilated, converted and utilized are understood and properly managed only when their chemistry is understood. Now, much of the chemistry of life is understood reasonably well, and most nutritional deficiencies could be easily diagnosed and successfully treated. Nutritional requirements and food energy values are well known, and the public is becoming better informed about these and vitamin and mineral needs, as well. Still, the interrelationships among food constituents, particularly as related to metabolism, and the delicately balanced chemistry of the body place a thorough understanding of nutrition well beyond our present capabilities. Therefore, although the mechanisms for studying specific aspects of nutrition are presently available (e.g., calorimetry for determining food energy values), many aspects of nutrition can be studied only by observing the organism's total response (e.g., long-term animal feeding studies for determining if a food additive might have adverse effects on the consumer).

Knowledge, in the field of nutrition, is relatively new, especially regarding the vitamins and some trace elements, and medical practitioners are not always as well informed in nutrition as they should be. Consequently, while information on the early symptoms of nutritional

deficiency is often available, some cases of dietary inadequacies may g
unnoticed to the detriment of those involved.

The need to maintain a nutritionally adequate diet has spurred th
development of data relating to dietary needs. The Food and Nutritio
Board of the National Research Council has published a table o
recommended dietary allowances which are considered adequate fo
optimum body functioning (see Table 2.1). When the dietary intake i
insufficient for short periods, body reserves may be substituted, especial
ly for energy needs, but eventually, the body must be replenished wit
essential food components. When the dietary intake is insufficient fo
long periods, diseases resulting from inadequate nutrition develop, a
described in Table 2.2.

By gross analysis, in terms of food components, man has been reportec
to consist of approximately 18% protein, 0.6% carbohydrate, 15.5% fat
3% minerals, 0.000001% vitamins, and the rest (about 63%) water. It i
considered by many experts in nutrition that the body requires daily
helpings of foods from certain basic food groups that include: (1) meats
poultry, fish, eggs, and beans; (2) green and yellow vegetables; (3) milk
cheeses and other dairy products; (4) breads and cereals; and (5) fruits

While water has no nutritional value, it nevertheless plays an impor
tant role in nutrition. As a major component of the body's transpor
system (blood and lymph), it is instrumental in the distribution of energy
components to the points of need and in the collection and removal o
metabolic waste products through the kidneys, the sweat glands and the
lungs. Therefore, the body must maintain an adequate amount of water
(about 2/3 of the body weight) in order to function properly. Water enters
the body as drinking water and as a major component of most foods anc
beverages. It is estimated that for proper body functioning, a person
eliminates, and therefore requires, more than 2 qt of water each day
Over 1/2 of the intake should be in the form of drinking water.

With modern technological advances it is possible to ensure sufficient
vitamin and mineral intake through the availability of capsules that
contain concentrated amounts of these nutrients. However, it is desirable
that nutrients be derived from foods rather than pills, since foods contain
many additional components, the benefits of which may not be fully
understood. Studies have shown that it is better to obtain vitamin C
requirements from citrus fruits than from vitamin tablets, since oranges,
for example, contain other nutritional components, such as bioflavonoids,
that serve an important function in the biological processes of the body.
Thus, if vitamin C tablets are used in place of citrus fruits, the body may
sustain a lack of components that may have adverse effects on the
physiological functions of the body without giving any clue as to the

auses of undesirable body responses. The same has been suspected for he substitution of synthetic vitamin A for fish liver oil. It is now believed hat the fish liver oil contains beneficial nutrients that are not present in he synthetic vitamin A formulations. However, it must be remembered hat the content of any of the nutrients in foods is limited and that herapeutic needs for specific nutrients are sometimes so great that they an only be filled through the use of concentrated formulations.

An insight into nutrition can be gained by a consideration of the mportant food substances (proteins, carbohydrates, fats, vitamins and iinerals) and we will facilitate their presentation by classifying them iainly on the basis of their chemical characteristics and/or general ehavior.

ROTEINS

Proteins are the chief organic constituents of muscles and other ssues. Proteins are major components of the enzymes which regulate nd carry out the general metabolism and functional processes of living hings. Proteins are part of the intracellular and extracellular structure f animals; they make up the structure and composition of many ormones and antibodies (disease-resisting components), and are con-erned with many other factors involved with body functions. Proteins ontain nitrogen, carbon, hydrogen, oxygen, and sometimes sulfur and hosphorus. All proteins contain nitrogen at a level of about 16%. The nalysis for proteins is determined indirectly by analyzing for protein itrogen, then multiplying the results by 6.25 to determine the actual mount of protein analyzed. All proteins are composed of amino acids aving the general formula

$$R-CH \cdot COOH$$
$$| $$
$$NH_2$$

here R could represent any one of a variety of chemical structures. In he simplest amino acid (glycine), R represents one hydrogen atom— us the formula for glycine is

$$H-CH \cdot COOH$$
$$|$$
$$NH_2$$

In larger amino acids, the R could represent a complex structure such s in methionine

$$(CH_3 \cdot S \cdot CH_2 \cdot CH_2 \cdot CH \cdot COOH)$$
$$|$$
$$NH_2$$

TABLE 2.1. RECOMMENDED DIETARY ALLOWANCES, REVISED 1980[a]. FOOD AND NUTRITION BOARD, NATIONAL ACADEMY OF SCIENCES–NATIONAL RESEARCH COUNCIL

Designed for the maintenance of good nutrition of practically all healthy people in the U.S.A.

Category	Age (Years)	Weight (kg)	Weight (lb)	Height (cm)	Height (in.)	Protein (g)	Fat-Soluble Vitamins Vitamin A (μg R.E.) [b]	Vitamin D (μg) [c]	Vitamin E (mg αT.E.) [d]	Water-Soluble Vitamins Vitamin C (mg)	Thiamin (mg)	Riboflavin (mg)
Infants	0.0–0.5	6	13	60	24	kg × 2.2	420	10	3	35	0.3	0.4
	0.5–1.0	9	20	71	28	kg × 2.0	400	10	4	35	0.5	0.6
Children	1–3	13	29	90	35	23	400	10	5	45	0.7	0.8
	4–6	20	44	112	44	30	500	10	6	45	0.9	1.0
	7–10	28	62	132	52	34	700	10	7	45	1.2	1.4
Males	11–14	45	99	157	62	45	1000	10	8	50	1.4	1.6
	15–18	66	145	176	69	56	1000	10	10	60	1.4	1.7
	19–22	70	154	177	70	56	1000	7.5	10	60	1.5	1.7
	23–50	70	154	178	70	56	1000	5	10	60	1.4	1.6
	51+	70	154	178	70	56	1000	5	10	60	1.2	1.4
Females	11–14	46	101	157	62	46	800	10	8	50	1.1	1.3
	15–18	55	120	163	64	46	800	10	8	60	1.1	1.3
	19–22	55	120	163	64	44	800	7.5	8	60	1.1	1.3
	23–50	55	120	163	64	44	800	5	8	60	1.0	1.2
	51+	55	120	163	64	44	800	5	8	60	1.0	1.2
Pregnant						+30	+200	+5	+2	+20	+0.4	+0.3
Lactating						+20	+400	+5	+3	+40	+0.5	+0.5

[a] The allowances are intended to provide for individual variations among most normal persons as they live in the United States under usual environmental stresses. Diets should be based on a variety of common foods in order to provide other nutrients for which human requirements have been less well defined.

[b] Retinol equivalents: 1 retinol equivalent=1 μg retinol or 6 μg β-carotene.

[c] As cholecalciferol: 10 μg cholecalciferol=400 I.U. vitamin D.

[d] α-tocopherol equivalents: 1 mg d-α-tocopherol=1 α T.E.

[e] 1 N.E. (niacin equivalent)=1 mg niacin or 60 mg dietary tryptophan.

[f] The folacin allowances refer to dietary sources as determined by Lactobacillus casei assay after treatment with enzymes ("conjugases") to make polyglutamyl forms of the vitamin available to the test organism.

ESTIMATED SAFE AND ADEQUATE DAILY DIETARY INTAKES OF ADDITIONAL SELECTED VITAMINS AND MINERALS[a]

Category	Age (Years)	Vitamins Vitamin K (μg)	Biotin (mg)	Pantothenic Acid (mg)	Trace Elements[b] Copper (mg)	Manganese (mg)
Infants	0–0.5	12	35	2	0.5–0.7	0.5–0.7
	0.5–1	10–20	50	3	0.7–1.0	0.7–1.0
Children	1–3	15–30	65	3	1.0–1.5	1.0–1.5
and	4–6	20–40	85	3–4	1.5–2.0	1.5–2.0
Adolescents	7–10	30–60	120	4–5	2.0–2.5	2.0–3.0
	11+	50–100	100–200	4–7	2.0–3.0	2.5–5.0
Adults		70–140	100–200	4–7	2.0–3.0	2.5–5.0

Source: from Recommended Dietary Allowances, Revised 1980. Food and Nutrition Board, National Academy of Sciences-National Research Council, Washington, D.C.

[a] Because there is less information on which to base allowances, these figures are not given in the main table of the RDA and are provided here in the form of ranges of recommended intakes.

Niacin (mg N.E.) e	Vitamin B6 (mg)	Folacin f (µg)	Vitamin B12 (µg)	Calcium (mg)	Phosphorus (mg)	Magnesium (mg)	Iron (mg)	Zinc (mg)	Iodine (µg)
		Water-Soluble Vitamins				Minerals			
6	0.3	30	0.5 g	360	240	50	10	3	40
8	0.6	45	1.5	540	360	70	15	5	50
9	0.9	100	2.0	800	800	150	15	10	70
11	1.3	200	2.5	800	800	200	10	10	90
16	1.6	300	3.0	800	800	250	10	10	120
18	1.8	400	3.0	1200	1200	350	18	15	150
18	2.0	400	3.0	1200	1200	400	18	15	150
19	2.2	400	3.0	800	800	350	10	15	150
18	2.2	400	3.0	800	800	350	10	15	150
16	2.2	400	3.0	800	800	350	10	15	150
15	1.8	400	3.0	1200	1200	300	18	15	150
14	2.0	400	3.0	1200	1200	300	18	15	150
14	2.0	400	3.0	800	800	300	18	15	150
13	2.0	400	3.0	800	800	300	18	15	150
13	2.0	400	3.0	800	800	300	10	15	150
+2	+0.6	+400	+1.0	+400	+400	+150	h	+5	+25
+5	+0.5	+100	+1.0	+400	+400	+150	h	+10	+50

The RDA for vitamin B_{12} in infants is based on average concentration of the vitamin in human milk. The allowances after weaning are based on energy intake (as recommended by the American Academy of Pediatrics) and consideration of other factors, such as intestinal absorption.

The increased requirement during pregnancy cannot be met by the iron content of habitual American diets or by the existing iron stores of many women; therefore, the use of 30 to 60 mg supplemental iron is recommended. Iron needs during lactation are not substantially different from those of non-pregnant women, but continued supplementation of the mother for 2 to 3 months after parturition is advisable in order to replenish stores depleted by pregnancy.

	Trace Elements [b]				Electrolytes		
Fluoride (mg)	Chromium (mg)	Selenium (mg)	Molybdenum (mg)	Sodium (mg)	Potassium (mg)	Chloride (mg)
0.1−0.5	0.01−0.04	0.01−0.04	0.03−0.06	115−350	350−925	275−700
0.2−1.0	0.02−0.06	0.03−0.06	0.04−0.08	250−750	425−1275	400−1200
0.5−1.5	0.02−0.08	0.02−0.08	0.05−0.1	325−975	550−1650	500−1500
1.0−2.5	0.03−0.12	0.03−0.12	0.06−0.15	450−1350	775−2325	700−2100
1.5−2.5	0.05−0.2	0.05−0.2	0.1−0.3	600−1800	1000−3000	925−2775
1.5−2.5	0.05−0.2	0.05−0.2	0.15−0.5	900−2700	1525−4575	1400−4200
1.5−4.0	0.05−0.2	0.05−0.2	0.15−0.5	1100−3300	1875−5625	1700−5100

[b] Since the toxic levels for many trace elements may be only several times usual intakes, the upper levels for the trace elements given in this table should not be habitually exceeded.

TABLE 2.2.   DISEASES DUE TO MALNUTRITION

Name of Disease	Cause	Manifestations
Kwashiorkor and marasmus	Protein and calorie deficiency	Growth retardation, impaired mental development, edema, enlarged liver, pigment changes i skin, low serum proteins, enzyme deficiency
Xerophthalmia and keratomalacia	Vitamin A deficiency	Night blindness, infection of eye with loss of vision, skin changes (hyperkeratosis)
Anemias	Deficiency of iron, folic acid or vitamin B-12	Pallor, weakness, heart failure, low hemoglobin and red cell count
Endemic goiter	Iodine deficiency	Thyroid enlargement, cretinism
Beriberi	Thiamin deficiency	Peripheral neuritis, central nervous system disturbances, heart diseas
Ariboflavinosis	Riboflavin deficiency	Cheilosis, glossitis, seborrheic dermatitis
Pellagra	Niacin-tryptophan deficiency	Dermatitis, mental changes, diarrhea, inflammation of gastro-intestinal tract
Scurvy	Vitamin C deficiency	Hemorrhages, improper growth of bone and supporting tissue, anemia
Rickets	Vitamin D deficiency	Improper bone growth, abnormal calcium and phosphorus metabolism, skeletal deformities

In proteins, amino acids are chiefly linked together to form peptides b a peptide bond (—CO—NH—). The bond links the carboxyl grou (COOH) of one amino acid with the amino group (NH_2) of another amin acid with the release of one molecule of water.

Because proteins contain carbon, they can be used as fuel since part o the molecule can be oxidized, sometimes involving deamination, t supply energy.

Proteins are required by humans for growth (protein synthesis) and fo repair and maintenance of cells. Since mature adults have, in essence ceased to grow, their protein requirement is less, per unit weight, tha that of those who are still growing.

While the human requires proteins, all proteins are not of a suitabl composition to supply the needs of the body, especially that of th growing child. Generally, animal proteins are complete proteins or of composition which can supply all the body requirements, while vegetabl proteins are not complete. Some animal proteins are more complete tha others. Proteins are complete or not complete, depending on the amin acids (the prime components of proteins) which they contain. Ma requires a source of ten amino acids (the essential amino acids). Ther are many amino acids not required by man. The chief difference betwee essential and nonessential amino acids is that man cannot synthesize th essential amino acids in his body—these he must obtain through his diet

Most proteins are evaluated against egg albumin, considered to be a complete protein. However, fish, meats, poultry and milk are considered to supply the essential protein components for growth and repair of cells. If proteins are to be obtained from vegetable sources alone, the human must have a diet which includes a variety of different vegetable foods.

The ten amino acids essential for man are leucine, isoleucine, lysine, methionine, cystine, phenylalanine, tyrosine, threonine, tryptophan and valine. Among these it should be noted that the body can utilize phenylalanine to form tyrosine but cannot form phenylalanine from tyrosine. Also, the body can form cystine or cysteine from methionine but cannot form methionine from cystine or cysteine.

Certain nonessential amino acids have some utilization by the body:

Glycine (nonessential) is utilized by the liver to detoxify certain components of foods, such as benzoic acid. It may also be involved in the synthesis of several body components, such as the bile acids. Glutamic acid (nonessential) may act as a source material for the synthesis of other amino acids.

Histidine (nonessential) is needed for growth and for the repair of human tissues and is converted to a substance which stimulates the secretion of hydrochloric acid in the stomach to facilitate gastric function.

Proline and hydroxyproline (nonessential amino acids) contain a structure found in hemoglobin (blood pigment) and in the cytochromes (compounds which are essential for oxidation and reduction reactions in the body).

Arginine (nonessential) is required for the detoxification of ammonia and amines resulting in the production of urea. Arginine is classified as nonessential only because it can be synthesized from other amino acids in the body.

The functions of certain essential amino acids are as follows: phenylalanine and tyrosine are used by the body to make the hormones, adrenaline and thyroxine, and are also involved in the formation of melanin, a pigment which is present in the skin, hair and parts of the eye.

Tryptophan is an amino acid from which a substance involved in the constriction of blood vessels is formed, and is present in components of blood involved in clotting.

Cystine, cysteine and methionine are sources of a part of the structure of insulin and the keratin of hair and are involved with oxidation-reduction reactions in the body.

While proteins eaten in excess of that required for growth or cellular repair may be utilized as a source of energy, it is not considered that they are efficiently utilized for this purpose.

CARBOHYDRATES

In order to carry out its day-to-day physiological functions and maintain a constant body temperature (invariably in an environment of changing temperatures, usually less than that of body temperature), the body requires a constant source of energy. Beyond its continuing main-tenance needs for energy, the body periodically needs relatively larger amounts of energy to do work or to engage in other vigorous physical activities. Man derives his energy mainly from carbohydrates (55-65%), although he can also utilize fats and proteins for this purpose.

The carbohydrates are a class of chemical compounds that consist of carbon, oxygen and hydrogen. The carbohydrates that are important in nutrition include the sugars, the starches, the dextrins, and glycogen. Cellulose, pectin and other carbohydrates are not important nutri-tionally.

Sugars

Sugars, important in nutrition, consist of monosaccharides, having the general formula $C_6H_{12}O_6$, and disaccharides, having the general formula $C_{12}H_{22}O_{11}$. Although the monosaccharides consist of 3-carbon sugars (trioses), 4-carbon sugars (tetroses), 5-carbon sugars (pentoses) and 6-carbon sugars (hexoses), only the latter are important in human nutri-tion as sources of energy.

Glucose, a 6-carbon sugar, is one of the simplest carbohydrates found in foods. While many foods contain traces of glucose, it is found in significant amounts only in fruits, such as grapes. Fructose, also a 6-carbon sugar, is found in fruits and honey. Both of these sugars can be utilized by the body as a source of energy.

Sucrose (the ordinary table sugar derived from sugar cane and beets) is a 12-carbon sugar which is broken down in the intestine to glucose and fructose, hence utilized as a source of energy.

Lactose, the 12-carbon sugar present in milk, is broken down in the intestine to glucose and galactose (6-carbon sugars), both of which can be used as sources of energy.

Starches

Starches are carbohydrates which are storage materials in the seeds and roots of many plants. Corn, wheat, rice and other grains, as well as potatoes and other root-like vegetables, contain significant amounts of starch. Starch is made up of many units of glucose linked together in

ifferent forms. In the intestine, starch is broken down to glucose and
tilized as a source of energy. Cooking (moist heat) causes starch grains
o swell and rupture, thus converting starch to a form which is readily
igested.

In the body, much of the glucose may be utilized directly as a source
f energy, but some of it is converted into fat, the muscles utilizing fatty
cids indirectly as fuel for energy. Excess carbohydrates, not required for
nergy, when ingested (eaten) will be stored in the body as fat.

Dextrin

Dextrin is an intermediary breakdown product of starches. It is
roduced in the body by the action of saliva and pancreatic juice
n starch, and its presence in the intestine is considered beneficial to
he digestive process.

Glycogen

Glycogen is produced in the liver from glucose (the end product of
arbohydrate digestion), and it is stored in the liver, as well as in the
nuscles where it is available for immediate use as energy. Both the liver
nd the muscles can store only a limited amount of glycogen; therefore,
vhen an excess of carbohydrates is ingested, there will be a tendency to
levelop an excess of glycogen. The excess carbohydrates will then be
onverted to fat and stored in the body as fat. The body maintains an
quilibrium between glucose, the energy producing sugar, and glycogen,
vhich can be converted to glucose as the glucose in the blood is used up
o produce energy. The production of energy from glucose involves
xidation of the sugar with the release of water and carbon dioxide which
re easily removed from the body.

$$\text{Glucose} + \text{oxygen} \rightarrow \text{energy} + \text{carbon dioxide} + \text{water}$$

$$C_6H_{12}O_6 + 6O_2 \rightarrow E + 6CO_2 + 6H_2O$$

FATS

Fats are glyceryl esters of fatty acids (see Chapter 24). Fats, as do
carbohydrates, contain the elements carbon, oxygen and hydrogen, but
he proportion of oxygen in fats is less, and it can be said that fats are fuel
oods of a more concentrated type than are the carbohydrates.
Carbohydrates and fats are interchangeable as fuel foods, but it can be
hown, by calorimetry, that fats produce more than twice the heat energy

produced by carbohydrates. One gram of fat yields 9 Cal, while 1 gram c carbohydrate yields 4 Cal. An additional advantage of fat from th viewpoint of energy availability is that it stores well in large amounts i adipose tissues. Thus fat, considered to be a reserve form of fuel for th body, is an important source of calories. Paradoxically, this is nc advantageous in affluent societies where the problem is not the avail ability of food for energy, but rather the health hazard of obesity.

Fats may occur in foods as materials which are solid at roon temperature or as oils which are liquid at room temperature. Solid fat contain comparatively small amounts of fatty acids with two or mor groups of adjacent carbons which are not fully saturated with hydrogen That is, these carbons could accept another hydrogen: $-CH_2-CH_2-$ (saturated), $-CH=CH-$(unsaturated).

Vegetable and marine fats (fish oil, whale oil, etc.) tend to contai unsaturated fatty acids, hence are oils and liquid at room temperature whereas the fats of most land animals (cattle, hogs, poultry) contai comparatively large amounts of saturated fatty acids, and therefore ar solid at room temperature.

Small amounts of three fatty acids, linoleic acid: $CH_3-(CH_2)_4-CH=$ $CH-CH_2-CH=CH-(CH_2)_7-COOH$; linolenic acid: CH_3-CH_2- $CH=CH-CH_2-CH=CH-CH_2-CH=CH-(CH_2)_7-COOH$; and ara chidonic acid: $CH_3-(CH_2)_4-CH=CH-CH_2-CH=CH-CH_2-CH=$ $CH-CH_2-CH=CH-(CH_2)_3-COOH$ are considered to be essential t life and health. Vegetable oils (excepting coconut and olive oils) contai considerable amounts of linoleic and linolenic acids, and the huma body is able to synthesize arachidonic acid from the former two fatt acids.

There are a number of phospholipids which are similar to fats in tha two of the alcohol groups $(-OH)$ of glycerine are esterified to fatty aci while the third alcohol group is esterified to a side chain containin phosphorus and nitrogen. While these compounds are found in animals their exact function is not known.

Another substance, sphingomyelin, is an important constituent c nerves and brain tissues. This is a kind of lipid in which glycerol i replaced by a long-chain alcohol containing nitrogen.

There are a number of sterols which have important functions in th body. These are very complex chemical compounds containing a alcohol group to which fatty acids can be esterified. The sterol cholesterc is involved in the composition of bile salts which play a role in th emulsification of fats in the intestine, hence, in the digestion of fats Ergosterol, another sterol, may be converted to vitamin D in the bod under the influence of sunlight or ultraviolet light.

When fats are ingested, they are either hydrolyzed to glycerine and fatty acids by the enzyme lipase in the small intestine and reformed into fat in the intestinal wall, or they are emulsified and absorbed as such. If fats are to be utilized for energy, they will be oxidized to carbon dioxide and water through a complex process involving a number of enzymes, while small amounts of fat may be excreted as waste. Excess fats not required for energy, when eaten, are eventually deposited as such in the body.

VITAMINS

There are a number of vitamins required in small amounts by the human body for sustaining life and good health. Some are fat-soluble, others are water-soluble.

Vitamin A

Vitamin A is a fat-soluble vitamin. It is found only in animals, although a number of plants contain carotene, from which vitamin A can be produced in the body once the plants containing carotene are eaten. Vitamin A may be formed in the body from the yellow pigments (containing carotene) of many fruits and vegetables, especially carrots. Vitamin A is also found in the fats and especially in the liver oils of many saltwater fish. Vitamin A is required for vision. Epithelial cells (those cells which are present in the lining of body cavities and in the skin and glands) require vitamin A. This vitamin is also required for resistance to infection. Deficiency of vitamin A may cause impairment in bone formation, impairment of night vision, malfunction of epithelial tissues and defects in the enamel of teeth. The daily recommended requirement for vitamin A is about 5000 I.U. (I.U. = international unit; 5000 I.U. of vitamin A = about 1.5 mg vitamin A crystalline form.)

Vitamin D

Vitamin D (calciferol or activated ergosterol) is fat soluble. This vitamin is necessary for normal tooth and bone formation. Deficiencies in vitamin D result in rickets (deformities of bone, such as bow-legs and curvature of the spine) and teeth defects. Fish oils, and especially fish liver oils, are excellent sources of vitamin D. The human body is also able to synthesize this vitamin from components of the skin through exposure to ultraviolet light or sunlight. The daily requirement for vitamin D is 00 I.U. or about 15 μg.

Vitamin E

Vitamin E, of which there are four different forms (the tocopherols), is fat-soluble.

The four tocopherols have the same name except with the prefixes alpha-, beta-, gamma- and delta- (the first four letters of the Greek alphabet). The four compounds are closely related, with some differences in the molecular weights and in the position and number of certain molecular constituents.

This vitamin is an antioxidant which serves to prevent the oxidation of some body components, such as unsaturated fatty acids, and is necessary for reproduction. Almost all foods contain some vitamin E, although corn oil, cottonseed oil, margarine and peanut oil are especially good sources of this vitamin.

While the symptoms for vitamin E deficiency in man are not clearly established, experiments with various animals have shown that vitamin E deficiency has an adverse effect on reproduction with apparent irreversible injury to the germinal epithelium. Other symptoms noted in animal studies include injury to the central nervous system, growth retardation, muscular dystrophy and interference with normal heart action. There is some question as to the significance of vitamin E deficiency in the human diet, and there are no known, substantiated recommendations as to the minimum daily requirement for this vitamin. In vitamin preparations, it is added in amounts of a few milligrams. (A milligram is approximately 0.000035 oz.)

Vitamin K

Vitamin K is also fat-soluble. It is essential for the synthesis of prothrombin, a compound involved in the clotting of blood. Cabbage, spinach, cauliflower and liver are especially good sources of vitamin K, although moderate amounts are found in many other vegetables, as well as in cereals.

The significant symptom of vitamin K deficiency in humans and in animals is the loss of the ability of the blood to clot which is, of course, a dangerous condition that can result in death whenever bleeding from cuts, etc., occurs.

The minimum required amount is not yet established, and it is believed that humans ordinarily receive adequate amounts of vitamin K in the diet.

Vitamins

The B vitamins are water-soluble. Thiamin—vitamin B-1—is involved in all bodily oxidations which lead to the formation of carbon dioxide. It is necessary for nerve function, appetite and normal digestion. It is also required for growth, fertility and lactation. The symptoms of vitamin B-1 deficiency are retardation of growth, palpitation and enlargement of the heart, blood pressure effects and wet beriberi (generalized edema or accumulation of fluids). The various effects of a disturbance of the nerve centers such as forgetfulness, difficulty in thinking, etc., are other manifestations of vitamin B-1 deficiency.

This vitamin is often lacking in the diet because much of the naturally occurring amounts of it in food are destroyed during the processing of the food. The adult requirement of vitamin B-1 is related to the food (calorie) intake and is about 1.0 mg per day. Fresh pork is an excellent source of vitamin B-1, and the heart, liver and kidneys of pork, beef and lamb are fair sources.

Riboflavin—vitamin B-2—is water-soluble. This vitamin makes up a part of enzyme systems involved in the oxidation and reduction of different materials in the body. Deficiency of riboflavin generally results in growth retardation and may result in vision impairment, scaling of the skin and lesions on mucous tissues. Neuritis is another deficiency effect. The minimum intake of riboflavin for an adult is about 2.0 mg per day. The liver and kidney of pork, beef and lamb are excellent sources of riboflavin, and the heart of these animals is a good source. Fair amounts of riboflavin are found in the muscular tissues of pork, beef and lamb, while more is found in veal.

Niacin (nicotinic acid) is another B vitamin. This compound is part of an enzyme system regulating reduction reactions in the body. It is also a compound which dilates blood vessels. Deficiency of niacin causes pellagra (a disease that causes diarrhea, dermatitis, nervous disorders and sometimes death). The requirement for niacin is about ten times that for thiamin. Beef, hog and lamb livers are excellent sources of niacin. Other organs and the musculature of these animals are good to fair sources.

Pyridoxine (vitamin B-6) is part of the enzyme system which removes CO_2 from the acid group (COOH) of certain amino acids and transfers amine groups (NH_2) from one compound to another in the body. It is also needed for the utilization of certain amino acids. Deficiency manifestations are dermatitis around the eyes, eyebrows and angles of the mouth. There may also be a sensory neuritis, and a decrease in certain

white blood cells and an increase in others. The daily requirement fo
pyridoxine has not been established, but bananas, barley, beef and bee
organs, cabbage, raw carrots, yellow corn, lamb and the organs of lamb
malt, molasses, peanuts, pork and the organs of hogs, potatoes, rice
salmon, sardines, tomatoes, tuna, wheat bran and germ, flour and yam
are good to excellent sources of pyridoxine.

Biotin is reported to be a coenzyme in the synthesis of aspartic acid
which plays a part in a deaminase system and in other processe
involving the fixation of carbon dioxide. Deficiency of this compound i
unusual, but can be demonstrated by the feeding of raw egg white whic
contains the substance, avidin, which ties up biotin. Deficiencies o
biotin cause scaling skin, skin lesions and a deterioration of nerve fibers
Due to the production of biotin by the microbial flora of the intestines
the requirement for this compound is not known. Liver is an excellen
source of biotin, and peanuts, peas, beans and whole cooked eggs ar
good sources.

Pantothenic acid, a vitamin required for normal growth, nerve de
velopment and normal skin is a component of enzyme systems involve
in metabolism (e.g., acetylation processes). It is believed, and there i
evidence, that pantothenic acid is intimately related to riboflavin i
human nutrition. Deficiency symptoms can be successfully treated wit
either compound. Deficiencies of this vitamin cause degeneration o
nerve tissues with resulting muscular weakness, numbness, malaise, et
Scaling skin and dermatitis, diarrhea with bloody stools and ulceratio
of the intestine are also deficiency symptoms. The requirement fo
pantothenic acid is about 10.0 mg per day. The organs of animals (live
heart, brain, kidney), and eggs, whole wheat products and peanuts ar
excellent sources of pantothenic acid. The muscular tissues of animal
cheese, beans, cauliflower, broccoli, mushrooms and salmon are ver
good sources of this vitamin.

Folic acid is required for the formation of blood cells by the bon
marrow and is involved in the formation of the blood pigmer
hemoglobin. It is also required for the synthesis of some amino acid
Deficiency symptoms involve pernicious anemia. The requirement fo
folic acid is approximately 10.0 mg per day. Nuts, dried beans, turnip
lentils, corn and shredded wheat are good sources of this vitamin, whil
liver and wheat bran are excellent.

B-12 (cobalamine) is a very complex chemical compound. This vita
min is required for the normal development of red blood cells, and
deficiency in it causes an acute pernicious anemia. The exact require
ment of vitamin B-12 is yet unknown, since some B-12 is synthesized b
bacteria in the intestine. The organs of animals are excellent sources o

itamin B-12 and the muscles of warm-blooded animals and fish are good
ources.

scorbic Acid

Ascorbic acid or vitamin C is required for the formation of intercellular
ubstances in the body, including dentine, cartilage and the protein
etwork of bone. Hence, it is important in tooth formation, the healing of
roken bones and the healing of wounds. It may be important to
xidation-reduction reactions in the body and to the production of
ertain hormones. Deficiency of vitamin C causes scurvy (spongy gums,
ose teeth, swollen joints, hemorrhages in various tissues, etc.) and
npaired healing of wounds. The daily requirement for vitamin C is
pproximately 30 mg. Orange juice is an excellent source of vitamin C.
omato juice, if it has been processed properly, is a fair source of this
itamin. Green peppers, cabbage, broccoli and Brussels sprouts are
xcellent to good sources of vitamin C, while other vegetables, such as
eas, spinach and lettuce are good to fair sources. Many fruits contain
air amounts of vitamin C.

MINERALS

A number of minerals or elements are required for normal body
unctions. Iron is required, since it is an essential part of both the blood
igment, hemoglobin and the muscle pigment, myoglobin. Some body
nzymes also have a composition which includes iron. Deficiencies of iron
ause anemia. Liver, animal muscle tissues, eggs, oatmeal, wheat flour,
ocoa and chocolate are good sources of iron. Approximately 10.0 mg of
on are required daily.

Iodine is required by all vertebrate animals, including the human,
nce it is a component of the hormone, thyroxine, produced by the
hyroid gland. This hormone regulates metabolic levels. Deficiency of
odine leads to low level metabolism, lethargy and goiter. Requirements
f iodine are believed to be about 0.1 mg daily. Sea food and saltwater
sh are the best sources of iodine. In areas where the water is known to
e deficient in iodine, iodized table salt may be used in place of regular
able salt.

Sodium is required by the human, as it is part of all of the extracellular
uids of the body. Since table salt is used by essentially all people, there
little likelihood of deficiencies except in diseases involving prolonged
omiting or diarrhea.

Potassium is present in body cells and is associated with the function of muscles and nerves and with the metabolism of carbohydrates. Deficiency of potassium is unusual except in cases of disease involving prolonged diarrhea. Sources of potassium include meats, eggs, oranges and bananas.

Phosphorus, an important component of bones and teeth, is also associated with essential body lipids, and its intake should be in a ratio 1–1.5/2 calcium/phosphorus. Sources of phosphorus are meats, fish, eggs and nuts.

Calcium is required for bone and tooth structure and is necessary for the function of nerves and muscles. Calcium is also required for the clotting of blood. Deficiencies of calcium may lead to weak and flacci muscles. Adults require a daily intake of about 750 mg of calcium. Milk cheese, sardines and canned salmon are especially good sources of calcium.

Magnesium is a minor component of bones and is present in soft tissu cells. Deficiency of magnesium is unusual, since most vegetables, cereal and cereal flours, beans and nuts contain adequate amounts to take car of daily requirements.

All body proteins contain sulfur, for it is a component of some amino acids. Certain vitamins also contain sulfur, required for the function of some enzyme systems. Meats, fish, cheese, eggs, beans, nuts and oatmea are all good sources of sulfur.

Fluorine is present in body tissues in trace amounts and helps to prevent tooth decay. Drinking water is the chief source of fluorine and fish is also a good source of this element. In high concentrations, fluorine is poisonous.

Copper is required for some body enzyme systems and is present in trace amounts in all tissues. Like fluorine, copper is poisonous in high concentrations. Fruits, beans, peas, corn, flour, rye, oats, eggs, liver, fish and oysters are adequate sources of copper. Dietary deficiency of copper is unknown.

Cobalt is a component of vitamin B-12, the only compound present in the body known to contain this element. Very small amounts of this mineral are, therefore, required. Sufficient amounts of cobalt are present in most foods, and some may be absorbed from cooking utensils. Excessive amounts of cobalt cause toxic effects. Dietary deficiency of cobalt is unknown.

Zinc and manganese are present in all living tissues. Most human diets contain 10–15 mg of each metal per day. Both metals are important components of a number of enzyme systems. Dietary deficiency of either zinc or manganese is uncommon because of their ubiquity in food

However, there are recorded cases of the effects of deficiencies of each element. A deficiency in zinc has been attributed to dwarfism, gonadal atrophy and possible damage to the immune system. A deficiency of manganese in experimental animals resulted in bone disorders, sexual sterility, abnormal lipid metabolism, and even toxicity.

Selenium, molybdenum and nickel are all found in trace amounts in the body. Selenium helps depress the symptoms of vitamin E deficiency, as well as muscular dystrophy in animals. Nickel has a role in physiological functions. Molybdenum is involved in oxidative and catabolic enzyme reactions. In excessive amounts, it results in copper deficiency. The molybdenum/copper balance can be restored by treatment with sulfur. Deficiencies of these metals are not common for man.

Vanadium deficiencies for man are not known; however, deficiencies of this element in birds and animals result in growth retardation, deficient lipid metabolism, impairment of reproductive function and bone growth retardation.

Silicon is found in unpolished rice and grains and is quite prevalent in beer. Certain diseases involving connective tissue are believed to result when it is not present in adequate quantities. The required amounts for man are not known.

Tin, occurring naturally in many tissues, has been found necessary for the growth of rats. It is believed essential to the structure of proteins and possibly other biological components. As it is present in most foods, tin deficiencies should not occur except possibly in foods that undergo refinement processes.

Chromium plays a physiological role thought to be related to glucose metabolism—perhaps by enhancing the effectiveness of insulin. While it is a normal body component, its content decreases with age.

Aluminum, boron and cadmium are also found in trace amounts in the human body, but neither their roles nor the effects of their deficient or excessive amounts are known.

3

Sanitary Handling of Foods

To refer to the opening remarks in Chapter 6, the factors that contribute to food-borne diseases are related to ignorance of proper food-handling procedures or to the unwillingness of food handlers to comply with the guidelines for handling foods sanitarily. It is unfortunate that we are unable to reduce the large number of food poisonings that occur in the United States each year, as there is no insurmountable reason why it could not be done. While it is not possible to provide an accurate figure, it has been estimated that at least 1 million persons are made ill from food poisoning each year in the United States with nearly all illness due to bacteria.

From what we know about bacteria that are involved in food-borne diseases, we can control their growth by controlling the temperature of the foods. Thus, foods held at temperatures below 38°F (3.3°C) or above 145°F (62.8°C) will not undergo a buildup of pathogens. Just as important is keeping the contamination levels of pathogens quite low by following sanitary guidelines in the handling of foods. To do this effectively, sanitary procedures must be practiced in all phases of food handling up to the moment it is consumed, and two prime needs in any sanitation program are (1) the appointment of a person to take the responsibility for the program and (2) the institution of measures to ensure the continuation of the program. It must also be remembered that sanitary handling practices are of no meaningful use unless the food itself is safe and unadulterated when it is procured.

It is often said that good sanitation is good economics. Actually, good sanitation is an additional cost of production. However, the good economics are evident in the long run through customer satisfaction, continuing sales, minimized losses from spoilage and minimized probability of damaging lawsuits that could arise from food poisonings.

PERSONAL HYGIENE

It has been said that it isn't the careless few who serve thousands in restaurants and institutions, but the potential sickness that they carry that should be our concern. Thus, the responsibility of *all* food handlers is a serious one, and the preparation and handling of foods in restaurants and institutions, and even in the home, should not be entrusted to any but knowledgeable and reliable people. Preferably, food handlers should have had at least one course in sanitation, and they should have been made aware of intentional (e.g., preservatives) and unintentional (e.g., pesticides) food additives. It is surprising that the curricula of primary and secondary schools do not include compulsory training in sanitation and personal hygiene as they pertain to food handling. After all everybody must sometimes eat foods prepared by others and everybody sometimes prepares or handles foods that are to be consumed by others Therefore, each one of us has the responsibility of knowing and practicing food sanitation and above all, of personal cleanliness. Clean people prepare clean wholesome foods; unclean people prepare unclean food which can easily transmit disease. Without question, sanitary programs are no more effective than the people connected with them. This includes not only those who handle food, but also managers upon whose attitude the effectiveness of sanitary programs depends. When management imposes and enforces strict sanitary measures, employees recognize and adopt them. On the other hand, when management fails to demonstrate a strong sanitary policy, both the morale and the sanitary attitudes of the employees are easily eroded.

Rules of personal hygiene which must be strictly observed by food handlers are reasonable and require little more than common sense and awareness.

(1) Persons with communicable diseases, including skin infections should never be allowed to handle foods to be consumed by others. Obviously, this means that all food handlers must undergo periodic physical examinations to establish this qualification

(2) Food handlers should observe physical cleanliness and should wear clean (preferably white) work uniforms (no jewelry).

(3) They should keep the head covered, the fingernails short and clean (no nailpolish).

(4) When possible, gloves should be worn, but whether or not gloves are worn, the hands should be washed and dipped in a disinfectant prior to handling food. The hands should be rewashed any time they are used for anything else prior to handling food again.

(5) When handling foods, the hands should not touch the mouth, nose or other part of the body, especially body openings, since these are possible sources of pathogens. Remember that the hands are involved in most instances of personal contamination of foods.

(6) During work, food handlers should neither smoke, drink nor eat in the work area.

(7) Pets and other animals do not belong in the food-processing area.

(8) Sneezing and coughing should be confined to a handkerchief, and in fact, the individual should leave the work area when either is imminent. After using a handkerchief, the hands are to be rewashed.

(9) Cloths should not be used for cleaning.

(10) Foods that appear to be unwholesome, or that may contain unacceptable contaminants, should not be handled.

SANITATION IN THE HOME

The educational importance and the impact of the home on the adult and his attitude are neither widely recognized nor socially encouraged. This is unfortunate, because the home is where the bad virtues (prejudice, deception, selfishness, greed, etc.) or the good virtues (tolerance, honesty, generosity, cleanliness, etc.) are moulded into the character. Thus, the institution that can do most to increase the number of people having healthy attitudes and clean habits is the home.

Personal Habits

The transmission of communicable diseases among humans is facilitated by many of our social habits, such as kissing, shaking hands, etc., as well as through foods. The bacterial flora on the body can be controlled to a degree by bathing and by changing to fresh clean clothing regularly, but the bacteria emerging from the openings in the body, especially to the respiratory, alimentary and urinogenital systems cannot be kept at very low levels for any length of time because of the constant outflow of bacteria from these systems.

All these sites are accessible to the hands and are, from time to time, reached or closely approached by the hands, which are, of course, used also to handle foods! Therefore, it is imperative that following a trip to the bathroom, which children should be taught to use properly, the hands should be washed with soap and water and wiped on a clean towel. Also, prior to handling foods, whether to prepare the foods or to eat them, the hands must be carefully washed with soap and water and wiped with a clean towel. Remember, while your own personal habits are sanitary, you have no assurance that the hand you shook in the office was clean by your standards. In fact, it may have had a residual of pathogens from a contaminating source unknown to you. Children should be taught to use handkerchieves rather than fingers to clean their noses—remember that the concentration of bacteria, viruses, etc., in the nose is considerable. Children should be taught to use handkerchieves to confine coughing and sneezing and not to cough or sneeze in the direction of foods and people. Children's hair should be kept clean and neat. They should be informed of the need to stay healthy and to respect that right for others. They should be taught not to pick at pimples with white heads, and boils, as these are "germ mines" holding great concentrations of dangerous staphylococci. Their nails should be kept short and clean. If children can be made to develop healthy attitudes and healthy habits, the rest should come easily.

The Home Environment

The home should be kept clean by periodic cleaning and by setting rules of conduct for members of the household such as discouraging litter, encouraging use of ash trays (if discouraging smoking is unsuccessful), etc. The surfaces should be kept dust free, preferably by vacuum cleaning, and surfaces that contact foods should be of easily cleaned material, such as plastics and stainless steel. Tableware, such as knives, and pots and pans should have handles of metal or plastic rather than wood. Tableware and utensils should be cleaned as soon after use as possible and should not be left uncleaned for long periods either in the sink or on the counter, since bacteria will grow and may become airborne to contaminate other food. Tableware and utensils should be washed in hot water and detergent and rinsed, then immersed in hot water (that has been heated to a minimum of 170°F (76.7°C) for at least $1/2$ min.). Automatic dishwashers may be used with confidence, since they clean effectively and since the temperature of the water can be raised to a higher level than can be tolerated in hand washing. Keep the refrigerator and freezer clean and free from odors. Keep a clean house and enforce the

abits that keep it clean, and members of the household will be
prompted to help keep it clean. The home should be kept free from
odents and other pests, such as roaches and flies, by a strict preventive
program and by a continuing check for indications of the presence of
hese undesirable elements. When they are present, concerted efforts to
eliminate them should not be spared.

Care of Food

Care of food in the home involves precautions in a number of areas.

Food Shopping.—When shopping for foods, shop at clean stores where
employees observe the rules to be found earlier in this chapter. Shop for
he perishable foods (e.g., milk, meat, fish) last, and do not delay going
ome once the shopping has been completed. Remember that bacteria
which spoil foods) grow to very large numbers in just hours while in a
warm car. Once home, the perishable foods should be unpacked immediately and transferred quickly to the refrigerator or freezer.

Food Storage.—Perishables should be kept in a refrigeraor held at as
ow a temperature (above freezing) as possible. Remember that the most
mportant deterrent to food spoilage is low temperature. The refrigerator
should be kept at temperatures in the range 32°–38°F (0°–3.3°C), preferably on the lower side of the range. Frozen foods should be held at 0°F
− 17.8°C) or below. Neither the refrigerator nor the freezer can lower the
emperature of bulk foods quickly enough, and spoilage occurs during the
period of cooling; therefore, when the bulk of foods can be reduced it
should be done. Large pots of stew, bulk hamburger, large cuts of meat,
etc., could all be divided and put in small containers. Packaging of the
oods is important, especially in impervious films, to minimize freezer
burn, oxidative deterioration, dehydration, etc. Refrigerated foods
should be covered, except for ripe fruits and vegetables. All cooked foods,
meats, fish, cold cuts, bacon, frankfurters, should be packaged before
refrigerating. Nuts should be packaged and refrigerated to slow down the
oxidation of their fats which leads to rancidity. Greens (e.g., spinach)
should be refrigerated unwashed. It should be remembered that fresh
oods should be used as soon as possible. Fresh foods, with some
exceptions, deteriorate rapidly. Some foods are best held at room
emperature and these include baked goods (to hold for more than two
days, they should be frozen), unripe fruit and bananas. When meats,
eggs, poultry, fish and other perishable foods are to be held, it should be
remembered that the safe temperature ranges are those below 38°F

(3.3°C) and above 145°F (62.8°C). Therefore, these foods should not be held at temperatures in the range 38°–145°F (3.3°–62.8°C).

Home-prepared Meals.—Meals that are consumed following their preparation are not likely to be responsible for food poisoning. On the other hand, meals that are consumed hours after preparation, such as is customary at picnics and outings, should be supervised by knowledgeable people. This is especially true of salads containing eggs, chicken or turkey, or products made from them. Prepared foods that are not to be used for long periods should be immediately refrigerated and reheated, if necessary, just prior to use. Some products such as puddings, custards and eclairs should be held under refrigeration at all times. Home freezers are not of sufficient capacity to flash-freeze stuffed poultry, chicken and especially turkey, and, therefore, poultry should not be stuffed prior to freezing. Instead it should be stuffed prior to cooking. Leftovers should be used as soon as possible; this is especially true of salads, chicken and other perishables. If, due to odor or other indicator, there is any doubt about the safety of a food, it is best to discard it unless advice can be obtained from a food scientist.

Extreme care must be taken with foods to be taken on a picnic or outing. Insulated containers should contain enough ice to keep the sandwiches, etc., refrigerated until used.

It is best to refrain from handling pets while preparing foods.

FOOD SERVICE SANITATION

It has been reported that about 80 million meals per day are served in approximately 335,000 restaurants in the United States. The U.S. Public Health Service has reported that about $^2/_3$ of all reported food poisonings result from meals served in restaurants. In terms of numbers, no one can be sure, since only a small percentage of all food poisonings is ever reported. One thing is known, and that is the number is very high, with some estimates indicating that it is about one million per year. Thus, sanitation in restaurants requires considerable improvement, since it is known that nearly all food poisonings are avoidable. The main reason for food poisoning is the poor attitude of some workers and the ignorance of others. Therefore, it must be the responsibility of the managers and owners of restaurants to employ only those with healthy attitudes and to make sure that they have been properly educated. Attitude has to be the most important criterion for employment, since the education part is simply a matter of a little time and effort. On the other hand, no amount of education in sanitation can improve an unhealthy attitude. People with unhealthy attitudes *must* be removed from food handling responsibilities. The problem is aggravated by the fact that restaurants are visited by large numbers of people over short periods (taxing the cleaning and food handling efforts of the facilities) and by the fact that some of the customers are bound to have communicable diseases.

Whereas restaurants outnumber other feeding establishments (e.g., hospitals, school and industrial cafeterias, etc.), in the latter, about 70 million meals are served daily.

The number of people involved in serving meals in restaurants and institutions is estimated to be 3.7 million. Whether food is to be prepared and served in restaurants or in institutional cafeterias, etc., safety precautions and sanitary procedures are very similar.

In general, the floor plan of an area where people are fed need not be special except that it can be easily cleaned and kept clean. The surrounding area should be pleasant, the floor should be carpeted and of a type that is easy to clean thoroughly.

It should be mandatory to use only potable (drinkable) water in foodserving establishments. If nonpotable water is used for cooling refrigeration units in walk-in refrigerators, for example, this water should not be connected with the potable water system, and the pipes should be painted with some identifying color.

In the food preparation and utensil-cleaning areas, the floors should be constructed of acid-resistant unglazed tile or of epoxy or polyester resin on a suitable base material. Also, the floor should be sloped to drains to facilitate cleaning and prevent the accumulation of water. Drains should be separate from toilet sewer lines to a point outside of the building and should be so constructed as to prevent the possibility of back-up into the building. The walls of food preparation and utensil cleaning areas should be constructed of smooth glazed tile to distances equivalent to splash height, since this type of construction makes for easier and more effective cleaning. The junction of the walls and floor should be coved or curved, which also facilitates cleaning, their being no angled corners where food materials can lodge.

The surfaces of benches and tables in food preparation areas should be of stainless steel or plastic, since such materials are impervious, noncorodible and easy to clean. For the same reason, cooking utensils, including steam kettles, should be constructed of stainless steel.

Areas in which steam cooking, steam cleaning or deep-fat frying is done should be provided with hoods and exhaust fans to the outside. Hoods should be constructed of stainless steel for ease in cleaning and should be equipped with traps to prevent condensed moisture from running back into foods which are being prepared.

Cutting boards should be constructed of plastic, preferably Teflon, as such material is easy to clean, does not absorb water and does not foster bacterial growth.

Walk-in refrigerators should have floors constructed of unglazed tile which are sloped to drains to facilitate cleaning. The drains should no empty directly into the sewerage system. Instead, they should empty into a sink or other container which empties in turn into a drainage system so there is no possibility of back-up into the refrigerator. The walls and ceilings of walk-in refrigerators should be constructed of glazed tile to facilitate cleaning.

Wash basins, soap, hot and cold water, and a container of disinfectant (preferably a solution of one of the iodophors) together with paper towel should be present in the food preparation, utensil washing and food exit areas to make certain that anyone having left his or her particular operation washes, disinfects and dries his hands prior to resuming work

The management of food-serving establishments should obtain raw food materials only from reliable sources. It should also determine that the water to be used for drinking, cooking and cleaning is potable.

The precautions to be used in cooking foods are no different from those which have been given for household food safety, but if foods, including gravies, are to be held on steam tables, the temperature of all parts of the food should never fall below 145°F (62.8°C), and preferably not below 150°F (65.6°C). Also, any food container held on steam tables should be emptied, removed and replaced with a new container of the particular food instead of being partially emptied and refilled with more food.

If leftover cooked foods are to be refrigerated, they should be placed in covered impervious containers (plastic or metal) and labeled. A card catalog of leftover foods should be maintained so that such material may be thrown away if held for periods of more than four days at temperatures of 38°F (3.3°C) or above.

If cream-filled pastries, such as eclairs or pies, or salads, such as potato, tuna fish, crab meat, chicken, etc., are to be held in the refrigerator or in display cases, the temperature of such storage area should be 38°F (3.3°C) or below.

Personnel with boils or pus-producing infections on their hands should not be allowed to handle foods or to clean utensils or equipment. Any personnel known to have had a recent intestinal ailment should be excused from work until such a period as it can be determined that he or she is not infective.

All garbage and waste materials should be held in leakproof metal or plastic containers with tight-fitting covers when held on the premises of the food-serving establishment. Rubbish should also be held in this manner. After emptying, each container should be cleaned inside and outside, and the water used for such cleaning disposed of in the sewerage system. Such containers should be washed in an area well separated from that used for washing utensils; containers of garbage and trash should be stored in vermin-proof rooms, well separated from the food preparation and serving areas. The floor of such storage rooms, and the walls up to distances of splash height, should have an impervious easy-to-clean surface. Storage areas of this type should be cleaned periodically. Care should be exercised to see that rodents and insects do not become established in waste storage areas.

An effective method of sanitizing glasses, chinaware and utensils is to wash in water at least at 120°F (48.9°C) with detergent followed by rinsing, then immersing in clean water at 170°F (76.7°C) for 0.5 min. Immersion for 1 min in a solution containing 50 ppm available chlorine or 12.5 ppm of an iodophor, in each case at a temperature not below 75°F (23.9°C), may be substituted for the hot water dip for purposes of sanitizing utensils and tableware.

PLANT SANITATION

Plant sanitation is necessary, first because it is a law (see sections 402(a)(3) and 402(a)(4) of the Food, Drug and Cosmetic Act), and second because it is good ethics, good economics, and we all expect it.

An important part of plant sanitation is the establishment of a strict quality control over the incoming raw materials. No amount of plant sanitation can remain effective if incoming materials are allowed to bring in contaminants.

Many of the factors concerned with food safety in plants manufacturing food products are the same as those indicated for food preparation and serving establishments. There are some additional considerations, however.

Plant Exterior

The surroundings for food plants should be neat, trim and well landscaped. There are several reasons for this. Nice surroundings have a good psychological effect upon those who work within. If the environs are well kept, the personnel working there are much more apt to try to keep things neat and clean on the inside. If the surroundings are dirty or cluttered, those working in the plant are apt to become careless in matters concerned with general sanitation. All parking spaces, roadways, walks, etc., should be paved so that dust contamination of the air will be minimized, and contamination, such as animal droppings, will be washed away with each rain rather than be soaked into the ground to be airborne during dry spells. The area surrounding a food plant, including platforms, should not be used for storing crates, boxes, machinery, etc., since these materials may become a harborage for rodents which may eventually find their way into the plant. There should be no area around the plant where the landscaping allows potholes or depressions of any kind in which water may accumulate and become a breeding place for insects which then may become established within the plant. Food materials, ensilage piles or other organic wastes should not be present in any exposed area near the plant, since they attract and become breeding places for insects, especially flies, which are difficult to control in food plants, even in the best conditions. There should be no neighboring plants, such as chemical, sewage, poultry, tanneries, etc., that may transfer bacteria, chemicals, etc., to the food plant.

Plant Construction

Food manufacturing plants are best constructed of brick or concrete, since wood is difficult to maintain in a clean and sanitizable condition and is more vulnerable to invasion by rodents, birds and other pests. If the plant is constructed of wood, the foundation should be "rat-stopped" (constructed of cement to a distance of several feet below and above the ground). The walls and roof junction of the food plant should be weatherproof and impenetrable by insects. In food processing or utensil washing areas, the junction of the wall and floor should be curved and have no angled corners, in order to facilitate cleaning. Window ledges should be slanted to prevent their use by personnel for storage of materials. The floor should be made of acid-resistant unglazed tile or of epoxy resin material, an epoxy tile grout laid on cement, for instance. The floor should be sloped to drains so that water does not accumulate. A cement surface is undesirable, since it tends to become pitted, leaving areas where water and food scraps accumulate and where bacteria may grow to large numbers, thus becoming sources of bacterial contamination and putrid odors, as well. The walls should be covered with glazed tile at least up to distances of splash height, in order to facilitate cleaning.

Raw materials should be separated from areas producing the finished product by a solid nonleaking wall (no openings, no doors) or by using separate buildings. The boiler room must also be separate and closed off. Each entryway to the area producing the finished product should be equipped with a shallow pan containing a disinfectant, so that those who enter the area must step into the pan, thus disinfecting their shoes or boots.

Equipment

Food-processing equipment should be so designed that all surfaces contacting foods are smooth, relatively inert, nonabsorbent, easily reached for cleaning, and of materials that are easily cleaned and sanitized. Moreover, it is desirable that as much of the equipment as possible can be cleaned without disassembling, but that disassembly to the required degree is possible and easily done. These specifications are desirable because they ensure that sanitation of equipment is possible and because sanitation can be accomplished quickly, effectively, and inexpensively. They are also desirable because they do not contribute to delaying tactics on the part of employees.

Assistance on the sanitary aspects of equipment is available from federal, state, county, and municipal agencies. One considered to be outstanding is the "Guide for the Sanitary Design and Construction of Food Equipment" (General Code: G P) issued by the Bureau of Food and Drugs of the New York City Health Department, 125 Worth St., New York, NY 10013.

The equipment, benches and machinery used in food processing and utensil cleaning areas should be of such materials and design as to make cleaning as easy as possible. Surfaces should be smooth (about 150 grit). Almost never should such equipment be constructed of wood. Even cutting boards, knife handles, shovel handles, etc., should be of hard plastic or other materials impervious to water. Black and cast irons and ordinary steel may be used for equipment which has no contact with foods, such as retorts, can sealers, etc. However, since these materials have rough surfaces and are subject to corrosion and are difficult to clean, their surfaces should not come in contact with food materials. While the surface of new galvanized iron is corrosion resistant, the zinc covering soon wears off, exposing the iron which can corrode. Also, zinc may cause a discoloration of certain foods. For these reasons, galvanized iron is not acceptable as a surface for contact with foods. Copper is used in steam kettles for the manufacture of certain foods like jams and jellies, because it is a good conductor of heat. When used in processing other foods, it must be kept scrupulously clean, however; otherwise, oxides accumulate, accelerating the destruction of vitamin C and the oxidation of fats. Alkaline materials used on copper equipment may cause a discoloration of foods. It is generally undesirable to use copper in food processing equipment, even though it is among the best conductors of heat available. Monel metal, an alloy consisting mainly of nickel and copper, is suitable for food processing equipment but is expensive. Aluminum conducts heat well but is subject to corrosion when contacting alkaline materials or fruit acids. Glass is used for piping for the transportation of liquids, such as milk, and such piping can be cleaned in place without dismantling. It is not suitable for lining metal equipment, nor is enamel, being subject to chipping, thus exposing the metal which may then corrode. Glass or enamel chips may also become incorporated in the food. Rubber is used for conveyor belts which transport materials, but rubber is not easily cleaned and belts lined with Teflon or a suitable metal are preferable to those lined with rubber. Stainless steel is doubtless the most suitable material for the construction of equipment which contacts food. It is noncorrosive and easily cleaned and sanitized. A special stainless steel may be required for areas contacted by chlorides, such as those occurring in sea water or brine. Piping and pumps should have no threaded joints where food can accumulate. Sanitary design for such equipment calls for flush joints held together by clamps, allowing thorough cleaning and sanitizing. Pipes that carry food materials should have no dead ends which cannot be cleaned and where food materials can accumulate and decompose. With such construction, surges on the line cause decomposed material to enter the mainstream of the food material passing through the pipes. Pipes should not be joined to tanks, hoppers, etc., so that the pipe end extends into the tank itself. In such cases, when the liquid falls below the level of the pipe, some food remains in the pipe end where it may decompose and eventually contaminate new material entering the tank. Tanks, flumes, thermometer wells, pots, pans, etc., should have only curved corners and junctions of sides and bottom in order to facilitate cleaning.

Personnel Facilities

Locker rooms should be provided, separate for male and female personnel, and a sufficient number of lockers should be available to provide one for each worker so that outside clothing may be stored. These locker rooms should be kept clean and tidy.

Separate toilet rooms with self-closing doors (that do not open directly to processing area) should be provided for men and women workers. Toilet rooms should have wash bowls with hot and cold running water, soap dispensers, paper towels and containers for refuse. The minimum number of toilets and wash basins is given in Table 3.1. Urinals may be substituted for toilets in the men's rooms on a one to one basis, but the number of toilets must never be fewer than ⅔ of the number given in Table 3.1. The hand washing units should be of the foot-activated type.

If personnel are to eat lunches at the plant, a room separate from other rooms, including locker rooms, should be provided for this purpose. This room should be kept clean and sanitary. Drinking fountains should never be located in toilet rooms.

All personnel working in food processing or utensil cleaning areas should be provided with clean outer uniforms daily.

In processing or utensil cleaning areas, there should be wash basins, a container of disinfectant (preferably a weak solution of an iodophor) and paper towels so that those in charge can make sure that workers wash and disinfect their hands before returning to work once having left their particular area of operation. This is of great importance to good food plant sanitation.

Food-processing plants must have adequate light to ensure that employees can perform their duties effectively. Table 3.2 cites the minimum light requirements to carry out different operations in the plant.

Food-processing plants should have good ventilation with filtered air, to reduce the atmospheric humidity, and hence the amount of condensation, which if allowed to occur, promotes the growth of bacteria and molds on the surfaces of walls, ceilings, floors, equipment, utensils, foods, etc.

TABLE 3.1.   REQUIRED LAVATORY FACILITIES

Number of Employees	Minimum Number of Toilets	Minimum Number of Wash Basins
1–9	1	1
10–24	2	1
25–49	3	2
50–74	4	3
75–100	5	4
>100	5 + 1/30 additional employees	4 + 1/50 additional employees

TABLE 3.2.   MINIMUM LIGHT REQUIREMENTS FOR FOOD PROCESSING PLANTS

Operation	Minimum Light (Ft-Candles)
Sorting, grading, inspection[1]	50
Processing, active storage	20
Instrument panels, switchboards	10
Toilet rooms, locker rooms, etc.	10
Dead storage	5

[1] Local lighting for inspection may have to be as high as 100–150 ft-candles, depending on the type of inspection performed.

Storage

Dry materials such as breadings, flour, etc., to be stored, should be held in a room constructed of materials, such as brick or cement, which do not allow the entrance of insects and rodents. Such rooms should be refrigerated to about 50°F (10°C) to prevent the hatching of eggs and the development into adult insects, since insect eggs are almost invariably present in such materials.

Cleaning the Plant

All floors, walls, benches and tables, conveyors, hoppers, fillers, kettles, utensils, etc., used for processing foods should be thoroughly cleaned and sanitized at least once, and preferably twice, per 8-hr working shift. Large plants should have cleaning crews with a foreman. A list of approved cleaning compounds was published by the U.S. Department of Agriculture in 1968, and it is available from that agency. The water used for cleaning should be potable and have a temperature of about 130°–140°F (54.4°–60.0°C). The type of detergent used in cleaning should be suitable to remove the type of soil which will be encountered. Depending on the application, detergents should have certain properties. In general, detergents should not be corrosive. When used in hard water, they should not form precipitates. They should have good wetting, and in many cases emulsifying, properties. They should be good solvents for both organic and inorganic soils. They should saponify fats and have good dispersal and deflocculating properties. They should not form residual films on surfaces. High-pressure water and high-pressure steam can be used to flush hard-to-reach places with detergent.

Chemical agents used for controlling microbes include bacteriostats, which prevent the growth and spread of bacteria, and bactericides, which not only stop bacterial growth, but also destroy the bacteria. Some agents are bacteriostatic in small amounts and bactericidal in large amounts. Effective sanitation programs should be concerned mainly, if not completely, with bactericides which include compounds in the following chemical classes: halogens, phenolics, quaternary ammonium compounds, alcohols, carbonyls and miscellaneous others. The halogens, chlorine and iodine, are considered to be the most important sanitizing agents known. Such common uses as chlorination of drinking water and iodine treatment for cuts, etc., make these compounds familiar. Phenolics, such as cresol, are considered to be also very good sanitizers; however, they have some disadvantages, among which are high irritation characteristics and relatively high cost. Quaternary ammonium compounds, while effective in the control of algae and some bacteria, are relatively ineffective against a variety of microbes that are not resistant to the halogens and phenolics. Alcohols are not as effective as generally believed, and while some carbonyls, such as formaldehyde, are effective, they are also hazardous to use. There are other miscellaneous sanitizers, but they are generally low in effectiveness and high in cost.

After cleaning and rinsing with hot water, equipment should be immmersed for about 1 min or rinsed with a solution of an iodophor containing 12.5 ppm of available iodine. Iodophors, or "tamed iodines," are combinations of iodine and surface active agents which have the sanitizing advantages of iodine with minimized disadvantages of iodine. Iodine by itself is not very soluble in water, has a high vapor pressure, is corrosive and leaves a stain. In combination with surfactants, these undesirable properties are minimized. A solution containing 50 ppm of available chlorine at a temperature not below 75°F (23.9°C) may be substituted. All equipment so constructed as to hold liquids should be thoroughly drained after cleaning and sanitizing and containers, such as pans, should not be nested after cleaning and sanitizing, since this prevents draining and evaporation of moisture and thus provides moisture in which bacteria may grow. For a summary of the properties of sanitizing agents see Table 3.3.

Water Supply

The water supply should be adequate for filling the plant's needs. All water that may contact either foods or surfaces that may be contacted by foods must be of potable quality (safe for drinking). The water used for cleaning should be of adequate temperature and pressure and supplied via a plumbing system of adequate capacity and in conformance to building codes.

TABLE 3.3. ADVANTAGES AND DISADVANTAGES OF THREE IMPORTANT CLASSES OF SANITIZERS

Hypochlorites (Liquid)	Iodophors	Quats[1]
	Advantages	
Inexpensive	Stable	Stable
	Long shelf-life	Long shelf-life
Active against all micro-organisms	Active against all micro-organisms except bacterial spores and bacteriophage	Active against many microorganisms, especially the thermoduric types
Unaffected by hard water salts	Unaffected by hard water salts	Form bacteriostatic film
Water treatment	Noncorrosive	Prevent and eliminate odors
Active against spores	Not irritating to skin	Nonirritating to skin
Active against bacterio-phage	Easily dispensed and controlled	Noncorrosive
Easily dispensed and controlled	Acid nature prevents film formation	Stable in presence of organic matter
Not film forming	Concentration easily measured by convenient field test	Easily dispensed and controlled
Concentration easily measured by convenient field test	Visual control (color)	Stable to temperature changes
	Good penetration qualities	Good penetration qualities
	Spot-free drying	May be combined with nonionic wetting agents to formulate detergent sanitizers
	Disadvantages	
Short shelf-life	Not as effective against spores and bacterio-phage as hypochlorites	Incompatibility with common detergent components
Odor	Expensive	Germicidal efficiency varied and selective
Precipitate in iron waters	Should not be used at temperatures exceeding 120°F (48.9°C)	Slow in destruction of coliform and Gram-negative psychrophilic bacteria (like *Pseudomonas*)
Adverse effect on skin	Staining of porous and some plastic surfaces	Not effective in destruction of spores and bacteriophage
Corrosiveness on some metals	Germicidal action adversely affected by highly alkaline water or carryover of highly alkaline detergent solutions	Expensive
		Slow to dissipate (residual problem)
		Objectionable film on surfaces treated
		Foam problem in mechanical application
Use concentration 200 ppm Cl_2	25 ppm I_2	200 ppm quat

Courtesy of Klenzade Products Division, Economics Laboratory, Inc.
[1]Quaternary ammonium compounds.

ewage Disposal

Sewage disposal must be through a public sewerage system or through system of equal effectiveness in carrying liquid disposable waste from ıe plant. The sewerage system must conform to building codes and in no ay be a source of contamination to the products, personnel, equipment r plant. Drains must be sufficient to ensure the rapid and complete ˙ansfer of all wash water, spilled liquids, etc., to the sewerage system.

ANITATION IN RETAIL OUTLETS

Most regulations applicable to foods in retail stores are the same as ıose which are applied in food manufacturing plants, but there are some recautions which are especially applicable to retail outlets. Fresh uncut ıeats should be stored in a walk-in refrigerator with walls and ceilings of lazed tile to facilitate washing and cleaning. The floor should be ɔnstructed of unglazed tile and sloped to drains. The temperature of the ıeat storage room should be held at 32°–37°F (0°–2.8°C). Carcasses or des should be hung on hooks attached to rails. Sawdust should not be sed on the floor since this creates dust which is, to some extent, a source f contamination.

The room where meat is cut into retail portions should have the same all, ceiling and floor construction as does the meat storage area. enches used for cutting meat should have surfaces of stainless steel, and ıtting boards should be made of a plastic material, preferably Teflon, ˙hich is impervious and easily cleaned and sanitized. The temperature of ıis room should be about 50°F (10°C), since personnel find it difficult ɔ work at lower temperatures. However, a low relative humidity must be ıaintained in the cutting area, otherwise the meat will condense ıoisture (sweat) which will facilitate the growth of spoilage bacteria. lso, neither cut nor uncut meat should be allowed to accumulate in this rea but should be moved back into the storage area or into display cases s soon as possible. The meat cutting area should be cleaned and ınitized at least once per 8-hr working period and should follow ssentially the same methods as indicated earlier for food manufacturing lants.

In the grinding of meat, such as hamburger, separate grinders (or ˙inder heads) should be used for beef and pork. The reason for this is ıat pork may contain an infective roundworm which causes the disease ıown as trichinosis in humans. In pork, this is usually taken care of by ɔoking to a temperature which destroys the worm cysts. Hamburger, however, is sometimes eaten undercooked or only lightly heated, in which case the cysts would not be destroyed. If, therefore, the same grinder is used for beef and pork, small pieces of pork containing cysts may contamininate the ground beef. It seems reasonable that a grinder head be used for pork, then washed and used for beef without any hazard. However, there is always the chance that the grinder head will not be thoroughly washed.

Retail display cases for cut meats, chicken, cooked and fresh sausage products, bacon, cold cuts, etc., should be held at temperatures of 32°–38°F (0°–3.3°C) and should be of the open top variety. Also, personnel in charge of display cases should make certain that products move more or less in the order of "first in first out" and that no item remains in the display case for long periods of time.

Large or small canned hams of the type requiring refrigeration should be held, at all times, at temperatures of 38°F (3.3°C) or below. They should *never* be displayed in general store areas, aisles, or windows where there is no refrigeration. Unfortunately, retail outlets do not always observe safe practices.

Fresh produce is oftentimes poorly handled in retail outlets. The cells of such foods continue to respire after they are harvested, and the higher the temperature at which they are held the faster the rate of respiration. Respiration brings about chemical changes in fresh produce which cause a deterioration of quality. Loss of sweetness, loss of succulence, toughening, and the development of off-flavors are some of the changes which may take place in fresh produce because of respiration. For instance, sweet corn on the cob, freshly picked, then held at 35°F (1.7°C) is perfectly good in taste and texture after 15 days of storage. At high temperatures, its quality will be lost in a few hours. Storage areas for fresh produce should be clean and held at 32°–37°F (0°–2.8°C). Potatoes, turnips and cabbages should be held at about 50°F (10°C), since cabbage and turnips stand up well at this temperature, and potatoes convert starch to sugar at 40°F (4.4°C) or below—becoming sweet. Most fresh produce should be displayed in an area or cabinet held at 32°–37°F (0°–2.8°C) or partially surrounded with ice, since such temperatures maintain quality. A possible exception is lettuce which, if held in ice, must be held in tempered ice [temperature brought up to 32°F (0°C)]; otherwise, the produce may freeze, causing the leaves to wilt.

Fish and shellfish, such as shucked clams, oysters, scallops, and shrimp should be held in a display case surrounded by ice. This provides a temperature of about 33°F (0.6°C) and is the best way to maintain this low temperature, without freezing, for products which are extremely perishable.

Milk, cream, soured cream, cheeses and butter should be held in an open top display case, the temperature of which is held at 32°–37°F (0°–2.8°C).

Frozen foods, which are not indefinitely stable, are rarely handled under satisfactory conditions in retail stores. At –30°F (–34.4°C), frozen foods deteriorate at an extremely slow rate. At 0°F (–17.8°C), many foods will have a storage life (no noticeable loss of quality) of at least 6 months, and some foods have a storage life of at least 1 year at this temperature. As the storage temperature is raised above 0°F (–17.8°C), for each 5 degrees F (2.8 degrees C), the rate of deterioration is approximately doubled. Thus, a product which has a storage life of 6 months at 0°F (– 17.8°C) will have a storage life of only 3 months at 5°F (–15°C). When frozen foods are delivered to the retail store, they should not be allowed to stand on platforms or in a room at high temperature but should be immediately placed in the frozen storage room. The frozen storage room should always be held at a temperature of 0°F (–17.8°C) or below, preferably at –20°F (–28.9°C). Display cases used for holding frozen foods in the retail area should be of the open top type, or of the enclosed shelf type. Shelf-type display cases for frozen foods are not suitable unless they are enclosed by doors. The reason for this is that cold air is heavier than warm air, hence, in the open top case the cold air tends to remain in the area where the foods are held. Shelf-type frozen food cases must be refrigerated by blowing cold air out and over the product. Since this air is heavier than warm air, it tends to flow outward and downward into the room. In display cases of this kind it is, therefore, difficult to maintain temperatures of 0°F (−17.8°C) or below around the entire product. Those items in front where warm air has access are usually surrounded by air temperatures much higher than 0°F (−17.8°C), and hence subjected to an accelerated rate of deterioration.

Canned foods require some consideration in retail handling. To begin with, the buyer of canned foods for a retail outlet should have the knowledge, or employ personnel with the knowledge, to determine whether or not these products have been sufficiently heat processed for safe consumption. That means that the foods must be heated to the point where all spores of the bacterium known as *Clostridium botulinum* have been destroyed. Actually, in order to prevent spoilage (not disease) by other bacteria, canned foods should be heated beyond the point at which all disease-causing bacteria will have been destroyed.

An adequate backlog of canned foods must be available to the retail store. This means that there must be a warehouse where canned foods are stored. Such warehouses should be held at temperatures not above 75°F

(23.9°C), because bacteria known as thermophiles which are difficult to destroy by heat, may be present as spores in an occasional can of food. These bacteria do not cause disease, but they can cause spoilage of canned foods. Since they are so heat resistant, it would take excess heating to reach the point where all of the thermophilic spores would have been destroyed should they be present. It is fortunate that these bacteria grow only at high temperatures [usually well above 75°F (23.9°C)]. For this reason, the temperature of the warehouse where canned foods are stored should be regulated as previously stated. Nor should the storage warehouse for canned foods be held at temperatures below 50°F (10°C), for if this is done, when higher temperatures are reached in this area or when the cans are placed in the retail outlet at higher temperatures, moisture may condense on the surface (the cans may sweat) and cause rusting of the outside surface which discolors the label and is otherwise unsightly, and it may eventually weaken the can to permit microbial invasion of the contents.

Canned foods, both those stored in the warehouse and those held in retail stores, should be handled on a first-in, first-out basis. The reason for this is that while most canned foods have a relatively long storage life, they are not indefinitely stable. The usual type of deterioration after long periods of storage is due to internal corrosion of the container which results in a swelled can or in leakage. Deterioration of this type is most often encountered in acid foods, such as tomato products, in which case internal corrosion produces hydrogen gas causing the can to swell. The food in this case may be perfectly edible, but the consumer would be running a risk to eat the food from a swelled can since he cannot be sure that the cause of the swelling was not due to gas produced by some disease-causing bacterium. Also, swelled cans sometimes burst, causing the product to be spread over a wide area, creating a considerable mess and bad odors.

4

Regulatory Agencies

A major evolution in modern societies is the widespread use of food which is produced and often preserved in areas remote from the consumer. Since the consumer doesn't know how the food was handled, he doesn't know if it's safe to eat. (In ancient times, slaves and animals were sometimes compelled by their masters to eat food of questioned safety before it was eaten by their masters.) Thus, as the entire population presently requires protection, it is a proper government function to determine the wholesomeness and purity of foods and to protect the consumer against economic fraud as well as health hazard. Yet government seldom assumes this responsibility spontaneously, and protective regulatory legislation is passed and enforced only after consumer-oriented persons or groups stimulate broad public support for government action.

In the United States, the basic regulation covering the safety and legitimacy of commercial foods was not enacted until several states had already passed laws to protect consumers against adulterated and misbranded food and drugs within their own states. It was not until 1906 that the Federal Food and Drug Act was passed, due to the efforts of dedicated people, such as Dr. Harvey W. Wiley. Dr. Wiley is given credit for the enactment of the first federal regulation covering safe and pure foods and drugs. Subsequent additions to federal regulations have followed.

There are several federal agencies for regulating foods and food products sold in the United States, but only three have enforcement authority. These are the Food and Drug Administration (Department of Health, Education and Welfare) and the Meat Inspection Division and

Poultry Inspection Service, both of which come under the Department c
Agriculture.

THE FOOD AND DRUG ADMINISTRATION

The Food and Drug Administration (FDA) is probably the mos
important enforcement agency since it regulates all our foods excep
meat and poultry and, in some instances, can even regulate thes
products. There are two general regulatory categories: adulteration an
misbranding. A food is adulterated if (1) it is filthy, putrid or decom
posed, (2) it is produced in unsanitary conditions, (3) it contains an
substance deleterious to health. A food is misbranded if (1) it is a food fo
which standards of identity have been written and it fails to comply wit
these standards, (2) it is wrongly labeled, (3) it fails to meet th
regulations for fill of container.

Adulteration

Adulteration is not difficult to determine since there are tests that ca
be made to detect sources of contamination such as rodents (hairs
pellets or urine), insects, dirt and other detritus. Also, if a food is putrid
this can be detected by the ordinary human senses, a fact which is wel
known and accepted. However, the detection of decomposition is no
easy and often scientists do not agree on what constitutes decompositio
of a particular food. Citations based on the decomposition of a food
therefore, frequently have to be settled in court.

Regarding additives which may be present in foods, the administratio
and the industry know that certain chemicals are toxic and cannot b
added to foods at all. The FDA has a GRAS (generally regarded as safe
list which specifies which chemicals may be added to foods and, in man
instances, how much may be added to a particular food. Many com
pounds on this list come under what is called the Grandfather Clause
these chemicals having been used in foods for years with no apparent il
effect. For some chemicals which can be added to foods and for any nev
chemical which will be added, tests have been or will be made by feedin
several kinds of animals (e.g., rats, dogs, mice, guinea pigs) a die
containing up to 100 times the level at which the chemical would be used
over a period of several generations. The results of such tests ar
determined by observations on the weight and general health of th
animals, as well as their ability to breed, and on autopsies and chemica

ests for specific enzyme activities, etc. Time, trained personnel and special facilities are required for testing a new food additive. This is a very expensive process, requiring, as a rule, the outlay of several hundred thousand dollars, and no producer of such a new compound is apt to initiate such testing, which must satisfy the FDA, unless he is certain that the new additive will provide specific advantages and have great utility.

It is difficult to enforce the section of the law prohibiting substances that may be deleterious to the health of the consumer. For example, pathogenic bacteria, such as the *Salmonella* organisms often present in food, can cause disease and even death. The FDA and food scientists know that poultry and other foods generally contain these organisms but are unable to control the situation. If enforcement were attempted, for instance, in the case of poultry, the whole industry would be shut down, since approximately 25% of the product contains salmonellae. The testing for this group of organisms requires several days to obtain results, another factor which serves to impede enforcement. This regulation, therefore, should probably be changed or reworded, since laws which cannot be enforced are useless and lead to unsatisfactory practices.

Foods found to be adulterated are seized by the FDA and destroyed.

Misbranding

If a standard of identity has been set up for a particular food, the food can only contain the ingredients specified in the standard. If the food is found to contain other ingredients or additives, it will be seized and destroyed. As an example, sulfur dioxide is allowed in some foods and might be used to provide good color in ketchup. However, there is a standard of identity for ketchup in which sulfur dioxide is not included. Therefore, if ketchup were found to contain sulfur dioxide, it would be seized and destroyed. Foods for which a standard of identity has been established and which do not conform to these standards are destroyed on the premise that they have no identity. Standards of identity have been set up for some bakery products, cacao products, cereal flours and related products, alimentary pastes, milk and cream, cheeses, processed cheese, some cheese foods and spreads, some canned fruits, fruit preserves and jellies, some canned shellfish, eggs and egg products, oleomargarine, some canned vegetables, canned tomatoes, tomato products and other foods. A food which is misbranded may be wrongly labeled as to weight, portions, etc., or ingredients if no standard of identity has been established for it. If wholesome, wrongly labeled products need not be destroyed. Instead, they can be relabeled to comply with ingredients, weight, etc., and sold. If the product fails to meet the "fill of container" requirement, it may be relabeled to specify this fact and sold under the new label.

It should be noted that, theoretically, the FDA can regulate only those foods which are shipped interstate (from one state to another), but there are ways to get around this. For instance, if a company ships one food interstate but not another, both foods may come under the jurisdiction of the FDA.

The U.S. Public Health Service

The U.S. Public Health Service has regulatory authority over the sanitary quality of drinking water and foods served on interstate and international carriers (airlines, trains, etc). This agency carries out research and surveillance on food-borne diseases (infections and intoxications) and sanitary processing and shipping of foods. It is now a part of the FDA. In cooperation with state regulatory agencies, it sets up standards for coastal waters from which bivalve shellfish, such as oysters and clams, may be harvested for consumption. It also sets up specifications for waters from which bivalves may be harvested and then subjected to purification procedures or relayed in approved waters. It also helps set up standards for the bacteriological quality of the bivalve shellfish and for shellfish-growing waters. Shellfish dealers are registered and must keep records indicating the area from which bivalves were harvested, from whom they were purchased, and to whom they were sold. The state and the Public Health authorities keep a list of approved shellfish dealers. It is up to the state authorities to make sanitary surveys of the bivalve growing areas and bacteriological tests of the waters of shellfish-growing areas and of the shellfish. If a particular dealer does not comply with regulations, he is taken off the approved list and any product which he ships interstate will be seized by the FDA. It is also probable that state authorities would seize his product shipped intrastate. Also, if the state program for shellfish sanitation does not meet the requirements of the Public Health Service, all producers within the state are taken off the approved list and no bivalve product can be shipped interstate.

The Public Health Service also sets up standards for milk and cream together with state authorities. This includes the control of disease in dairy herds, the management and milking of herds, the bacteriological quality of raw milk and cream and certified milk, pasteurization procedures including time and temperatures, and the bacteriological quality of pasteurized milk and cream.

The FDA has the power to inspect any food processing or food handling lant and to close any plant which it considers to be unsanitary or to be dulterating foods in any way. FDA inspectors inspect some food plants ut do not have the personnel to inspect them all even annually. Most nforcement results from chemical or bacteriological analyses of some roduct. If according to their analysis a product is found to be dulterated, it will be seized and destroyed, either with or without court earings. In a case where the FDA learned that a certain boat, without eezer capacity, had been fishing for an extraordinarily long time prior) landing its catch, the FDA inspectors waited until the product was lleted, packaged under a particular label, and frozen, then seized the roduct. This resulted in a court case in which the FDA convinced the dge that the product was decomposed. The product was destroyed and e producers fined.

HE MEAT INSPECTION BUREAU

The Federal Meat Inspection Act, made law in 1906, is administered y the United States Department of Agriculture (USDA) through its leat Inspection Bureau, a branch of the Agricultural Research Service. his law regulates the safety of meats (beef, pork, lamb) entering into terstate commerce. The Meat Inspection Bureau is also an enforce- ent agency. It differs from the FDA in that most of the regulation is rried out by inspectors stationed at the plant processing the food. They al with any food which contains more than a small percentage of meat. hen cattle, hogs or sheep are slaughtered, one or more inspectors (who e veterinarians) must be stationed at the slaughtering plant if any rts of such products are to be shipped interstate. The animals may be spected prior to slaughter, and if diseased, are destroyed. However, the ain inspection comes after slaughter. As the animals are slaughtered, e carcass, entrails and organs are tagged and identified with a rticular carcass. The veterinarian inspectors test the viscera to de- rmine whether or not the animals were diseased. Diseased animal rcasses including organs and entrails are covered with a dye and must ː disposed of as tankage and not used for human consumption. 'ankage is slaughterhouse waste that is heat processed and dried and ed as fertilizer.)

The Meat Inspection Bureau also has "lay" inspectors. These inspec- rs have some training but are not veterinarians. Lay inspectors deal imarily with those plants or areas of plants where meat is cut into rtions for further processing or for shipment as fresh cuts; where sausages (fresh sausage, frankfurters, bologna and other cooked sausage dried sausage, etc.) are produced; and where hams, shoulders and bacon are cured, smoked, and so on. Lay inspectors determine that processing rooms and equipment are clean and sanitary before work is started, and if not satisfactory, the room is tagged and cannot be used for processing until cleaned and sanitized to the satisfaction of the inspector.

The Meat Inspection Bureau maintains lists of approved ingredients for cured and processed products, and it is part of the inspectors' job to determine that nothing is added to processed products which is not on the approved list. Inspectors also check the amounts of some approved materials which are added to processed products. The Meat Inspection Bureau has laboratories which make some analyses. For instance, a certain amount of water can be added to some cooked sausages or to cured hams. The amount of water allowed is specified (e.g., 10% of the finished product in frankfurters). In frankfurters, the added water gets into the product as ice during the cutting of the meat, one of the reasons for adding water in this case being that of keeping the ingredients (the meat emulsion) cool during cutting, which improves the quality of the finished product. The inspector cannot tell exactly how much water is being added during the cutting, but if he becomes suspicious he will take samples of the finished product, ship them to a bureau laboratory and have them analyzed for added water. In any case, samples for analysis may be taken periodically from processing plants. The same kind of inspection and analysis may be used for other processed products including hams.

States also have inspectors who regulate the slaughter and processing of meats which are used intrastate and not shipped interstate. However such inspection in the past has been far from adequate. In recent years federal authorities decided improvement in the inspection and regulation of local slaughtering and meat-processing plants was essential, and this improvement is presently being implemented.

The Poultry Inspection Service

The Poultry Inspection Service, similar in scope and operation to the Meat Inspection Bureau, is responsible for the inspection of poultry products. An agency within the USDA, it ensures that all poultry sold in interstate commerce is processed in U.S. government-inspected plants and is wholesome. Although poultry inspection is mandatory, the grading of poultry products is voluntary on the part of the processor. Poultry is inspected prior to dressing, during evisceration, during packing and after packing.

)THER REGULATORY AND/OR INSPECTION AGENCIES

There are some inspection agencies the activities of which are of a
egulatory nature but which are not necessarily enforcement agencies.
ne of these is the Voluntary Inspection Service of the U.S. Department
f Commerce. It is administered by the Department's National Marine
isheries Service, National Oceanic and Atmospheric Administration. If
 fish processor so desires, he may have an inspector stationed in his
lant on a permanent basis to assess the premises for sanitary conditions,
etermine that wholesome ingredients are used, and grade the product.
s previously stated, this is an entirely voluntary service, and the
rocessor must pay whatever costs are involved in such an inspection.
'ith this type of inspection, the product is evaluated and graded as A, B
r C in quality, the last grade being the lowest acceptable quality. The
acker can label the product accordingly. Continuous inspection is
vailable for certain canned and frozen fishery products. While the
ispectors do not inspect every fish, they do inspect a sufficient number
o that all lots of fish are sampled, and since each lot is relatively uniform
 quality, unacceptable fish are bound to be identified. This is also true
r canned fish. The inspection of frozen fish products does not permit
ie same degree of confidence, since frozen products are manufactured
om blocks of frozen fish or shellfish, usually prepared and frozen
itside the United States. It is true that the inspection service can and
>es sample the finished product to determine quality, but this can only
e a spot check which is not adequate to evaluate the quality of the whole
ick. There is another reason why grading canned fish is more effective
an grading frozen fish. Once the canned product has been processed,
e quality does not change to any extent. This is not the case with frozen
ods. Frozen fishery products are not indefinitely stable even at
F $(-17.8°C)$. At higher temperatures they deteriorate at a much faster
te. It is unfortunate that while manufacturers tend to freeze and store
hery products to a temperature of $0°F$ $(-17.8°C)$ or below, those who
ansport such products and especially retailers, during the handling and
splay of fishery products, tend to hold products at temperatures well
ove $0°F$ $(-17.8°C)$. For this reason a frozen fishery product labeled
ade A may actually be grade C or substandard by the time that it
aches the consumer.

The USDA has a grading service for fruits, fruit juices and vegetables,
th canned and frozen. Again, for canned fruits and vegetables, the
ading is quite effective, but for frozen products the same objections are
plicable as to frozen fishery products. What is needed to make the

grading of frozen foods effective is an enforcement agency to regulate th
temperatures at which these products are shipped, the temperatures a
which they are handled and stored at the retail level, and especially th
temperatures at which frozen foods are held during display at the reta
level.

The Environmental Protection Agency (EPA) authorizes and regulate
the use of pesticides and other environmental contaminants, and
monitors compliance and provides technical assistance to the states.

The Internal Revenue Service (IRS) enforces, with FDA collaboration
the Federal Alcohol Administration Act and other pertinent regulation
that control the commerce of alcoholic beverages (e.g., whiskey, win
beer, brandy).

The National Bureau of Standards (NBS) is responsible for setting th
official standards for units of weights and measures for all commerci
products, including foods.

The Office of Technical Services (OTS) issues voluntary "Simplifie
Practices Recommendations" in order to limit types and sizes of pacl
ages, bags, jars, etc., used as food containers.

The Federal Trade Commission (FTC) enforces the provisions of th
FTC Act which prevents unfair and deceptive advertising and unfair ar
deceptive trade practices.

The regulation and standardization of foods in international trac
represent a prodigious and nearly impossible task. Yet the benefits to I
derived are of such magnitude that they merit the necessary efforts
achieve them. The Codex Alimentarius Commission, an internation
organization, has been formed by over 90 countries to establish for
standards. Its importance is recognized when we are told that the U.
FDA detains about 40% of the imported foods that it checks. What of th
food shipments that escape inspection? What proportion of the import
foods is mislabeled? What proportion contains harmful substances or
contaminated or adulterated? It is because of these considerations th
the FDA detains so much of the imported foods that it checks. T
member countries of the Codex Alimentarius send their experts to t
international meetings, held in Rome, to help formulate the quali
standards that are more strict in some aspects than those of mai
countries. For example, the International standards require the listing
all ingredients in food formulations. The work done by this internation
body, slow though it is, should facilitate trade among the memb
countries, and the risks of food-borne illness and deceptive practic
should be substantially reduced.

5

Food Additives

DEFINITION OF FOOD ADDITIVES

Food additives may be defined as chemical substances which are deliberately added to foods, in known and regulated quantities, for purposes of assisting in the processing of foods, preservation of foods, or in improving the flavor, texture or appearance of foods.

The term does not include chance contaminants. An additive may be reactive or inactive; it may be nutritive or nonnutritive; it should be neither toxic nor hazardous. Some substances, such as pesticides, packaging components, etc., are added to foods unintentionally, and these are, of course, undesirable, and may be hazardous to health. Because of their toxicity, their presence is closely regulated by strict government tolerances.

Many food additives are classified as GRAS ("Generally Regarded As Safe") additives. Additives are classified as GRAS when they have been used without apparent harm for long periods, long before regulations were put into effect.

PHILOSOPHY OF FOOD ADDITIVES

Foods are made entirely of substances that, in the pure form, can be described as chemicals or chemical compounds. It is important to note that our knowledge of the composition of foods, because of its complexity, is by no means complete. For instance, it is reported that one of the

most important of man's natural foods, human milk, contains well ov
100 chemical compounds.

Unfortunately, the interpretation of the word chemical is too oft
inaccurate. Thus, some consumers are apprehensive about purchasing
food that is preserved by treating it with a chemical with which they a
unfamiliar. However, a number of foods may be preserved with tab
salt, which is a chemical. Consumers are not apprehensive about usin
salt as a preservative, because they are familiar with it, at least f
adding taste and sometimes for bringing out the flavor in foods, yet tab
salt is definitely a chemical, having the name, sodium chloride, and tl
formula, NaCl. Refined sugar, vinegar, spices and other substances th
are routinely added to foods are also chemicals or mixtures of chemica
and we do not question the use of these, either. The characteristics
chemicals which we use with confidence are: (1) familiarity and (
frequent use. The characteristics of chemicals that arouse skepticism
consumers are that they are uncommon and unfamiliar.

A large number of chemical additives are unfamiliar, and there is
need for regulatory agencies to question their use from the standpoint
safety. Obviously then, we should not fear the use of chemicals, but v
do need to screen them for safety when their effects on human health a
not known. Some lessons have been learned along these lines. F
example, indiscreet use of certain additives used for coloring candy a
popcorn was reported to have caused diarrhea in children resulting in t
removal of these dyes from the FDA approved list of additives. There a
a number of related ideas which must be remembered when dealing wi
food additives: (1) all foods are composed of chemical compounds, ma
of which can be extracted and added to other foods, in which case th
are classified as additives; (2) any additive or chemical compound can
injurious to health when particularly high levels of that compound a
added to foods; (3) any additive or chemical compound can be safe to u
when particularly low levels of that compound are added to foods; (4)
is necessary to evaluate each additive for its usefulness and toxicity in
sensible, scientific way, regardless of how safe its proponents say it is a
how toxic its opponents say it is.

The use of radiation for preserving foods has been declared an additiv
and whether or not it should be approved by the FDA makes it the prir
example of extreme opposition and extreme favor. Quite often, t
tendency to take a strong position for the use of an additive might ma
a proponent overlook or rationalize undesirable investigative facts cc
cerning the additive. On the other hand, opponents tend to ma
irrational demands of investigators to prove the safety of an additive; 1
example, opponents to the use of radiation for preserving foods ha

ggested that radiation should not be approved for preserving foods ntil all possible chemical effects of the process have been identified. his, without going into detail, is an impossible task. It would be just as possible to identify all the chemical effects of frying food and of baking od.

Given present capabilities, our most reasonable evaluation of an dditive for safety can be made through conventional animal feeding udies. The overall physiological effects that an additive may have on imals of two or three different species over a specified number of nerations is the most comprehensive, as well as the most reliable, way evaluate the safety of a food additive.

It should be remembered that chemical materials cannot be added to ods unless their use, in the quantities added, has been approved by the A. Moreover, additives are tested for toxicity by regulatory agencies concentrations much greater than those allowed in foods. It should o be remembered that most food additives are components of natural ods and that without these additives the quality of many foods would greatly inferior to those to which we have become accustomed. The elf-life or availability of many foods would also be greatly limited were additives to be eliminated from foods. Food additives are difficult to assify mainly because they overlap each other in numerous combina- ns of effects. It should be remembered, therefore, that the following assification is not a precise one.

NTISPOILAGE AGENTS

Although foods can be sterilized (as by heat processing) and contained such a way as to prevent contamination by microbes during storage, it ill is often necessary in some cases to forego sterilization, thus making necessary to take other steps to prevent microbial degradation of the od. Foods can be protected against microbial attack for long periods nonths to years) by holding them at temperatures below freezing (see hapter 13). They can be preserved for shorter periods (several days) by lding them in ice or in a refrigerator at temperatures in the range °–46°F (0°–7.8°C) (see Chapter 12). Foods can also be preserved by tering them to make them incapable of supporting microbial growth. rying is an example of this type of preservation. Foods must also be eserved against color and texture changes.

Quite often it is either impossible or undesirable to employ conven- nal preservation methods, and a large variety of food additives is ailable for use, alone or in combination with other additives or with mild forms of conventional processes, to preserve foods. Usually, preservatives are used in concentrations of 0.1% or less. Sodium diacetate and sodium or calcium propionate are used in breads to prevent mold growth and the development of bacteria which may produce a slimy material known as rope.   Sorbic acid and its salts may be used in bakery products, cheeses, syrups and pie fillings to prevent mold growth. Sulfur dioxide is used to prevent browning in certain dried fruits and to prevent wild yeast growth in wines used to make vinegar. Benzoic acid and sodium benzoate may be used to inhibit mold and bacterial growth in some fruit juices, oleomargarines, pickles and condiments. It should be noted that benzoic acid is a natural component of cranberries.

Salt is an excellent microbial inhibitor, mainly due to its suppression of the water activity (see Chapter 11) of the material to which it is added. Its effectiveness is enhanced when the food is also dried or smoked or both. Smoking also imparts a partial preservative effect.

Weak acids, such as sorbic acid, or salts of weak acids, benzoates, propionates, nitrites, certain chelating agents (chemicals which tie up metals and prevent the catalytic action of metals), and other chemical additives are effective preservatives. Natural spices also have antimicrobial properties. Antibiotics, relatively new antimicrobial agents, have been used as food additives and are still used to preserve animal feeds and human foods in some countries. Their use in human foods is banned in the United States and in some other countries.

Since many antimicrobial agents are generally toxic to humans, their use must be regulated not to exceed established levels beyond which they are hazardous to human health.

Nitrites, proven inhibitors of *Clostridium botulinum*, and nitrates are added to cured meats, not only to prevent botulism, but also to conserve the desirable color as well as add to the flavor of the products.

ANTIOXIDANTS

Antioxidants are food additives used, since about 1947, to stabilize foods which by their composition would otherwise undergo significant loss in quality in the presence of oxygen. Oxidative quality changes in food include the development of rancidity from the oxidation of unsaturated fats resulting in off-odors and off-flavors and discoloration from oxidation of pigments or other components of the food.

Although it would seem relatively simple to prevent oxidation of food by proper packaging and precautions during handling, the facts are: (1 that oxygen is difficult to exclude from food systems, especially since it is often closely associated with the food and (2) that only minute amounts of oxygen are sufficient to degrade the food.

There are a large number of antioxidants, and although they may function in different ways, the net effect of each is to prevent, delay or minimize the oxidation of foods to which they are added. One of the ways by which some antioxidants function involves their combination with oxygen. Others prevent oxygen from reacting with components of the food. When only a limited amount of oxygen is present, as in a hermetically-sealed container, it is possible for some antioxidants to use up all of the available free oxygen, because they have a relatively great affinity for it. Some antioxidants lose their effectiveness when they combine with oxygen; therefore, there is no advantage to using this type of antioxidant unless the food is enclosed in a system from which oxygen or air can be excluded. In the use of antioxidants, it should be kept in mind that other precautions are necessary to minimize oxidation, since heat, light and metals are prooxidants, that is, their presence favors oxidative reactions. Many of the antioxidants used in commerce occur naturally in foods (e.g., vitamin C, vitamin E, citric acid, amines and certain phenolic compounds). However, the amines and the phenolic compounds can be toxic to humans in low concentrations; therefore, they and the synthetic antioxidants require strict regulation of their use in foods. It should be pointed out that the potency of the naturally occurring antioxidants is not as great as that of the commonly used synthetic antioxidants. The antioxidants that are considered to be the most effective and therefore are most widely used are butylated hydroxyanisole (BHA), butylated hydroxytoluene (BHT) and propylgallate. These are usually used in formulations that contain combinations of two or all three of them, and often in combination with even a fourth component, very often citric acid. The main purpose in adding citric acid is for its action as a chelator (a chelator ties up metals which thereby prevents metal catalysis of oxidative reactions).

Fats and shortenings, especially those used in bakery goods, fried foods, etc., are subject to oxidation and the development of rancidity after cooking. To prevent this, chemical antioxidants in concentrations up to 0.02% of the fat component may be added.

The use of antioxidants is regulated by the Food and Drug Administration and is subject to other regulations, such as the Meat Inspection Act and the Poultry Inspection Act. Their use is limited so that the maximum amount that can be added is generally 0.02% of the fat content of the food; there are some exceptions to and variations of that rule.

NUTRIENTS

The need for a balanced and ample nutrient intake by the human body is well known. Although nutrients are available in foods, losses of fractional amounts of some of them through processing, and increasing frequencies of improper dieting, have led to the practice of adding minimum daily requirements, or sizeable fractions of minimum daily requirements, of a number of nutrients to popular foods, such as breakfast cereals, baked goods, pasta products, low-calorie breakfast drinks, etc. Nutrient additives include mainly vitamins, proteins and minerals.

Vitamin D is an exceptional example of the value of the food additive concept. The major source of vitamin D for humans lies in the existence of a precursor compound lying just under the skin which converts to the vitamin form when we are exposed to the radiant energy of the sun. However, in many cases, exposure to the sun is sporadic and insufficient especially in areas where there is normally insufficient sunshine or in cases where outdoor activities are of insufficient duration. Thus, vitamin D is added to nearly all commercial milk in a ratio of 400 U.S.P. units per 0.95 liter (1 qt). (Vitamins A, C and some of the B vitamins are added to some foods.)

It might interest the reader to know that the vitamin D used as an additive in milk is produced by exposing a precursor compound, a sterol (for example, 7-dehydrocholesterol), to radiation from a source of ultraviolet rays. This is especially interesting since the production of vitamin D within our bodies is due to the action of ultraviolet radiation from the sun.

The addition of protein concentrate (produced from fish or soybean) to components of the diet of inhabitants of underdeveloped countries has been used successfully to remedy the high incidence of protein malnutrition. It should be noted that soybean protein is incomplete and requires the addition of some amino acids in which it is deficient. Children especially, succumb in large numbers to the disease, *kwashiorkor*, that results from insufficient protein intake.

Among minerals, iron has received major attention as a food additive mainly because of its role in preventing certain anemias.

FLAVORINGS

Flavorings are compounds, many of which are natural, although there are also many synthetic ones, that are added to foods to produce flavor r to modify existing flavors. In the early days of man's existence, salt, ugar, vinegar, herbs, spices, smoke, honey, and berries were added to oods to improve their taste or to produce a special, desirable taste. The ange of natural and synthetic flavoring available to the modern food echnologist is very large. Essential oils form a major source of flavorings. Essential oils are odorous components of plants and plant materials that re the characteristic odors of the materials from which they are xtracted. Because of the large production of orange juice, quantities of ssential oil of orange are produced as by-products. For this reason, there s little need for the production of synthetic orange flavoring.

Fruit extracts have been used as flavorings, but these are relatively veak when compared to essential oils and oleoresins. An oleoresin is a olvent extract of spices from which the solvent, usually a hydrocarbon, as been removed by distillation. Because of their weak effects, fruit xtracts may be intensified by combining them with other flavorings.

Synthetic flavorings are usually less expensive and more plentiful than atural flavorings. On the other hand, natural flavorings are often more cceptable. However, they are quite complex and difficult to reproduce ynthetically. In fact, one of the problems with natural flavorings is that hey may vary according to season and other uncontrollable variables. ynthetic flavorings, however, can be reproduced quite accurately. Many rtificial flavors, such as amyl acetate (artificial banana flavor), enzaldehyde (artificial cherry flavor) and ethyl caproate (artificial ineapple flavor) are added to confectioneries, baked products, soft rinks and ice cream. These flavorings are added in concentrations of .03% or less.

FLAVOR ENHANCERS

Flavorings either impart a particular flavor to foods or modify flavors lready present. Flavor enhancers intensify flavors already present, specially when the desirable flavors are relatively weak. Monosodium lutamate (MSG) is one of the best known and most widely used flavor nhancers. This compound occurs naturally in many foods and in a ertain seaweed that was used for centuries as a flavor enhancer for soups nd other foods. It is only within the last hundred years that the reason or the effectiveness of the seaweed was discovered to be MSG. The way n which MSG enhances flavor is not yet understood. While it is effective t relatively low levels (parts per thousand), there are other compounds alled flavor potentiators which also enhance flavors but are extremely owerful, effective in parts per million and even per billion. These compounds have been identified as nucleotides, and their effect is attributed to their synergistic properties (properties which intensify the effect of natural flavor components).

ACIDULANTS

The pH of a solution is defined as the logarithm of the reciprocal of the hydrogen ion activity in the solution. In pure water, at ordinary temperatures, a small amount of the water molecules is dissociated to hydrogen ions that are positively charged and hydroxyl ions that are negatively charged.

HOH	$\rightleftarrows$	H^+	+	OH^-
Water		Hydrogen		Hydroxyl
molecule		ion		ion

It can be seen that each of the dissociated water molecules yields one hydrogen ion (H^+) and one hydroxyl ion (OH^-). In pure water, the dissociation is such that the hydrogen ion concentration is 1×10^{-7} and the hydroxyl ion concentration is also 1×10^{-7}. The product of these two values is $(1 \times 10^{-7}) \times (1 \times 10^{-7}) = 1 \times 10^{-14}$. All values are in moles per liter. In water solutions, the product of the concentration of the hydrogen ion and the concentration of the hydroxyl ion is always 1×10^{-14} (the ionization constant for water). When an acid, such as hydrochloric acid (HCl) is added to water, it results in an increase in the hydrogen ion concentration because the number of hydrochloric acid molecules that dissociate is much greater than the number of water molecules that dissociate. (The Cl^- has no direct effect on the pH.)

HCl	$\rightarrow$	H^+	+	Cl^-
Hydrochloric		Hydrogen		Chlorine
acid		ion		ion

In a solution made up of water and hydrochloric acid there will then be more hydrogen ions than there are in pure water, depending on the amount of acid added. For example, there might be 1×10^{-2} H^+ in the acid/water solution. To determine the pH of water and of the acid solution cited previously:

The pH of water = logarithm of the reciprocal of the hydrogen ion concentration

$$= \log 1/[H^+]$$
$$= \log 1/10^{-7}$$

$= \log 10^7$ (When a number from the denominator is brought to the numerator, the sign of the exponent must be changed.)

$= 7$

The pH of the acid solution $=$ logarithm of the reciprocal of the hydrogen ion concentration

$= \log 1/[H^+]$

$= \log 1/10^{-2}$

$= \log 10^2$

$= 2$

We can see from the previous equation that as the hydrogen ion concentration is increased, the pH is lowered. Because we know that $(H^+) (OH^-) = 1 \times 10^{-14}$, we can see that in the preceding equation, the OH^- concentration must be $(1 \times 10^{-14})/(1 \times 10^{-2}) = 1 \times 10^{-12}$. This means that some of the hydroxyl ions (OH^-) combined with the H^+ that were made excessive in number due to the addition of acid. On the other hand, if we add an alkaline compound, such as sodium hydroxide (NaOH) to water, there will be an increase in the concentration of OH^- in the resultant solution. The reason for this is that the number of NaOH molecules that dissociate is far greater than the number of water molecules that dissociate. (The Na^+ does not have a direct effect on

NaOH	$\rightarrow$	Na^+	$+$	OH
Sodium hydroxide		Sodium ion		Hydroxyl ion

the pH.) When the number of OH^- is excessive, some will combine with some of the H^+ to form water, thus causing a reduction in the concentration of H^+, for example to 1×10^{-10}. The pH of this solution then, is:

pH $=$ the logarithm of the reciprocal of the hydrogen ion concentration

$= \log 1/H^+$

$= \log 1/10^{-10}$

$= \log 10^{10}$

$= 10$

We can see from all this that when the pH of a solution is less than 7, the solution is acidic; when the pH of a solution is greater than 7, the solution is alkaline; and when the pH of a solution is exactly 7, it is neutral, as it is in pure water. Compounds such as water and solutions of neutral salts do not affect the hydrogen ion concentration. When acids are combined with bases they tend to neutralize each other's effect on the hydrogen ion concentration, as follows:

$$\text{HCl} \quad + \quad \text{NaOH} \quad \rightarrow \quad \text{NaCl} \quad + \quad \text{HOH}$$

HCl	NaOH	NaCl	HOH
Increases H^+ concentration (lowers pH value)	Decreases H^+ concentration (raises pH value)	Does not affect H^+ concentration	Does not affect H^+ concentration

From the root word, acid, in acidulants, one can conclude that this class of compounds tends to lower the pH of any food in which the compounds are incorporated. They also enhance desirable flavors, and in many cases, such as in pickled products, are the major taste component. Vinegar (acetic acid, CH_3COOH) is added to relishes, chili sauce, ketchup and condiments as a flavor component and to aid in the preservation of these products. Since the microbial spoilage of food is inhibited as the pH of a food is lowered, acidulants are used for that purpose in many cases. Many acidulants occur naturally in foods (e.g., citric acid in citrus fruits, malic acid in apples, acetic acid—the major component of vinegars; figs contain all three acids). Tartaric acid is widely used to lend tartness and enhance flavor. Citric acid is widely used in carbonated soft drinks. Phosphoric acid is one of the very few inorganic acids used as an acidulant in foods. It is widely used, comprising 25% of all the acidulants used in foods. Citric acid accounts for 60% of all acidulants used in foods.

In addition to their preservative and flavor enhancing effects, acidulants are used to improve gelling properties and texture. Acidulants are also used as cleaners of dairy equipment.

Acidulants may be used in the manufacture of processed cheese and cheese spreads for the purpose of emulsification as well as to provide a desirable tartness.

Acid salts may be added to soft drinks to provide a buffering action (buffers tend to prevent changes in pH) to prevent excess tartness. In some cases, acid salts are used to inhibit mold growth (e.g., calcium propionate is added to bread to prevent mold growth).

ALKALINE COMPOUNDS

Alkaline compounds are compounds that raise the pH. Alkaline compounds, such as sodium hydroxide or potassium hydroxide, may be used to neutralize excess acid which can develop in natural or cultured fermented foods. Thus, the acid in cream may be partially neutralized prior to churning in the manufacture of butter. If this were not done, the excess acid would result in the development of undesirable flavors.

Sodium carbonate and sodium bicarbonate are used to refine rendered fats. Alkaline compounds are also added to chlorinated drinking water to adjust the pH to high enough levels to control the corrosive effects of chlorine on pipes, equipment, etc. Sodium carbonate is also used in conjunction with other compounds to reduce the amount of hardness in drinking water. Sodium hydroxide is used to modify starches and in the production of caramel. Sodium bicarbonate is used as an ingredient of baking powder which is used for baked products. It is also a common household item used in a variety of cooking recipes. Alkaline compounds are used in the production of chocolate.

It is important to note that some alkaline compounds, such as sodium bicarbonate, are relatively mild and safe to use, while others, such as sodium hydroxide and potassium hydroxide, are relatively powerful reagents and should not be handled by inexperienced people.

SWEETENERS

Sweetening agents are added to a large number of foods and beverages. Table sugar (sucrose), the most commonly used sweetener in this country, and corn sugar and syrup, are covered in Chapter 23 and therefore will not be described in any detail here. Sweeteners include other sugars, as well as an abundance of natural and synthetic agents of varying strengths and caloric values.

Many sweeteners are classified as nonnutritive sweeteners. While this classification might imply a lack of nutritional value, the implication is correct only in a relative sense. That is, the caloric value of a nonnutritive sweetener, like aspartame, for example, is about 4 Cal/g as it is for sugar; however, since it takes only 1 g of aspartame to provide the same sweetness level as about 180 g of sugar (sucrose), it can be seen that the caloric contribution of aspartame is only about 0.5% that of sucrose. It is on this basis that nonnutritive sweeteners are classified.

The presently approved sweeteners are: saccharin, fructose, glycyrrhizin, xylitol, mannitol, sorbitol and thalose. A group of sweeteners that may have some probability of FDA approval in the future includes cyclamates, aspartame and neo-DHC (neohesperidin dihydrochalcone). A third group of sweeteners, having only a low probability of FDA approval, includes acetosulfam, D-6-chlorotryptophan, and a nondialyzable, water soluble extract of the tropical serendipity berry. The miracle berry (different from the serendipity berry) has been included in this last group, but there is no evidence of a sweet principle in the miracle berry.

Fructose

Of the other natural sugars used by man, fructose (also known as levulose), a monosaccharide ($C_6H_{12}O_6$), is the sweetest (nearly twice as sweet as the table sugar, sucrose); and it is the most water soluble of the sugars. It is hygroscopic, making it an excellent humectant when used in baked goods. The value of a humectant in baked goods is that it retards their dehydration. Solutions of fructose have a low viscosity which results in lower "body" feel than sucrose but in greater flexibility of use over a wide range of temperatures. Because of its greater solubility and more effective sweetness than sucrose, fructose is a better alternative to sucrose when very sweet solutions are required, as fructose will not crystallize out of solution, whereas sucrose will. Fructose has sometimes been called the fruit sugar, since it occurs in many fruits and berries. It also occurs as a major component in honey, corn syrup, cane sugar and beet sugar. In fact, sucrose, a disaccharide, is composed of glucose and fructose. Of these two components, the glucose moiety cannot be metabolized by diabetics, and it is for this reason that the ingestion of sucrose cannot be tolerated by diabetics. Fructose, on the other hand, does not require insulin for its metabolism and can, therefore, be used by diabetics with no concern. Its use also appears to reduce the incidence of dental caries. When used with saccharin, it tends to mask the bitter aftertaste of saccharin. As it apparently accelerates the metabolism of alcohol, it has been used to treat those suffering from overdoses of alcohol. It has been recommended as a rapid source of energy for athletes and, in combination with gluconate and saccharin, as an economic effective, safe, low-calorie sweetener for beverages. Despite the many advantages cited, the use of fructose has been limited, mainly because of its relatively high cost of production.

Molasses

Molasses can be considered a by-product of sugar production (see Chapter 23). The use of molasses as a sweetener in human foods is largely in baked goods that include bread, cookies and cakes. In addition to sweetening, molasses adds flavor and acts as a humectant. It is also used in baked beans and in the production of rum and molasses alcohol. (The greatest use of molasses, however, is in the production of animal feeds.) Molasses comprises about 60% sucrose, but the sucrose content can be lower, depending on the grade of the molasses and on the raw material from which it was produced. Thus, the sucrose content of cane blackstrap the final fraction of cane molasses) is only about 1/2 that of beet blackstrap (the final fraction of beet molasses). The fractions produced before the blackstrap are of higher grades and are those usually used for human consumption. Blackstrap generally is used for industrial purposes.

Honey

Honey, a natural viscous syrup, comprises mainly invert sugar. It is produced from the nectars of flowers, which is mainly sucrose, by the action of an invertase enzyme that is secreted by the honey bee. Honey is used as a direct sweetener, as an additive in a number of products, including baked goods, as well as other ways. It is relatively expensive.

Invert sugar, corn sugar and corn syrup are covered in Chapter 23 and will not be covered here.

Maple Sugar

Maple sugar is produced from the sap of the sugar maple tree. It comprises mainly sucrose and small amounts of other sugars, including invert sugar. Maple sugar is used in the manufacture of candies, fudge, baked goods and toppings. It is among the most expensive of sweeteners.

Lactose

Lactose ($C_{12}H_{22}O_{11}$), the sugar component of mammalian milks, is less sweet and less water-soluble than sucrose. While babies and young children generally are able to metabolize this sugar, some are unable to do so. The ability to metabolize the sugar appears to decrease with age. When a person is unable to metabolize lactose, the ingestion of milk may cause intestinal discomfort, cramps and diarrhea. The major source of lactose is whey, a cheese by-product. Because lactose is not as sweet as sucrose, larger amounts can be used in those foods whose texture benefits from a high solids content.

Maltose

Maltose ($C_{12}H_{22}O_{11}$) or malt sugar is produced during the malting process in brewing (enzyme conversion of starch). It is converted to alcohol by the action of yeasts through an intermediate conversion to dextrose. This sugar is much less sweet than sucrose, and it is used mainly in the manufacture of baked goods and infant foods.

Saccharin

Saccharin, the imide of o-benzosulfonic acid, is among the most important of the nonnutritive sweeteners used today, even though its safety is, to some measure, in doubt. It is used either as the sodium or calcium salt. Being about 300 times sweeter than sucrose (table sugar) it is a far more efficient sweetener than sugar. However, in cases where the texture of a product is dependent on the solids content, the use of saccharin is undesirable. In nearly every case, the organoleptic properties of candy, beverages and other sweetened foods are better in the sugar sweetened products than in the products sweetened with saccharin. In fact, saccharin has a slightly adverse effect on the flavor, as well as on the texture—it imparts a bitter taste, as well as reducing the desirable texture characteristic sometimes described as "body." Used in carbonated beverages, saccharin is believed to be responsible for the earlier release of the dissolved carbon dioxide than occurs when sugar is used as the sweetener. There is some evidence, and theory would support it, that microbial growth is higher in saccharin-sweetened foods because of the lower solids contents of such foods. On the other hand, the use of saccharin reduces the incidence of dental caries and it can be used by diabetics. It also lowers the overall caloric intake and is, accordingly, used by persons on weight reduction regimens.

The controversy surrounding the safety of saccharin is based on the development of bladder cancers in test animals used in the wholesomeness testing of this sweetener. However, it should be remembered that the test animals that developed the tumors were given extraordinarily high amounts of the sweetener (5% of the total diet). In tests where the amount was lower, there was no unusual incidence of bladder cancer. While the human intake of saccharin is substantially less than that fed to the test animals, and evidence presented at the National Academy of Sciences meeting, in March of 1975, was supportive of the continued use of saccharin, it remains a subject of inquiry. Cyclamate and aspartame were considered to be equally safe.

Cyclamate

Cyclamate had been used as a nonnutritive sweetener prior to its removal from the FDA approved list of additives. In many cases, it was used together with saccharin. Cyclamate is not as sweet as saccharin by weight, but it is about 30 times sweeter than sucrose. Like saccharin, it has been used as the sodium or calcium salt.

Aspartame

Aspartame is the common name for aspartyl-phenylalanine. It is a combination of the two amino acids from which its name is derived. First produced in 1969, it is reputed to be about 180 times sweeter than sucrose. Like cyclamate, it was approved and later banned by the FDA, even though exhaustive evidence of its safety has been presented by animal testing and by definition of its metabolic fate in both animals and humans.

Unlike saccharin and cyclamate, aspartame leaves no bitter aftertaste. It is quite expensive, about 200 times more so than sucrose, but as it is about 180 times sweeter than sucrose, its cost for obtaining a given unit of sweetness is not much more costly.

Xylitol

Xylitol is a polyhydric alcohol having the formula $C_5H_7(OH)_5$. It is presently used in chewing gum, mainly because of its noncariogenic property (it has not been found to cause tooth decay). It occurs naturally as a constituent of many fruits and vegetables, and is a normal intermediary product of carbohydrate metabolism in man and in animals. Commercially, it is produced by the hydrolysis of xylan (which is present in many plants) to xylose, which is then hydrogenated to produce xylitol. The xylitol is then purified and crystallized. Xylitol imparts a sweet taste, which also appears to have a cooling effect. As it is not metabolized by many microorganisms, it is quite stable.

Like cyclamate and saccharin, xylitol has been found to be carcinogenic in preliminary animal tests, and should additional tests corroborate these findings, its continued use as an additive would be in doubt. However, should xylitol be considered hazardous and discontinued as an additive, the fact that each person is reported to produce within the body about 10 g of the substance daily would somehow have to be reconciled with such a decision.

Sorbitol

Sorbitol is a polyhydric alcohol ($C_6H_8(OH)_6$) that is found in red seaweed and in fruits (apples, cherries, peaches, pears and prunes). It was first isolated from the sorb berries of the mountain ash; hence, its name. It is used as an additive because of its humectant property as well as its sweetening effect. It is used in cough syrup, mouthwashes and toothpaste. Another of its desirable properties is that it is not easily fermented by microorganisms. Because sorbitol is largely transformed to fructose by liver enzymes in the body, it is tolerated by diabetics, since fructose is not dependent on the availability of insulin for its metabolism. Sorbitol can be produced industrially by the electro chemical reduction or the catalytic hydrogenation of glucose.

Mannitol

Mannitol is a polyhydric alcohol having the formula $C_6H_8(OH)_6$. It is used in chewing gum, pharmaceuticals and in some foods. It is a naturally occurring sweetener in many plants, algae and molds. It occurs in the sap of the manna tree, an ash native to southern Italy, and can also be made by the reduction of either of the monosaccharides, mannose or galactose. Industrially, it is produced by electrochemical reduction or catalytic hydrogenation methods.While it is similar to sorbitol in many respects, it is less soluble than sorbitol.

Serendipity Berry

The serendipity berry contains the most intense natural sweetener known. It is the fruit of the plant *Dioscoreophyllum cumminsii,* in digenous to Africa. While the berry is too sweet to have found use among the natives (it is about 1000 times sweeter than sucrose), parts of the plant are eaten. The water soluble sweet principle occurs in the pulp of the small fruit (about 1/2 in. or slightly more than 1 cm in diameter) There is no known use of this naturally occurring sweetener at present Its chemical structure has not been identified as yet. It appears to be associated with the fruit's protein fraction, but it is believed to be other than proteinaceous.

Miracle Fruit

Miracle fruit, a berry from the plant *Synsepalum dulcificum,* is also indigenous to Africa. While the sweetness value of this berry is in question because it does not always appear to impart sweetness, it has been studied for its sweetness value. In our own experiment with the extract of miracle fruit, performed by placing the extract in the mouth, we found that lemons tasted sweet and not at all acidic but that the extract did not sweeten coffee or other unsweetened foods. We also found that sweetening lemons required about 1/2 hr before taking effect and that the effect lasted for only about 3 hr. Thus, adding this agent to lemon juice will not make the juice sweet. The juice will only taste sweet bout 1/2 hr after the mouth has been coated with miracle fruit extract. t has been proposed that the effect of this berry's extract is not to weeten but to numb the ability of the taste buds to detect acidity, llowing the sugar in the lemon to come through.

Dihydrochalcones

Dihydrochalcones are intensely sweet compounds obtained by the ydrogenation of chalcones found in naringin and neohesperidin, two lavanones occurring naturally in grapefruit and oranges. The sweetness evels of the dihydrochalcones vary, but the average is estimated to be bout 1000 times sweeter than sucrose. This class of compounds imparts cooling effect as well as a lingering sweetness effect, desirable charac- eristics for use in chewing gum. However, there may be a time lag before he sweetness of these compounds is felt; therefore, saccharin is used in ombination with them to provide the immediate sweet taste.

Other Sweeteners

A relatively new sweetener called SRI oxime V is prepared from erillartine, an aldoxime. It is reported to be over 400 times sweeter than ucrose and to have no adverse aftertaste, as does saccharin. It would ppear to have broad application as a sweetener, although there are no eports of any use thus far.

Sucaryl is a nonnutritive sweetener produced as a salt of sodium or alcium (the latter for the benefit of those who must exclude sodium rom their diets).

Glycyrrhizin, a natural sweetener, is 50 times sweeter than sucrose. It s a flavor potentiator on the GRAS list since 1973. A triterpenoid glycoside from the licorice root, it has a licorice flavor.

Thalose is not a sweetener, but it enhances the sweetness of sucrose. Therefore, with its addition, the amount of sucrose required to arrive at he desired level of sweetness can be lowered by about 10%.

Acetosulfam is a synthetic sweetener that tastes similar to saccharin ut has only 1/2 the sweetening effect of saccharin.

D-6-Chlorotryptophan is reported to be about 1300 times sweeter than ucrose. It has no aftertaste or toxicity.

Stevioside is a naturally occurring sweetener having a sweetening ffect 300 times greater than sucrose.

Dulcin is several hundred times sweeter than sucrose, but it may never e used because of its toxicity. It is produced by heating p-phenetidine vith urea or by reacting p-phenetidine hydrochloride with potassium yanate.

STARCHES

Although starches differ from each other somewhat, depending on the plant from which they are extracted, they are sufficiently similar chemically to be often classified together as starch. The two basic starch polymers are amylose and amylopectin. Starch is used as a source of carbohydrate, and because it is relatively inexpensive, is often used as an extender. Its properties also make it useful as a thickening agent. The major source of starch is corn, but some starch is also produced from sorghum, potatoes and wheat.

GUMS

Gums, a class of complex polysaccharides, are defined as materials that are dispersible in water and capable of making the water viscous. Many gums occur naturally in certain land and sea plants. Examples are gum arabic and agar. Many gums, such as the cellulose derivatives, are modified or semi-synthetic, and some gums, such as the vinyl polymers, are synthetic. Gums are used to stabilize ice cream and desserts, thicken certain beverages and preserves, stabilize foam in beer, emulsify salad dressings and form protective coatings for meat, fish and other products. Gums add "body" and prevent settling of suspended particles in chocolate milk, ice cream and desserts. They may also prevent the formation of large ice crystals in frozen desserts.

ENZYMES

Enzymes occur naturally in foods, and their presence may be beneficial or detrimental, depending on the particular enzyme (see Chapter 8 for information on enzymes). When the presence of enzymes is undesirable, steps are taken to inactivate them. When their presence is desirable, either the enzymes or sources of them are intentionally added to foods. Thus, the enzyme papain (from the papaya fruit) is added to steak to tenderize it. Many of the useful enzymes used in food processing are produced by microbes; consequently those microbes producing the desired enzyme(s) may be added intentionally to food materials. For example, specific yeasts are intentionally added in the production of bread, beer or cheese.

The use of enzymes as food additives presents no problem from the standpoint of safety, since enzymes occur naturally, are nontoxic and are easily inactivated when desired reactions are completed. Enzymes called amylases are used together with acids to hydrolyze starch in the production of syrups, sugars and other products.

Invertase

Certain enzymes, such as invertase, split disaccharides, such as sucrose (table sugar), to lower sugars (glucose and levulose). Invertase has many applications, and is used, for example, to prevent crystallization of the sucrose that is used in large amounts in the production of liqueurs. Without invertase, the liqueurs would appear cloudy.

Pectinase

Pectinases are enzymes that split pectin, a polysaccharide that occurs naturally in plant tissues, especially those of fruit. Pectin holds dispersed particles in suspension, as in tomato juice. Since it is desirable to keep the thick suspension in tomato juice, the pectinases that occur naturally in it are inactivated by heat. On the other hand, products such as apple juice are customarily clear, and this is accomplished by adding commercial pectinase to the product which degrades the pectin in the apple juice resulting in the settling out of the suspended particles which are then separated from the clear juice. In the manufacture of clear jellies from fruits, it is first necessary to add pectinase to destroy the naturally occurring pectin in order to clarify the juice. Once the juice has been clarified, pectin has to be added to produce the thick consistency of jelly. The pectinase that was added in the first place must be inactivated as part of the process, otherwise the enzyme would also break down the pectin that is added to produce the thick consistency.

Cellulase

Cellulases are enzymes that can break down cellulose, said to be the most abundant form of carbohydrate in nature. Cellulose, the principal structural material in plants, is insoluble in water and is indigestible by humans and many animals. Ruminants are able to digest cellulose because of a cellulase (produced by microorganisms in the large stomach) contained in their gastric juice. Commercial applications of cellulases are not widespread at present. Cellulases are used for tenderizing fibrous vegetables and other indigestible plant materials for the production of foods or animal feeds. They have other minor uses, as well.

Protease

Proteases are enzymes that break down proteins, polypeptides an peptides. Peptides are the structural units of which polypeptides consist and polypeptides are larger structural units that make up the protein There is a large variety of specific proteases, and each of them attack protein molecules at different sites, producing a variety of end products Proteases are used to produce soy sauce from roasted soybeans, chees from milk and bread dough from flour. They are also used to chill-proc beer (untreated beer develops an undesirable haze when chilled) and t tenderize meats. Proteases are widely used in the food industry, but the are reported to have even wider use in nonfood applications.

Lipase

Lipases, the lipid (fat or oil) splitting enzymes, have limited com mercial application, with oral lipases having the widest. Lipases pre pared from oral glands of lambs and calves are used in a controlled wa in the production of certain cheeses and other dairy products, as well a lipase treated butter fat which is used in the manufacture of candies confections and baked products. Lipases are also used to remove fa residuals from egg whites and in drain cleaner preparations.

Glucose Oxidase

Glucose oxidase is an enzyme that specifically catalyzes the oxidatior of glucose to gluconic acid. This reaction is important in preventing nonenzymatic browning, since glucose is a reactant in the undesirable browning reaction. The most important application of this enzyme is ir the treatment of egg products, especially egg whites, prior to drying. Eggs treated with this enzyme before they are dried do not undergo non enzymatic browning during storage, since the sugar has been removed. Ir some cases, the enzyme is added to remove traces of oxygen to preven oxidative degradation of quality. Examples of this type of application are bottled and canned beverages (especially beer and citrus drinks) and mayonnaise.

Catalase

Catalases are used to break down hydrogen peroxide to water and oxygen. Therefore, catalases are used when the presence of hydrogen peroxide is undesirable or when hydrogen peroxide is used for specific purposes, such as in bleaching, but then must be removed from the stem. Examples of the latter case are the uses of hydrogen peroxide for reserving milk in areas where heat pasteurization and refrigeration are navailable and in the manufacture of cheese from unpasteurized milk.

EQUESTRANTS

The role of sequestrants is to combine with metals, forming complexes ith them and making them unavailable for other reactions.

$$M + S \rightarrow MS$$

here:

$$M = metal$$
$$S = sequestrant$$
$$MS = complex$$

Sequestrants, as many other additives used for enhancing specific roperties of foods, occur naturally in foods. Many sequestrants have ther properties; for example, citric, malic and tartaric acids are cidulants but they also have sequestering properties.

Since metals catalyze oxidative reactions, sequestrants can be considered to have antioxidant properties. Thus, they stabilize foods against xidative rancidity and oxidative discoloration. One of the important ses of sequestrants as additives is to protect vitamins, since these mportant nutrients are especially unstable to metal catalyzed oxidation. equestrants are used to stabilize the color of many of the canned roducts and they help stabilize antioxidants. Sequestrants are especial-y helpful in stabilizing the color and the lipids in canned fish and hellfish. Since fish and shellfish naturally contain relatively high oncentrations of metal, these products normally have poor color stabil-ty, and the lipids tend to rancidify during storage.

Sequestrants are also used to stabilize the flavors and odors in dairy roducts and the color in meat products.

POLYHYDRIC ALCOHOLS

Many polyhydric alcohols (also called polyols) are used to improve exture and moisture retention because of their affinity for water. Many olyols are present in foods naturally, with glycerine being the predomi-ant one. However, only four of the many polyols are allowed as food dditives. They are glycerine, sorbitol, mannitol and propylene glycol. ll but the last have a moderately sweet taste (see section on sweeteners), though none are as sweet as sugar. Propylene glycol has a somewhat undesirable bitter taste, but is not unacceptable in small amounts. Sorbitol imparts a cooling sensation. Glycerine, on the other hand, imparts a heating sensation.

Polyols are used in the production of dietetic beverages, candy, gum and ice cream to contribute to texture as well as to sweetness. These compounds have a less adverse effect on teeth than sugar, due to the fact that they are not fermented as quickly as sugar and are usually washed away before they can be utilized by microorganisms.

SURFACE ACTIVE AGENTS

Surface active agents affect the physical force at the interface of surfaces. Commonly called surfactants, they are present in all natural foods, since by their nature they play a role in the growth process of plants and animals. They are defined as organic compounds that affect surface activities of certain materials. They act as wetting agents, lubricants, dispersing agents, detergents, emulsifiers, solubilizers, etc. One use for wetting agents is to reduce the surface tension of materials to permit absorption of water by the material. An example of their use is chocolate mixes used in the home to prepare chocolate milk by adding water.

Dispersions of materials depend on the reduction of interfacial energy, and this can be accomplished by certain surfactants.

Surfactants are used in the production of foods to prevent sticking, such as in untreated peanut butter. Surfactants are also used in cleaning detergents used on food equipment, and they can stabilize or break down foams.

Emulsifiers, such as lecithin, mono- and diglycerides and wetting agents, such as a class of chemicals known as "tweens," may be added to bakery products (to improve volume and texture of the finished products and the working properties of the dough and to prevent staling of the crumb), cake mixes, ice cream and frozen desserts (to improve whipping properties). Except for the tweens, the chemicals cited above are natural components of certain foods.

COLORANTS

We are accustomed to specific colors in certain foods, and colors often provide a clue to the quality of the foods. Experiments have been conducted to show that we avoid foods that are not colored as we expect them to be. Thus, we would suspect something were amiss if we were given orange juice that had a brown color. Blue colored orange juice would arouse even more suspicion. Consequently, in the production of orange-flavored soft drinks, candies, desserts, etc., it is usually necessary to add orange color (either natural or artificial) to lend credibility to the orange flavor.

Many colorants (compounds that add color to foods) are natural, and these include the yellow from the annatto seed, green from chlorophyll, orange from carotene, brown from burnt sugar and red from beets, tomatoes and the cochineal insect.

Some colorants, however, are derived from synthetic dyes. The synthetic dyes in use have been approved by the FDA.

Some compounds are used to produce a white color. Thus, oxidizing agents, including benzoyl peroxide, chlorine dioxide, nitrosyl chloride and chlorine are used to whiten wheat flour, which is pale yellow in color, at the end of its production cycle. Titanium dioxide may be added to some foods, such as artificial cream or coffee whiteners.

LEAVENING AGENTS

Leavening agents are used to enhance the rising of dough in the manufacture of baked products. Inorganic salts, especially ammonium and phosphate salts, favor the growth of yeasts which produce the carbon dioxide gas that causes dough to rise. Chemical leavening agents which react to form carbon dioxide are also used in baked goods. When either sodium bicarbonate, ammonium carbonate or ammonium bicarbonate is reacted with either potassium acid tartrate, sodium aluminum tartrate, sodium aluminum phosphate or tartaric acid, carbon dioxide is produced. Baking powder is a common household leavening agent that contains a mixture of chemical compounds that react to form carbon dioxide, producing the leavening effect.

6

Food-borne Diseases

Almost all the factors that contribute to food disease result from ignorance of proper food-handling procedures or from the unwillingness of some individuals in food industries (including food processors) to comply with the basic guidelines for proper food handling. Thus, food-borne diseases will continue to occur at unnecessarily high rates as long as: (1) food handlers do not employ strict sanitation in both their personal habits and in the maintenance of their work area and equipment, (2) foods are not properly refrigerated, (3) foods are not adequately processed, (4) cross-contamination situations are not avoided and, (5) management does not realize the importance of preventing food-borne diseases. Proper food-handling procedures include rather simple techniques, such as holding at specified temperatures, but they also include complex procedures, such as those for calculating processing times and predicting certain biochemical reactions that might result from processing modifications. Ordinarily, therefore, at some level in the food-processing stage, the services of a professional food technologist are required. Once a food has been properly processed, the remaining handling it undergoes does not require a professional food technologist, but it does require periodic quality control checks by personnel qualified in microbiology and sanitation.

Food-borne diseases are mostly caused by several species of bacteria, although viruses, parasites, amoebas and other biological, as well as chemical, agents may be responsible.

As has been previously stated, the bacteria in foods sometimes have good effects, but sometimes are undesirable in that they may result in the deterioration of foods or may cause disease in humans and other animals. The disease-causing bacteria in foods, or their end product which cause disease, are transmitted through the eating of foods. Food borne diseases are classified as food infections and food intoxications.

FOOD INFECTIONS

Food infections are those in which the disease organism is carried through foods to the host (human or animal) where it actually invades the tissues and grows to numbers which cause disease.

Salmonellosis

Salmonellosis is caused when foods contaminated with *Salmonella* bacteria are eaten. At the present time, approximately 20,000 cases of salmonellosis are reported yearly. About 75 deaths annually are due to this disease. However, it is considered that only about 1% of acute digestive illnesses are reported in this country, so that actually there may be many more cases of this disease.

Typhoid fever, of which there are fewer cases than of salmonellosis, hence fewer deaths, is caused by an organism belonging to the *Salmonella* species, but this disease is usually not considered to be the ordinary salmonellosis for three reasons: (1) the ordinary *Salmonella* organisms will infect other animals as well as man, while the typhoid germ is known to infect only man; (2) typhoid fever is usually more severe than the ordinary salmonellosis; and (3) in healthy adults, several hundred thousand to several million ordinary *Salmonella* bacteria (cells) must be ingested (eaten) to cause salmonellosis, while the ingestion of even one typhoid cell may cause typhoid fever.

It is known that antibiotics, especially chloramphenicol or some of the modified penicillins, are effective in treating salmonellosis, but since this disease is oftentimes not called to the attention of a physician, it may cause a more severe illness than if it were treated.

The ordinary symptoms of salmonellosis are: abdominal pain, diarrhea, chills, frequent vomiting and prostration. However, there are instances in which much more severe symptoms may be encountered. The incubation period (time after ingesting the organisms until symptoms are evident) is 7–72 hr. In typhoid fever, the incubation period is 7–14 days.

Persons with salmonellosis often become carriers of the organism for a period of time after they have recovered from the disease. That is, they continue to discharge the organisms in their faeces. Because of this, carriers often contaminate their hands with these organisms which may not be removed completely even after thorough washing. Hence, if carriers handle foods which are to be eaten by others, they may contaminate them with these bacteria, and in this manner, transmit the disease to others. In most cases, the carrier stage does not persist longer than 12 weeks after symptoms with salmonellosis, and for shorter periods with typhoid fever. However, there are isolated cases in which the carrier stage lasts much longer than 12 weeks, and 2–5% of those ill with typhoid fever may become permanent carriers.

The *Salmonella* bacteria are rod-shaped; they do not form spores, and thus are not especially heat-resistant. They are motile (can move about in the water, in foods or other materials in which they are found) and will grow either with or without air (oxygen). At the present time, more than 1400 types of *Salmonella* bacteria are known, all of which are considered to be infective to man. Obviously, these organisms are very widespread.

Whereas it is considered that many *Salmonella* bacteria must be taken in to cause the disease in a normal adult, it is known that the very old and especially the very young may contract the disease after ingesting (eating) only a few of these organisms. Therefore, any food, especially a food which can be eaten without cooking, should be kept essentially free of these bacteria.

The *Salmonella* bacteria grow at temperatures near 95°F (35.6°C), but they will also grow (more slowly) at both higher and lower temperatures than this. It has also been found that many foods are suitable for the growth of these organisms and that in some foods they will grow slowly at temperatures as low as 44°F (6.7°C) or as high as 114°F (45.5°C). Moreover, since the destruction of bacteria by heat is a matter which involves both time and temperature as well as the degree to which a food protects bacteria, temperatures as high as 140°F (60°C) may be required to bring about marked decreases in these bacteria during the cooking of some foods.

Regarding the destruction of bacteria by heat, it should be explained that as they are heated and a temperature is reached at which they are destroyed, they are not all destroyed at once. For instance, if at 120°F (48.9°C), 90% of the organisms would be destroyed in a period of 5 min, it would take 10 min to kill 99% of the organisms, 15 min to kill 99.9% of the organisms, and so on. It should be noted that if some of these *Salmonella* organisms in foods survive whatever heating they receive during cooking, and the food is thereafter held at temperatures at which they will grow (44°–110°F [6.7°–43.3°C]), especially at room temperatures, the organisms may grow again to large numbers.

Some types of cooking are not sufficient to destroy all *Salmonella* bacteria which may be present in foods. Examples of cooked foods in which these organisms may survive are: scrambled, boiled or fried eggs, meringue, turkey stuffing, oysters in oyster stew, steamed clams and some meat dishes. Foods which are eaten raw or without further cooking such as clams, oysters, milk powder, cooked crabmeat, smoked fish, etc. may be infective should they be contaminated with *Salmonella* bacteria. These foods should be kept free from *Salmonella* bacteria, especially since they may be eaten by young people who are quite subject to infections of this kind.

Shellfish, egg products, prepared salads, and to some extent, raw and cooked meats have often been associated with the transmission of salmonellosis. Raw shellfish may be taken from waters which are contaminated with *Salmonella* bacteria and cooked shellfish meats may be contaminated by humans since they are usually removed from their shell by hand. Poultry of all types, beef cattle and hogs may have salmonellosis or be carriers of the organism causing the disease; hence under conditions of cooking in which these organisms are not destroyed they may be transmitted to humans. Since poultry may have salmonellosis, the shell of eggs may contain the causative organism which, under some methods of handling, can be transferred to the contents within the shell.

Pets, such as cats and dogs can have salmonellosis and be carriers of the causative organism. Since this is the case, children, especially the very young, who handle materials contaminated by pets and who are very subject to contracting the disease, may contract salmonellosis from pets.

It may be wondered why animals should have salmonellosis. The reason seems to be that their feeds, especially fish meal, meat meal and bone meal, fed to them as supplements, often contain *Salmonella* bacteria. This is also true of some of the dried types of food used for feeding pets.

Methods or procedures which would help to eliminate salmonellosis or greatly decrease the number of cases are listed in the following paragraphs.

(1) Good sanitation methods and procedures in food manufacturing plants, in restaurants and institutions serving foods and in the home are essential. This includes not only the cleaning and sanitizing of equipment and utensils, the elimination of insects and rodents and the cleaning and sanitizing of floors, walls, etc. but also the personal cleanliness of workers or those preparing and

serving foods. All personnel should wash and sanitize their hands prior to handling foods once having left their working stations for any reason.

(2) All foods should be held at temperatures of 40°F (4.4°C) or below when not being cooked, prepared for cooking or being served. This would not eliminate *Salmonella* organisms from foods, but it would prevent their growth in the foods so that their number would be too small to cause salmonellosis in healthy adults.

(3) Foods which can be eaten without further cooking should be produced under the best conditions of sanitation, and some foods of this type, such as milk powder, should be subjected to frequent bacteriological examination to determine that they are essentially free of *Salmonella* bacteria.

(4) Where possible, foods (poultry stuffings, etc.) should be cooked to temperatures (at least 150°F [65.6°C]) at which it can be assumed that all *Salmonella* bacteria have been destroyed.

(5) Egg products (dried or frozen) should be pasteurized (heated to 140°F [60°C] for 3–4 min), then cooled prior to drying or freezing.

(6) Flocks of poultry known to have salmonellosis (this can be determined by testing) should be eliminated as egg producers. This can be done without economic loss since such poultry may be used as food.

(7) The food, especially protein supplements, given to pets and other animals, should be treated to eliminate the *Salmonella* bacteria which infect these animals. This can be done, for instance, by pelletizing the food or supplements, a treatment which raises the temperature sufficiently to destroy many of the bacteria which might be present.

(8) Animals which have died from disease should not be eaten. (This rule is sometimes not observed on farms.)

Shigellosis

Shigellosis, sometimes called bacillary dysentery, is a food infection caused by bacteria of the genus (group) known as *Shigella*. Each year, there are almost as many cases of shigellosis reported as there are cases of salmonellosis. There is also a higher mortality rate with shigellosis than with salmonellosis.

The ordinary symptoms of shigellosis are diarrhea with bloody stools (faeces), abdominal cramps and some fever. In severe cases, the symp-

toms are much more complex. There are only about 10 known species o
types of *Shigella* bacteria as compared with more than 1400 fo
Salmonella. However, one of these organisms, *Shigella dysenteriae*
usually causes a much more severe disease than either the ordinar
Salmonella or *Shigella* bacteria. The incubation period (time afte
ingesting the bacteria before symptoms are evident) is said to be as lon
as 7 days with an average of 4 days.

The *Shigella* bacteria are nonmotile (do not move about in th
solutions in which they live). They are rod-shaped cells which will grov
in both the presence and absence of oxygen. They do not form spores.

As in the case of salmonellosis, people may become carriers of th
Shigella organism after they have had the disease, hence may become
source of contamination or infection to others who eat foods which the
have handled. However, the carrier stage with shigellosis is shorter tha
is the case with salmonellosis.

Shigellosis is transmitted chiefly through water or milk, but th
disease has also been transmitted through the eating of soft, moist food
such as potato salad. At the present time, it is believed that when food
become the source of the organism in *Shigella* infections they have bee
contaminated, directly or indirectly, with small amounts of huma
faecal discharges.

Much remains to be learned about shigellosis. It is not known
animals other than man may have shigellosis and if they may be a sourc
of infection. The relative numbers of organisms required to cause th
disease in normal adults are not known. No investigations have bee
conducted to determine whether or not the *Shigella* organisms will gro
in foods, although it may be assumed that some foods would be suitabl
for the growth of these bacteria. Finally, we do not know the minimu
temperature at which *Shigella* bacteria will grow, the temperatures i
the higher ranges required to inhibit growth or the temperature at whic
these organisms start to die off.

Regardless of our lack of knowledge about shigellosis and the o
ganisms which cause the disease, certain control methods may be note
based on the knowledge of how food and drink should be handled.

(1) Since water is a known source of the organism causing shigellosi
all water used for drinking, for adding to foods or for the cleanir
and sanitizing of equipment and eating utensils should be potab
(drinkable). The potability of water is determined by sanitar
surveys and by bacterial tests made on samples of the water. Ar
food-manufacturing or food-serving operation should use on
water known to be potable. Municipal water supplies should

checked periodically to determine that they are not polluted (contaminated with human or animal discharges). In some cases, water from deep wells from sources not connected with the municipal water supply is used for foods and cleaning. Also, sea water is sometimes used for cleanup in food serving or food-manufacturing operations. In such instances, management should arrange for bacterial tests to be made on such water supplies at frequent intervals to determine they they are not polluted.

(2) Since it is good practice in food handling to refrigerate foods not in use, another method of controlling shigellosis is to hold foods at 40°F (4.4°C) or below at all times when they are not being served or prepared for serving.

(3) It is known that humans may become *Shigella* carriers; therefore, in food-handling operations such as food manufacturing, the preparing of food for serving or the serving of food, personnel known to have had an intestinal disease should be excluded from any tasks which would bring them into direct contact with any food.

Vibriosis

Vibriosis is a disease caused by a bacterium known as *Vibrio para-hemolyticus.* This disease has only become known in recent years and was first identified in Japan where it has caused outbreaks involving the infection of many people. Whereas in recent years, a few cases of food infection have been attributed to this disease, the frequency of such outbreaks is not known. It may be that in the past some or many of the food-borne disease outbreaks, in which it was not possible to determine the cause, were due to this organism. Since the disease and the organism have only been recently defined, there was formerly no way of recognizing the disease or of testing for it.

The symptoms of vibriosis are abdominal pain, nausea, vomiting with diarrhea and occasional blood and mucus in the stools (faeces). A fever involving a 1°-2°F (0.55°-1.1°C) rise in temperature is experienced in 60-70% of the cases. The period of incubation after ingestion (eating) of the organism is 15-17 hr and the symptoms last for 1-2 days. Whether or not people who have had this disease become carriers is not known.

The organism causing vibriosis is a short, curved, rod-shaped, motile cell. It grows with or without oxygen and is believed to require 2-4% sodium chloride for growth. The organism is found naturally in the ocean, and, since it grows fastest at 86°-104°F (30°-40°C), is found in highest concentrations near the shoreline during the summer months.

Raw fish and molluscs (squid, octopus) have been the foods most ofte: involved in the transmission of vibriosis, but it is known to be presen sometimes in shellfish, such as clams and oysters, and has also bee: found in cooked crab meat. The extent to which the latter foods ma have transmitted the organism and caused the disease is not known.

The control of vibriosis would appear to involve the following precau tions:

(1) Since the *Vibrio* is found in high concentrations in seawater onl during the period in which coastal waters are quite warm, it woul appear to be preferable not to eat raw molluscs (squid, octopu: clams and oysters) during the months of July, August and Septem ber in temperate climates or in whatever months the coast; waters of a particular region are warmest.

(2) Since *Vibrio parahaemolyticus* is quite heat sensitive and would b destroyed by whatever cooking is necessary to remove crab mea from the shell, the presence of the organism in cooked crab mea must have been due to contamination after cooking. This indicate that good food plant sanitation is a method of controlling vibriosi: It should be noted that some plants which process marine foods us seawater for the cleaning of equipment, floors, etc. This should n be allowed; the use of potable fresh water which has been ade quately chlorinated for the cleaning and sanitizing of food plan and food-plant equipment is a method of controlling vibriosis.

Cholera

Cholera is rarely encountered in the United States, the only cas occurring in people coming from certain other countries in which th contracted the disease prior to entering this country. In the Near and F East, cholera causes, from time to time, much sickness and death.

The symptoms of cholera include diarrhea with an abundance watery stools, vomiting and prostration. Eventually, since the patient unable to retain water taken by mouth, dehydration becomes a wea ening factor in the disease. Due to the rundown condition of the patier secondary (other types) infection may set in.

The organism causing cholera is *Vibrio comma*. It is a short curved r which is motile. This bacterium is aerobic (requires oxygen to grow).

Cholera is ordinarily transmitted through drinking water, but it can ! transmitted through contaminated foods that have been washed polluted water or handled by persons having the disease.

The control of cholera appears to require mainly pure water for drinking, adding to foods and cleaning and sanitizing utensils and equipment used for the manufacture, preparation and serving of foods.

Trichinosis

Trichinosis is not a bacterial disease. It is caused by the microscopic roundworm, *Trichinella spiralis.* Approximately 100 cases of the disease are reported in this country each year, with an indeterminate number of unreported cases. There are also a few deaths attributed to this disease each year. Trichinosis is transmitted through the eating of pork, although some wild game, such as bear, has also been known to cause trichinosis. Many of the reported cases are due to pork processed by farmers or local butchers. However, some of the products of larger packers have also transmitted this disease.

The symptoms of trichinosis vary with the number of organisms ingested. If large numbers of larvae (stage of the life cycle of the worm) are eaten, nausea, vomiting and diarrhea develop 1–4 days after the food is eaten. If only a few organisms are eaten, symptoms may be absent. On the seventh day after they are taken in, the larvae (produced by the adult worms) migrate from the intestines to the muscles, causing an intermittent fever as high as 104°F (40°C) that can last for a few weeks. The upper eyelids may swell due to the accumulation of fluids. Once the larvae have located in areas between the muscle fibers, a cyst is formed which becomes calcified. In this state, the larvae remain dormant in the body over a period of years.

The trichina organisms in meat are present as the larval stage of a roundworm, and in the human intestine, develop into adult roundworms. The adult worms unite sexually and the females produce larvae which migrate from the intestine to the muscle tissues.

The following paragraphs describe a number of ways in which trichinosis can be controlled or prevented.

(1) Since the trichina larvae are destroyed by a temperature of 137°F (58.3°C), fresh pork should not be eaten unless all parts have been heated to this temperature or higher. It is a good general rule never to eat fresh pork which shows any pink or red color or which has not been heated to the point where no indication of uncooked meat fluids can be seen.

(2) The USDA has several rules governing the processing of pork sold as cured products. (a) The fresh pork shall have been frozen and held at 5° to − 20°F (− 15° to − 28.9°C) for a period of 6–30 days,

the time of holding depending upon the temperature at which it i held and the size of the portion of pork; (b) all parts of the cure product shall have been heated to at least 137°F (58.3°C) in ready to-eat products; (c) for dried, summer-type sausage (Italia salami, cervelat, etc.), the product shall have curing compound added, after which it must be held for at least 40 days a temperatures not lower than 45°F (7.2°C).

(3) The cooking of garbage which is to be fed to hogs is considered t be a method of controlling trichinosis since uncooked pork may b present therein. Such pork might contaminate the hog eating i thus causing it to be a source of infection to humans.

Amebiasis

Amebiasis, as trichinosis, is not a bacterial disease. A single-celle animal (an amoeba) causes amebic dysentery in humans. The organis: is called *Endamoeba histolytica*.

Amebiasis varies greatly in symptoms from patient to patient an periodically in severity in the same patient. Diarrhea is a commo symptom of the disease, and it may be persistent and severe, mild, occasional. Abdominal pain, fatigue and fever are sometimes encou: tered. Incubation lasts from 2 days to several months, but is usually 3 weeks.

The control of amebiasis is essentially a matter of good sanita procedures. (1) Drinking water, water added to foods and water used f washing and sanitizing of equipment and utensils and for irrigation crops should be potable. (2) Water to be used for foods or for equipme: or utensils which contact foods, taken from deep wells, lakes or pon should be tested periodically for bacteriological safety. Although this not a bacterial disease, it appears to occur when bacteria indicative pollution with human discharges are present. (3) Persons known to ha had amebic dysentery should be rigidly excluded from food handling any kind.

Other Food Infections

Tuberculosis (caused by *Corynebacterium tuberculosis*) a: brucellosis (caused by *Brucella melitensis* and also known as undula fever) are diseases which in the past were transmitted through milk b have been controlled in recent years by heat pasteurization of milk a: the testing of dairy herds which eliminated infected animals.

The ingestion of foods may be involved in infectious hepatitis wh infected food handlers who are careless in their personal habits a:

nvolved in preparing and/or serving food to others, or when con-
aminated shellfish are eaten raw or without adequate cooking.

Streptococcal infection is quite rare and can be prevented by
pasteurization.

Taenia solium (a tapeworm that may infest pork), *Taenia saginatta* (a
apeworm sometimes found in beef) and *Diphyllobothrium latum* (a
apeworm sometimes found in fish) all cause illness in man, but none of
hese pose any hazard when foods are thoroughly cooked. Small worms of
he genus *Anisakis* may also infect fish and cause illness in man when the
nfected fish is not properly cooked.

FOOD INTOXICATIONS

Food intoxications are those diseases in which the causative organism
rows in the food and produces a chemical substance in the food which is
oxic to man and other animals.

Staphylococcal Poisoning

The symptoms of staphylococcal poisoning are nausea, vomiting,
abdominal cramps, diarrhea and prostration. While the symptoms last,
suffering may be acute, but this is usually for a period of only a few hours.
Generally, the patient recovers without complications. The incubation
period after ingestion or eating food containing the toxin is from 1 to 7 hr,
usually 3 to 6 hr. Staphylococcal poisoning is often wrongly called
ptomaine poisoning. Ptomaines are produced by bacteria in some foods
when extreme decomposition occurs. Most ptomaines are not poisonous,
and it is unlikely that many people would eat foods decomposed to this
extent. It is probable, therefore, that ptomaine poisoning rarely occurs.

The bacterium that causes staphylococcus poisoning is *Staphylococcus aureus,* the same bacterium which causes white-head pimples,
nfections, boils and carbuncles. These cells are spherical or ovoid in
shape, nonmotile, and, in liquid cultures, arrange themselves in
grapelike clusters, in small groups, in pairs or in short chains. They grow
best in the presence of air (oxygen), but they will also grow in the absence
of air. They will grow in media or in foods which contain as much as 10%
salt (sodium chloride). When these organisms grow in foods they produce
a toxin which can be filtered away from the food and from the bacterial
cells.

Almost any food (except acid products) is suitable for the growth of
Staphylcoccus aureus, but certain foods have been most often the cause
of staphylococcal poisoning. The food types most frequently involved in

this disease are ham and ham products, bakery goods with custard fillings, chicken products and especially chicken salad, potato salad and cheddar cheese. The reason why ham products are frequently involved is that in their preparation they may become contaminated with *Staphylococcus aureus*, and since this product contains 2–3% salt, other bacteria which might grow and inhibit the growth of staphylococci are themselves inhibited by the salt. Also, people who handle ham and its products are apt to believe that such foods are not perishable. However ham and ham products are perishable and should always be held at 40°F (4.4°C) or below.

Products, such as eclairs with custard filling, have frequently been involved in staphylococcal poisoning since when the filling is cooked, the bacteria which otherwise might grow and inhibit staphylococci are killed while staphylococci may survive the cooking. Also, the filling may be contaminated with the organism after cooking. Again, people handling foods may tend to hold this type of product at room temperature. It should be held under refrigeration at all times.

The toxin produced by *Staphylococcus aureus* is not readily destroyed by heat. With most cooking methods, the organism itself would be destroyed, but if the organism has grown and produced toxin prior to cooking, the toxin would still be present. Therefore, such foods as milk powder have caused staphylococcal poisoning, yet no living staphylococci could be isolated from them. In the case of milk powder, the bacteria were destroyed during the heating required for drying the milk

Humans are the main source of the organism causing staphylococcal poisoning. It has been found that approximately 40% of normal human adults carry *Staphylococcus aureus* in their noses and throats. Thus, the fingertips and hands often become contaminated with the organism Also, any person having infected (containing pus) cuts or abrasions is a definite source of the organism. Cows may also be a source of *Staphylococcus aureus*, particularly if the animals have mastitis (an infection of the udder).

In foods, *Staphylococcus aureus* will grow at temperatures as low as 44°F (6.7°C) and as high as 112°F (44.4°C). In some foods (turkey stuffing), temperatures as high as 120°F (48.9°C) have sometimes been necessary to destroy the organism.

Staphylococcal poisoning usually occurs after a food has been held at temperatures which allow the organism to grow at relatively fast rates.

Methods of controlling staphylococcal poisoning are:

(1) Hold all foods, when not being eaten or prepared for eating, at temperatures of 40°F (4.4°C) or below.

(2) Prohibit all persons having boils, carbuncles or pussy abrasions or cuts on their hands from handling foods.

(3) Require all personnel handling food in food manufacturing or food serving establishments to wash and sanitize (with such solutions as chlorine or one of the iodophors) their hands prior to performing their particular tasks.

(4) Eliminate use of milk from cows with mastitis for human consumption.

Botulism

Botulism is an unusual disease occurring only rarely (about 15 cases per year in the United States) but with a high mortality rate (in the past, over 50%, but more recently, about 30%). The symptoms of botulism are vomiting, constipation, difficulty of eye movement, double vision, difficulty in speaking, abdominal distension and a red, raw sore throat. In severe cases, and this disease is usually severe, the breathing mechanism and eventually heart action are affected, often resulting in death.

There are seven types (A, B, C, D, E, F, G) of the bacterium *Clostridium botulinum* which may cause botulism in humans or other animals. The types that affect man most often are A, B and E. Man is most susceptible to the toxin of types A and E. The botulinum organisms are spore formers which have some heat resistance, and they grow only in the absence of oxygen (air) or under conditions in which oxygen is quickly taken up by chemical substances (reducing compounds) present in some foods. The toxin is produced in the food by the bacteria before the food is eaten. Whereas the various types (A through F) of *Clostridium botulinum* are classified mainly by the fact that different antitoxins are required to neutralize each of the different toxins, there are some differences in the growth (cultural) characteristics of the cells of the different types. The types of *Clostridium botulinum* have different minimum temperatures for growth. Types E, F, and some B will grow at temperatures as low as 38°F (3.3°C), although growth will be comparatively slow at this temperature, while types A and B will not grow at temperatures below 50°F (10°C). The maximum growth rate for all types is at 86°–95°F (30°–35°C).

Unlike the toxin produced by *Staphylococcus aureus*, that produced by *Clostridium botulinum* is readily destroyed by heat. In foods, all botulinum toxins present would be destroyed by bringing the temperature to that of the boiling point of water (212°F [100°C]). Destruction of the toxin starts at temperatures well below 212°F (100°C). This is

a fortunate circumstance, since most cases of botulism are caused b
home-canned foods, many of which are heated prior to serving. Thi
heating has no doubt saved many lives, since while the organism itse
may not be destroyed by such heating, the toxin is usually destroyed, an
it is the toxin, not the organism, which causes the disease.

Clostridium botulinum will not grow in acid foods (pH 4.5 or below
yet there have been some acid foods (pears, apricots, tomatoes) whic
have been involved in causing the disease. In such cases, it is believe
that some other organism (mold, yeast or bacterium) has first grown an
raised the pH of the food to the point where *Clostridium botulinu*
would grow.

Botulinum toxin affects the nerves associated with the automati
functions of the body (contraction and dilation of blood vessels, breath
ing, heart action, etc.). At least in the case of type E poisoning, eve
when symptoms are recognized, the patient may be treated with ant
toxin and eventually recover. With type A botulism, the situation is nc
so certain, and it is considered by some that once symptoms hav
occurred, the toxin is fixed, and antitoxin treatment does little goo
Since the toxin blocks the function of the nerve synapses, the voluntar
nervous system also becomes affected.

Since most cases of botulism are caused by home-canned foods, or
method of control would be to make sure that all vulnerable hom
canned foods are pressure cooked at times and temperatures suitable t
destroy all spores of *Clostridium botulinum* which may be presen
Pamphlets are available from the USDA which specify times an
temperatures for sterilizing different products in different size col
tainers.

Though extremely rare, there have been, recently, some commerciall
canned products which have caused botulism. All manufacturers of suc
products should have technical advisors who can determine that the hea
processes given their products are sufficient to destroy all spores
Clostridium botulinum that may be present.

Since types E and F *Clostridium botulinum* will grow at temperatur
as low as 38°F (3.3°C), all flesh-type foods, especially fish (since type
is often present), should be stored at temperatures below 38°F (3.3°C

A good control precaution is to heat all canned vegetables (especial
home-canned) to the boiling point prior to serving.

Perfringens Poisoning

Perfringens poisoning has sometimes been classified as a food infectic
and at other times as a food intoxication. Current research indicates th

when this organism grows in foods, it produces a substance (an enzyme or other compound) which causes an intestinal disturbance in humans.

Perfringens poisoning is caused by *Clostridium perfringens* which, like *Clostridium botulinum,* grows only in the absence of oxygen. Also, as in the case of the botulism organism, perfringens is a spore-forming bacterium, although the spores of this organism are not as heat resistant as those of some types of the botulinum spores. The symptoms of perfringens poisoning are diarrhea and abdominal pain or colic. The illness occurs 8–22 hr after the food has been eaten, and the symptoms are of short duration (one day or less). The number of people involved in an outbreak is quite often comparatively large. In perfringens poisoning, what ordinarily happens is that meat or poultry is cooked, then held at comparatively high temperatures, then served. The spores survive the cooking since they have some heat resistance, then when the meat or gravy is held at room temperature or on a steam table where the temperature is below 140°F (60°C), the spores grow out to large numbers, causing illness.

In order to control perfringens poisoning, meats and gravies should be refrigerated at temperatures of 40°F (4.4°C) or below shortly after cooking if not immediately eaten, or be held at temperatures not lower than 140°F (60°C) in preparation for serving.

Other Food Intoxications

Other diseases may occur from the accidental accumulation of toxins in foods exposed to unusual environmental concentrations of chemical or biological toxins from polluted areas. In such cases, the toxins are not easily removed or destroyed; therefore, we must rely on our regulatory and public health agencies to make periodic inspections of foods that are suspect, as well as the areas from which they are produced.

Section II

Causes of Food Changes

7

Microbial Activity

Foods are normally contaminated with microbes (or microorganisms). They are so small that we need a microscope to see them. Microbes include bacteria, yeasts, molds, algae, protozoans and others. However, the organisms that normally contaminate and spoil foods are the bacteria, with yeasts and molds of secondary importance. This chapter deals primarily with bacteria, yeasts and molds.

Under normal conditions, microbes feed on the food in which they live and reproduce. During their life cycles, they cause a variety of changes in foods, most of which result in a loss of the foods' quality. In some cases, the controlled growth of specific microbes can produce desirable changes, such as the change that results in the formation of sauerkraut from cabbage and wine from grapes. Microorganisms function through a wide variety of enzymes that they produce, and the changes in foods attributed to microorganisms are actually brought about by the chemical action of these enzymes.

Most scientists classify bacteria, yeasts and molds as members of the plant kingdom. On the other hand, these microbes have some characteristics associated with members of the animal kingdom. Some scientists postulate that these microbes do not classify as either plants or animals and should comprise a third kingdom of living things, the "Protista."

CHARACTERISTICS OF MICROBES

The physical and biological abilities, requirements and tolerances of microbes are factors that determine the effects of microbes on food.

Structure and Shape of Microbes

Bacteria, molds and yeasts have rigid cell walls enclosing the cell materials and the cytoplasm, but they differ greatly in their properties. The microbes most important in the spoilage and controlled changes in food materials have various shapes (see Fig. 7.1). Many are rod-shaped and occur either as single cells, as two adjoining cells or as short chains of cells. Other bacteria are spherical in shape (cocci). Some cocci exist mainly as a grape-like cluster of cells, the staphylococci, others as a cube-like cluster of spherical cells, the sarcina. Some cocci occur in groups of two, the diplococci, or in chains, the streptococci. Some disease-causing bacteria are included in the latter 2 groups.

Some bacteria are curved or comma-shaped rods, the vibrios, and others are long, slender, corkscrew-shaped cells, the spirochaetes. Other bacterial groups either are not important to changes in foods or are disease-causing types (see Chapter 6).

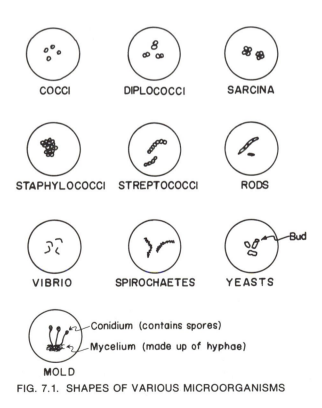

FIG. 7.1. SHAPES OF VARIOUS MICROORGANISMS

Molds are multicellular (bacteria and yeasts are single cells) and are made up of branched threads (hyphae), consisting of chains of cylindrical cells united end to end. Some of the hyphae serve to secure nutrients from the material in which they are growing while others produce the spores which provide for reproduction, or for new mold growth. Some molds produce a mycelium (mass of hyphae) which has cross walls (septa) while others do not have cross walls in the mycelium.

Yeasts form single cells or chains of cells which may be spherical or of various shapes between the spherical and the cylindrical.

One characteristic of bacteria, yeasts and molds is that they can exist either as active, "vegetative cells" or as spores. As vegetative cells, they metabolize, reproduce, cause food spoilage, and sometimes disease. They generally have a considerable effect on the environment. Their activity, however, is dependent on several environmental conditions, as we shall see later on. When conditions are unfavorable (such as high temperatures), the vegetative cells begin to die before they can reproduce, and soon all the vegetative cells die. Many microorganisms, however, also exist in the spore state, sometimes because it is a necessary step in reproduction (as in molds), so there are spores at any given time to permit the microbe to survive unfavorable conditions. In other cases, certain microorganisms, as some bacteria, exist in both the vegetative and the spore states so that when conditions become unfavorable, the spores survive.

In order for spores to survive the conditions that are destructive to their vegetative counterparts, they must have special properties. In general, spores are concentrated forms of their vegetative counterparts; thus spores contain less water, and they are more dense and smaller than the vegetative cells. The spore wall is thicker and harder than the cell wall.

The Size of Microbes

Bacteria are comparatively small. The single cell of many bacteria is about 40 millionths of an inch (1μ) in diameter. There are, however, some types of bacteria which may be 50 or more times larger than this. Due to their small size, bacteria cannot be seen with the naked eye, and when evidence of their growth can be seen, such as when slime forms on meat, the organisms will have multiplied to very large numbers, billions of cells to each square inch (6.5 cm^2) of the meat surface.

The single cell of the mold, although not visible without magnification, is much larger than the single bacterial cell, and any significant growth of molds on foods can be seen. The visible mold comprises mycelia with or without spore heads. The spore head carries the spore that will give rise to more growth.

Yeasts vary in size from that of the more common spherical bacteria to a form several times larger than this. Like bacteria, when grown in solution, the individual cells cannot be seen, but eventually, the cell numbers will accumulate to the point where the solution becomes cloudy. There are, however, some instances in which groups of cells will form visible clumps on foods or on the surface of solutions (colonies).

Reproduction in Microbes

Bacteria usually reproduce by fission, a transverse division across the cell to form 2 new cells. Under favorable conditions, this form of reproduction continues until there are billions of cells per ounce (29.6 ml) of solution or per square inch (6.5 cm^2) of surface of food material. (It should be noted that microorganisms must live in liquid material, and when they grow on food surfaces, they are actually living in the liquid available on or in the food.) There are some instances in which sexual reproduction occurs in bacteria, 1 cell uniting with another before division occurs, but this is not the usual circumstance and need not be discussed further here.

Under favorable environmental conditions, bacterial multiplication for some species occurs at an exceedingly rapid rate, a doubling of the number requiring about 20 min. From this standpoint, the number of cells with which a food is contaminated is very important, for if the number is high, only a few hours may be required for bacterial multiplication to reach a level that will cause food spoilage.

Molds usually reproduce by means of spores, each organism producing spores in great numbers. Mold spores are of 2 types: sexual spores produced by the fusion of 2 sex cells, or asexual spores which arise from the fertile hyphae. Most molds produce asexual spores. Asexual spores are formed on the sides or ends of hyphae threads (a conidium) or produced in a special spore case called a sporangium. Conidia contain many spores, and when a spore comes in contact with a suitable growth medium under favorable conditions, it grows to become the adult mold.

Yeasts reproduce by budding, by fission or by spore formation. In budding, a projection is formed on the original cell which eventually breaks off to form a new cell. Some types may undergo fission as do bacteria. One cell divides into 2 cells. Sometimes, yeasts reproduce by spore formations that may be sexual or asexual.

Motility in Microbes

Many types of bacteria are motile or are able to move about in the solutions in which they live by means of the movement of flagella, thin, protoplasmic, whip-like projections of the cell. Some bacteria, as the cocci, have no flagella and are not motile.

Molds are not motile, but they spread in vine-like fashion by sending hyphae outwards.

Yeasts are not motile; consequently, they exist in relatively compact clusters, unless distributed by agitation or by some other external dispersing force.

Effect of pH on Microbial Growth

Both the growth and the rate of growth of microbes are greatly affected by pH. (See Chapter 5 for discussion on pH.)

Thus, microorganisms have an optimum pH at which they grow most rapidly and a pH range above or below which they will not grow at all. Generally, molds and yeasts grow best at pHs on the acid side of neutrality, as do some bacteria. Many species of bacteria grow best at pHs which are at neutrality or slightly on the alkaline side. Some bacteria will grow at pHs as low as 4, while others grow at pHs as high as 11. At least part of the reason why fruits are usually spoiled by molds or yeasts and flesh-type foods (meats, fish, poultry and eggs) are usually spoiled because of bacterial growth, is because of the low pH (acidic) of fruits and the near neutral pH of flesh-type foods.

Nutritional Requirements of Microbes

Microorganisms, especially bacteria, vary greatly in nutritional requirements from species to species. In the presence of certain inorganic salts, some bacteria can utilize the nitrogen in air to form proteins and the carbon dioxide in air to obtain energy or to form compounds from which they can obtain energy. Others can utilize simple inorganic salts, such as nitrates, as a source of nitrogen and relatively simple organic compounds, such as lactates, as a source of energy. Nearly all yeasts can derive all their nitrogen from lysine, an amino acid. Some bacteria may require complex organic compounds for growth, including amino acids (the primary units of proteins), vitamins, especially those belonging to the B group, and traces of certain minerals.

It has been shown that, in some cases, not only are trace minerals necessary, but they need careful control to sustain an optimum growth rate. There is some evidence that demonstrates the ability of at leas some microbes to utilize substitute elements for required ones. Some times, 1 trace element may protect microbes from the toxic effects of th presence of other elements; thus, the presence of zinc has been reportec to protect yeasts against cadmium poisoning.

Molds and yeasts, like bacteria, may require basic elements (carbon hydrogen, nitrogen, phosphorus, potassium, sulfur, etc.) as well a vitamins and other organic compounds.

Although sugar is a nutrient important to microbes, some molds anc yeasts can grow well in concentrations that inhibit bacterial growth. I fact, yeasts grow extremely well in the presence of sugar.

Effect of Temperature on Microbial Growth

The optimum growth rate of microorganisms depends on the tem perature; they do not grow above or below a specific range of tem peratures. Again, growth temperatures vary with species, and bacteri are arbitrarily classified according to the temperatures at which the grow. Bacteria, classified as psychrophiles, grow fastest at about 68–77°F (20°–25°C), but some can grow, although more slowly, at temperature as low as 45°F (7.2°C), while others grow at temperatures as high as 86°F (30°C). Some will grow at temperatures as low as 19°F (− 7.2°C), as long as the nutrient solution for growth is not frozen. Bacteria classified as mesophiles grow best at temperatures around 98°F (36.7°C), but some will grow at temperatures lower than 68°F (20°C) and others at tem peratures as high as 110°F (43.3°C). Bacteria causing diseases of animals are mesophiles. Bacteria classified as thermophiles grow best at temperatures between 131°F (55°C) and 150°F (65.5°C), but some wil grow at temperatures as low as 113°F (45°C) and others at temperatures as high as 160°F (71.1°C) or slightly higher.

It should be pointed out that whereas microorganisms grow within a given range of temperatures, their growth rate decreases greatly at the low or high temperatures of that range.

When microbes are subjected to temperatures higher than those of their growth range, they are destroyed at rates that depend upon the degree to which the temperature is raised above the growth range. Some bacteria form heat-resistant spores, concentrated areas of protoplasm within the cell. The degree of heat resistance of spore-forming bacteria varies with the species. Some types may be destroyed by raising the temperature (in the presence of moisture) to 212°F (100°C) over a period of several minutes while others can survive boiling for periods of many ours. Some may even survive temperatures as high as 250°F (121.1°C) or many minutes. Bacterial spores are the most heat-resistant living hings known to man.

In general, there are bacteria that can grow and can survive under more extreme conditions than those tolerated by any of the molds or easts. Molds, as a class, can grow and survive under more extreme onditions than can yeasts. On the other hand, while the growth and urvival of the many species of bacteria cover a broad range of conditions, each species is highly selective, and the range of conditions under which it can metabolize is generally narrower than that of molds and easts.

Water Requirements of Microorganisms

Microorganisms grow only in aqueous solutions. A term, "water activity," (a_w), has been coined to express the degree of availability of water in foods. This term is applied to all food, the ordinary fresh-type ood having an a_w of about 0.99–0.96 at ambient temperatures. Low water activities, which limit the growth of microorganisms in foods, may be prought about by the addition of salt or sugar, as well as by the removal of water by drying. Under such conditions, the remaining water has been ied up by chemical compounds added to or concentrated in the food or pound to some food component, such as protein.

$$a_w = \frac{\text{equilibrium relative humidity}}{100}$$

Equilibrium relative humidity is reached in a food when the rate at which it loses water to its environment is equal to the rate at which it absorbs water from the envrionment.

Oxygen Requirements by Microorganisms

Some bacteria are aerobic, that is, they require oxygen (in air) for growth. Others grow best when the oxygen concentration is low (micro-aerophiles). Still others can grow either in the presence or absence of xygen (facultative aerobes or facultative anaerobes). A number of bacterial species will not grow in the presence of oxygen because it is oxic to them, and these are called anaerobes.

It should be pointed out that while oxygen is toxic to anaerobic bacteria, they sometimes can grow under conditions which appear to nclude the presence of oxygen. The explanation for this is that organic materials (animal and vegetable matter) contain compounds which themselves tie up the available oxygen. Also, in many instances, aerobic bacteria first grow and consume the oxygen and produce reducing compounds which also combine with available oxygen, thus producing conditions suitable for the growth of anaerobic bacteria. The corollary to this is that certain aerobic bacteria can utilize the oxygen in certain oxygen-bearing compounds, such as nitrates. In general, molds and most yeasts require oxygen for growth.

EFFECT OF MICROBES ON FOOD

As stated earlier, bacteria, molds and yeasts are the main causes of the spoilage of unpreserved foods, with bacteria playing the major role in the spoilage of meats, poultry, dairy and fish products. Molds and yeasts play the major role in the spoilage of fruits and vegetables. The changes in foods due to microbial action can be classified into 2 types: undesirable changes and desirable changes.

Undesirable Changes

Undesirable changes can be further subdivided into (1) those that cause food spoilage, not usually associated with human disease and (2) those that cause food poisoning whether or not the food undergoes observable changes.

Food Spoilage.—Food spoilage can be detected organoleptically. That is, we can either *see* the spoilage, *smell* the spoilage, *taste* the spoilage, *feel* the spoilage, or combinations of the 4 sensations. Quite often, the evidence of microbial growth is easily visible as in the case of slime formation, cotton-like network of mold growth, irridescence and greening in cold cuts and in cooked sausage, and even discrete large colonies of bacteria. In liquids, such as juices, microbial spoilage is often manifest by the development of a cloudy appearance or curd formation. The odors of spoiled protein foods are very objectionable and, in some cases, when they are intense enough, even toxic. Some of these obnoxious odors are common enough, and the compounds responsible for them are ammonia, various sulfides (as in the smell of clam flats), hydrogen sulfide (the typical smell of rotten eggs), etc. The tastes of spoiled foods range from loss of good characteristic taste to the development of objectionable tastes. Thus, when a pear or an orange spoils, the sweet characteristic taste of either is lost, and when milk spoils, it develops an acidic taste, sometimes also bitter. The feel of spoiled foods reflects the spoilage in different ways, depending on the type of food and the microbe involved in the spoilage, so some spoiled foods feel slimy, while others may feel mushy.

Food Poisoning.—Food poisoning may result from a variety of factors, but we are concerned with food poisoning resulting from the activity of microbes on the food. The subject is too broad to present here, and it is discussed in greater detail in Chapter 6 of this text. However, some remarks are appropriate here. Ptomaines result from the microbial decomposition of proteins and, in some instances, decarboxylation of amino acids. Theoretically, some amino acids can give rise to a ptomaine derivative plus carbon dioxide. A typical example is:

$$\text{histidine} \xrightarrow[\text{decarboxylation}]{\text{microbial}} \text{histamine} + \text{carbon dioxide}$$

where histamine is the ptomaine derivative said to be poisonous.

Ptomaine poisoning, which occurs only rarely, is caused by compounds that are formed in advanced stages of spoilage (the food is putrid), whereas most food poisonings are caused either by bacterial diseases or by toxins produced in foods through bacterial growth. In many cases, foods that can cause illness have no outward signs of spoilage.

In many instances, grain used for animal feeds becomes contaminated with mold growth during storage, the molds producing toxic materials which cause diseases in the animals to which the grain is fed.

Bacteria, and especially molds, cause many diseases of vegetables and cereals as they are grown, causing economic losses, but this subject is beyond the scope of our discussion.

Desirable Changes in Foods

While from the standpoint of food spoilage or food-borne diseases the effects of microorganisms are undesirable, many of their activities are beneficial and an absolute necessity to life. Were it not for the fact that microorganisms decompose both vegetable and animal wastes and convert them to a form which can be utilized by plants, there would be no life on earth due to a lack of basic elements. Moreover, were these wastes not decomposed and converted to soluble forms, waste materials would accumulate to the point where there would be no room for plants or animals.

During the development of civilization, man has learned how to utilize some of the products produced by the growth of certain microorganisms.

The bacterial products utilized by man are mostly produced by the bacteria which form lactic acid from sugars, although other types may form useful materials.

Cultured milks (soured cream, yogurt, buttermilk, etc.) are produced through the growth of the lactic acid bacteria.

In order to obtain the curd in the manufacture of cheese, milk is cultured with bacteria which produce lactic acid which in turn precipitates the casein, although this can be accomplished in a different way. The particular flavors and textures of many cheeses are attributed to bacterial growth during or after curd formation. Flavor and texture are influenced especially during aging or during the period in which the cheese is held in storage at a particular temperature for purposes of maturing.

The particular flavor of butter is due to the formation of small amounts of a chemical compound from sugars or citrates by the growth of lactic acid bacteria in the cream prior to churning.

Pickles and olives are at least partially preserved by acid formed by bacteria when the raw materials are allowed to undergo a natural fermentation.

The particular flavor and texture of sauerkraut are due to acid and other products produced by the growth of lactic acid bacteria in shredded cabbage to which some salt has been added.

The typical flavor of dried-cured sausage (Italian salami, cervelat etc.) is produced by the growth of lactic acid bacteria in the ground meat within the casing during the period in which the sausages are stored on racks and allowed to dry.

Acetic acid (vinegar) is produced from ethyl alcohol by the growth of *Acetobacter* which oxidizes the alcohol to acetic acid.

$$2(CH_3CH_2OH) \xrightarrow{O_2} 2HOH + 2(CH_3CHO) \xrightarrow{O_2} 2(CH_3COOH)$$

Ethyl alcohol   Water   Acetaldehyde   Acetic acid

Various types of bread and certain other bakery products are leavened (raised) by yeasts. The yeasts, in this case, not only produce carbon dioxide, a gas which causes the loaf to rise, but also produce materials which affect the gluten (protein) of flour, causing it to take on a form which is elastic. The elasticity of the gluten is necessary to retain the gas and to support the structure of the loaf.

All alcoholic drinks are produced through the growth of different species of yeasts. This includes beer, ale, wines and whiskeys. Whiskeys, brandies, gins, etc., are produced by distilling off the alcohol from materials which have been fermented by yeasts.

Molds are sometimes allowed to grow on sides or quarters of beef for purposes of tenderizing the meat and possibly to attain the development of particular flavors.

Some cheeses, such as Roquefort, Gorgonzola, blue and Camembert, etc., owe their particular flavor and texture to the growth of molds. Under present-day manufacturing procedures, these cheeses are inoculated with molds of particular species, and incubated to allow the molds to grow and produce the desired textures and flavors.

The utilization of bacteria, molds and yeasts to produce particular foods or beverages enjoyed by man, in many cases results from accidental discoveries in which foods or food materials have undergone natural fermentations or changes due to the growth of microorganisms. In many cases, these natural fermentations were first used as a method of food preservation, artificial refrigeration being then unavailable. Today, these foods have become special products valued for their own particular taste and texture, even though food handling and processing methods have done away with the need for preservation by natural fermentation, except in very primitive areas.

In general, it can be said that microorganisms, including bacteria, molds and yeasts, are essential to the existence of man and other animals on earth. Microorganisms can cause economic losses in that they cause diseases and the decomposition of animal foods. However, such decomposition is a part of the cycle which returns elements, especially nitrogen, carbon, hydrogen and oxygen, to a form which can be utilized by living things to form new organic materials.

Antibiotics, a group of compounds successful in combating microbial infections in man, are produced by other microbes.

8

Enzyme Reactions

THE NATURE OF ENZYMES

Enzymes, produced by living things, are compounds which catalyze chemical reactions. Reactions involving enzymes may be said to proceed in 2 steps. In step 1, $E + S \longleftrightarrow ES$ (where E = enzyme, S = substrate, and ES = unstable intermediate complex that temporarily involves the enzyme). In step 2, $ES + R \longleftrightarrow P + E$ (where R = a substance in the substrate that reacts with the complex, P = the final product of the reaction, and E = enzyme which is liberated from the complex). Their role is critical to life because they have the ability to catalyze the chemical reactions which facilitate life. Chemical reactions take place when the necessary reactants are present, but usually an energy input ("activation energy") is required to start a particular reaction. The analogy usually given to illustrate this concept is that of a boulder located at the top of a hill. The boulder has the potential energy for rolling down the hill, but must first be pushed over the edge (see Fig. 8.1). The potential energy of the boulder could be great, depending on its mass and altitude. It could start a rock slide or landslide involving a great amount of energy. Although the energy required to push the boulder over the edge is insignificant compared to the total energy involved in the rolling of the boulder down the hill, that initial energy (called "the activation energy") is nevertheless important, for without it there would be no landslide.

It is well known that the rate at which reactions take place is dependent on the temperature, among other things. The expression that is broadly accepted as the one that most nearly relates the rate of the

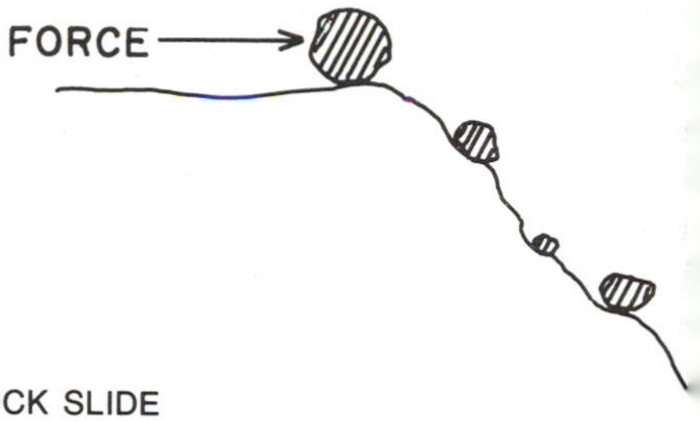

FIG. 8.1. FORCE NEEDED TO START ROCK SLIDE

reaction, the activation energy, and the temperature at which th reaction proceeds is the Arrhenius equation:

$$K = Ae^{-E/RT}$$

where: K = rate of reaction
A = constant
e = 2.718..., the base of the system of natural logarithms
E = activation energy
R = 1.99 cal/°C/mole, the gas constant
T = absolute temperature

For solution of the preceding equation, e and R are both known, K, E and T are values that may be known or need to be found. A is the orientation factor. It is associated with the probability that the reactive sites of reactants are properly oriented for reacting with each other.

Since reactions can proceed more rapidly at higher temperatures but quite often have to accelerate within a system of constant temperature as within our bodies, only by the action of enzymes can this occur Enzymes are produced by living things from the lowest single-celled members to the highest, most complex members of the plant or anima kingdoms, including man. All life depends upon enzymes to conver foods or nutrients to a form in which thay can be utilized and to carry ou cellular functions.

In composition, enzymes always contain a protein. They may also contain or require complex chemical compounds in order to become functional. These compounds are known as prosthetic groups (also called coenzymes) and are usually vitamins, especially those belonging to the E vitamin group. Some enzymes also require trace amounts of a metal such as copper, to function. Enzymes, therefore, consist of either pure proteins, proteins with a prosthetic group or proteins with a prosthetic group plus a metal cation.

Through their action, enzymes convert foods into less complex chemical substances which can be utilized for energy and for the building of cellular protoplasm. Proteins, fats and carbohydrates are thus broken down to less complex compounds by enzymes in order that they may be utilized by various organisms. More complex carbohydrates are broken down to glucose, a source of ready energy which can be absorbed and eventually converted to carbon dioxide and water or built up into fats which can be stored as a source of reserve energy. When reserve energy is needed, fats are first hydrolyzed (insertion of water to split the molecule) or broken down into glycerine and fatty acids. The fatty acids are then converted to acetates which can be utilized for energy. When the acetates are not completely used, those remaining may be recombined and deposited as fats, an energy reserve.

Proteins are broken down to their primary units (amino acids), in which form they may be incorporated into cellular protoplasm as proteins used for cell repair or for growth. In some instances, the nitrogen portion of the amino acid may be removed and the remaining compound oxidized to provide energy.

As previously stated, enzymes always contain a protein (chain of amino acids) in which the amino acids are combined in a particular sequence, the protein itself having a particular shape or configuration.

Amino groups

$$CH_3CHNH_2COOH + CH_2NH_2COOH \rightarrow CH_3CHNH_2CONHCH_2COOH + H_2O$$

Alanine	Glycine	Alanylglycine	Water
Amino acids		(A dipeptide)	

There are more than 22 known amino acids, some of which are much more complex than alanine and glycine. Proteins consist of a large number of amino acids combined in a particular sequence. Also, chains of amino acids are cross-linked one to another and the different proteins form special configurations and shapes. Proteins are also sometimes combined with carbohydrates, lipids, or phospholipids (fat-like compounds containing phosphoric acid as part of the molecule). These are called conjugated proteins. The number and sequence of amino acids in the chain, the shape or relationship of one protein chain with another and conjugation with carbohydrate or lipid all affect the characteristics (functional properties) and determine the manner in which a protein will react to physical and chemical energy.

Enzymes cause chemical reactions to occur at their fastest rates when the temperature is at some optimum level. Usually, this is in the range of 60°–150°F (15.6°–65.6°C), but some action may occur at temperature above or below the optimum range. Thus, some enzymes may reac slowly at temperatures well below that of the freezing point of water and others at temperatures above 160°F (71.1°C).

Since proteins are changed chemically and physically or are coagulated by high temperatures, especially when moisture is present enzymes are usually inactivated at temperatures between 160° and 200°F (71.1° and 93.3°C). There are some exceptions to this, however and at least one enzyme, which splits off fatty acids from fish phos pholipids, is known to remain active even after steaming at 212°F (100°C) for 20 min.

Enzymes have an optimum pH at which they cause reactions to occu at the fastest rate. Water solutions having a pH value less than 7 are said to be acidic; those having a pH value greater than 7 are said to be alkaline; and those having a pH value of 7 exactly are said to be neutral As in the case of temperatures, some action will occur at pHs above or below the optimum, although there are low and high limits beyond which a particular enzyme action cannot take place.

PROTEOLYTIC ENZYMES

Proteolytic enzymes or proteases cause a breakdown of proteins. They are present in plant and animal tissues. Proteases may be of the type which break proteins down into comparatively large chains of amino acids (polypeptides) or of the type which break polypeptides down into individual amino acids. The latter can be classified as polypeptidases Since most amino acids are water soluble, this means that tissues may in essence, be liquefied by proteases and polypeptidases.

In meat, such as beef, pork or poultry held in the eviscerated state (intestines and organs removed), the proteases present in the tissues are called cathepsins. The pH of meat tissues is on the alkaline side of the pH optimum for the cathepsins. Also, the temperatures and times under which these products are held prior to utilization are not such that extensive proteolysis can occur. Therefore, while there may be some tenderization of the tissues during holding, which may, in fact, be due to proteolysis, there is not extensive breakdown of the tissues.

In fish, proteolytic enzymes are much more active than in meats Even when fish is held in the eviscerated state in ice or under refrig eration, there may be sufficient proteolysis to cause softening of the tissues over a period of days. In fish held in the round (uneviscerated) proteolysis is accelerated due to a concentrated source of enzyme present in blind tubules (the pyloric caeca) attached to the intestines 'hus, even though fish in the round are refrigerated, within a few days, ifficient proteolysis may occur to dissolve the tissues of the abdominal all exposing the entrails. Members of the herring and mackerel families hich are handled in the uneviscerated state are quite subject to this 'pe of enzyme deterioration, especially if they have been feeding when aught. Such fish as flounders and ocean perch, also handled in the neviscerated state, seem not to be especially subject to this type of eterioration. Lobsters provide an especially good example of a de-riorative change which may take place through the action of pro-olytic enzymes. As long as the lobster is in the live state, autolytic roteolysis does not occur. However, if the lobster dies and then is held r some hours, even under refrigeration and especially at higher tem-eratures, proteolysis takes place to such an extent that the lower odominal portion will be partially liquefied. When such a lobster is ooked, the flesh will be soft and crumbly (short-meated) and part of the ail portion will have dissolved, leaving only a part of this section intact. or this reason, lobsters should never be held long after death prior to ooking. The present accepted practice is to cook lobsters from the live ate. Other crustaceans (shrimp and crabs) are also subject to enzyme roteolysis, although with shrimp, this usually is not extensive, especial-if the head portion (cephalothorax) is removed shortly after the shrimp caught. The relatively high activity of enzymes in marine species is ttributed to the low temperature conditions in the marine environment. hat is, in order to make reactions proceed at low temperatures, the ctivation energy system must be more efficient.

Plants also contain proteolytic enzymes, but these enzymes usually ontribute little to deterioration, especially as long as the tissues of fruits nd vegetables are not cut or damaged. Some plants provide an excellent ource of proteolytic enzymes. Bromelin is found in pineapple juice and so active that people who handle cut pineapple which has not been eated must wear rubber gloves; otherwise, the flesh of the fingers will be aten away. Papain is a proteolytic enzyme obtained from the latex of ie green papaya fruit. Ficin is a proteolytic enzyme obtained from the tex of certain fig trees.

Proteolytic enzymes from plants may be extracted and purified and ese enzymes may be employed, for instance, to tenderize meats.

XIDIZING ENZYMES

There are a number of oxidizing enzymes which bring about changes in ods that result in deterioration. In plants, peroxidases, ascorbic acid xidase, tyrosinase and polyphenolases may cause undesirable chemical

reactions to occur. Peroxidases may oxidize certain phenol-like com
pounds in root vegetables, such as horseradish, causing the prepare
product to become darker in color. This does not happen while the tissue
are intact but when the vegetable has been cut up or comminuted
Ascorbic acid oxidase, present in certain vegetables, oxidizes ascorbi
acid (vitamin C) to a form which is readily further oxidized by at
mospheric oxygen. The second oxidation product is not utilized by th
human as a vitamin. Therefore, the action of this enzyme may cause los
of the vitamin C content of foods. Peroxidases may also, indirectly, caus
a loss of vitamin C in vegetables. In this case, the compounds formed b
the action of peroxidase react with vitamin C. Phenolases are present i
some fruits and vegetables. These enzymes oxidize some phenol-lik
compounds, also present in plant products, causing brown or dark
colored compounds to be formed when the tissues are cut.

Tyrosinase oxidizes the amino acid tyrosine to form dark-colore
compounds. The molecule rearranges and further oxidizes to form a re
compound. Polymerization (combination of these compounds) results i
the formation of dark-colored melanin compounds.

The enzyme, tyrosinase, which is present in many fruits and vegeta
bles, may cause discoloration of the cut tissue and will also oxidiz
compounds related to tyrosine. This enzyme is also present in shrim
and some spiny lobsters and may cause a discoloration called "blac
spot." In shrimp, this often occurs as a black stripe on the flesh along th
edges of the segments of the tail or as a pronounced band where the she
segments overlap. It is not generally recognized, but tyrosinases are als
present in clams. Hence, shucked (deshelled) clams will darken at th
surface if oxygen is present and if the enzymes have not been inactivate
by heat.

These reactions occur only after the shrimp or clams die. In genera
oxidizing enzymes do not cause deteriorative changes in tissues which ar
intact. In fruits and vegetables, the tissues must be cut or bruised c

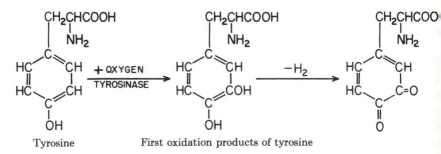

Tyrosine   First oxidation products of tyrosine

here must be a breakdown of cells by their enzymes before the action of oxidizing enzymes results in discoloration.

FAT-SPLITTING ENZYMES (LIPASES)

Fats are composed of glycerine and fatty acids. Glycerine is a polyhydric alcohol (3 alcohol groups), and fatty acids are short or long chains of carbon atoms to which hydrogen is attached, either to the fullest possible extent (saturated) or to a lesser extent (unsaturated), the latter resulting in reactive groups in the chain. At one end of the fatty acid chain there is an acid group.

In formation of fats, each 1 of 3 fatty acids combines with 1 of the 3 alcohol groups of glycerine, splitting off water in each case. On the other hand, where water and the enzyme lipase are present, fats are split into their component parts, glycerine and fatty acids.

The fatty acids in most fats that are found in nature consist of a chain of more than 10 carbon atoms, and these fatty acids have no particular flavor or odor. Hence, when lipase acts on most natural fats, no bad odors are generated. However, if fats or oils high in free fatty acid content are used for deep-fat-frying, the oil may smoke during heating, which is undesirable. There are some fats which contain short chain fatty acids, especially those fats present in the milk of cows or goats. These fats contain butyric (4 carbons), caproic (6 carbons), caprylic (8 carbons) and capric (10 carbons) acids. All these fatty acids have an odor and flavor. Butyric acid, especially, is pungent and considered to be distasteful. When lipase acts on butter, therefore, it splits off butyric acid which gives the butter a strong, undesirable taste (called rancid). Actually, butter is an emulsion of water in oil and contains about 16% water, the water being present as fine droplets. Butter becomes rancid by the action of lipase which is produced by bacteria that grow in the water droplets. The lipase acts on the fat surrounding the water droplets. Lipase rancidity in butter, therefore, is really a type of deterioration caused by bacteria.

Enzymes similar to lipase are called phospholipases. Phospholipases split phospholipids. Most phospholipids are similar to fats in that they contain glycerine, two alcohol groups of which are combined with fatty acids. The third alcohol group in this case is combined with a molecule containing phosphoric acid, a short chain of carbons and a nitrogen group with carbons attached to it. Lecithin is a typical phospholipid. Phospholipase splits off fatty acids from phospholipids. Such

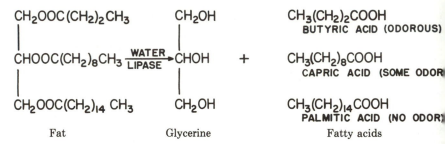

action may cause deterioration in foods in that it results in a destabiliza-tion of proteins which causes a toughening of the tissues and a loss c succulence (juiciness).

ENZYMES THAT DECOMPOSE CARBOHYDRATES (CARBOHYDRASES)

Fruits contain pectin which support the particular structure of th product. In processed fruit juices (for instance, tomato or orange juice) if the pectin is broken down, the solids tend to settle to the bottom leaving a clear serum on top. Pectin consists of a long chain o galacturonic acid molecules with the carboxyl groups partially esterifie with methyl alcohol. It has a high water-holding capacity. There ar pectic enzymes which will either break down the pectin molecule t smaller units or completely decompose the molecule to its primary unit galactose. The emulsifying properties of pectin may thereby be lost causing settling in fruit juices and softening in fruit. When the pectin i whole fruit breaks down, it may result in deterioration of the fruit due t the action of other enzymes or invasion of the tissues by microorganisms

In the sugar cane plant, there is an enzyme invertase which break down cane sugar (sucrose) having 12 carbon atoms to glucose an fructose, each having 6 carbon atoms. Before sugar cane is harvested therefore, a part of the plant must be removed to eliminate the source c the enzyme. Were this not done, there would be a loss of sucrose durin the processing of the cane. Many other carbohydrases, which break dow cellulose or starch or which break down more complex sugars to smalle units, exist.

APPLICATIONS

Whereas enzymes may cause a deterioration of foods, they may also b used in the processing of foods to produce particular products or t nodify the characteristics of particular products. Proteolytic enzymes obtained from plants may be used for tenderizing meat either by injecting animals with a solution of the enzyme prior to slaughter or by sprinkling the powdered enzyme on meat surfaces and allowing it to react, prior to cooking. In the manufacture of certain kinds of milk powder (e.g., to be used in chocolate), the lipases may be allowed to act on the milk-fat prior to drying to obtain a particular flavor in the finished product. The characteristic flavor of certain cheeses is due to the action of lipases on the milk-fat contained therein. In order to obtain the particular flavor of Roquefort, Gorgonzola or blue cheese, the milk-fat must first be broken down to fatty acids which can then be oxidized. The lipases which decompose the fats to provide fatty acids are produced by the molds allowed to grow in these cheeses. Enzymes produced by molds then oxidize the specific fatty acids that ultimately result in the unique flavors that characterize Roquefort and other exotic cheeses.

There are many applications of enzyme technology which involve the use of carbohydrate-splitting enzymes. In making malt, barley is germinated to obtain an enzyme which will convert starch to a sugar (maltose) that can be converted by yeasts to ethyl alcohol and carbon dioxide. By this means, various grains can be used as the source of sugar for fermentation. Carbohydrate-splitting enzymes are also used to modify starches used in foods and to modify starches used in sizing and laundering clothes.

There are many other applications of enzyme technology in the food and other industries and it is expected that the number of applications for enzymes will continue to increase. One of the factors which will serve to widen the use of enzyme technology is the development of immobilized enzymes. It has been found that enzymes can be fixed chemically to inert substances, such as glass beads. In this form, they can be packed into a column through which a solution or suspension of the material to be acted on (called the substrate) is allowed to pass. In this manner, the enzyme which is responsible for the conversion or change in the substrate is not lost or washed out with the substrate. Thus, the enzyme can be used for a number of substrate conversions. Moreover, in this form (immobilized enzyme), the active agent is much less subject to inactivation as, for instance, by high temperature.

9

Chemical Reactions

Foods sometimes deteriorate because of chemical changes not associated with microbial growth or enzyme-induced chemical reactions. Actually, certain components in foods are subject to reactions which involve either a combination with naturally occurring elements (such as oxygen) or with compounds present in the foods themselves.

OXIDATION

One of the changes that occur in foods is the oxidation of fats and oils. Fats and oils are chemically similar and are classified as lipids. Generally, fats are those lipids that are solid at room temperature (e.g., lard, suet), and oils those lipids that are liquid at room temperature (e.g., olive oil, corn oil). Fats and oils are glycerol esters of fatty acids, such that each molecule of fat or oil contains 3 fatty acids that may be the same as or different from each other. Generally, they are different from each other. The glycerol esters are formed by a condensation reaction which results in the formation of 1 triglyceride molecule and 3 molecules of water.

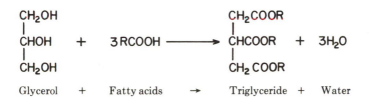

In the preceding formula, R is the symbol that represents one of man
different hydrocarbon chains. Usually, the Rs in the fatty acids wil
differ from each other, although they could be the same. However, eac!
of the Rs in the triglyceride must correspond to one of the Rs in the fatt
acids from which the triglyceride was formed. A hydrocarbon chai
simply means that it is made up of a number of carbon atoms bonde
together, as a string of beads, with one or two hydrogen atoms linked t
each of the carbon atoms. The carbon atoms on each of the 2 ends of th
string may sometimes have 3 hydrogen atoms attached to them.

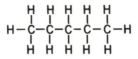

A hydrocarbon chain

Fats in nature usually contain long carbon chains, the carbons of whic
are mostly saturated with hydrogen with an acid (carboxyl) group at tł
end (e.g., butyric acid).

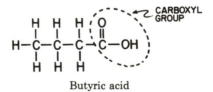

Butyric acid

The formula for butyric acid is usually written $CH_3 \cdot (CH_2)_2 \cdot COO$
Fats that are completely saturated with hydrogen atoms are calle
"saturated fatty acids." However, some fatty acids contain 1 or sever
groups of 2 adjacent carbon atoms in which the carbons are not ful
saturated with hydrogen (see the following formula).

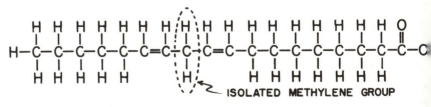

Linoleic acid

As can be seen, there are 2 sites of incomplete saturation shown by the double bonds (note that each of the unsaturated carbons has only 1 hydrogen atom attached to it). Fatty acids having 1 or more sites of incomplete saturation are called "unsaturated fatty acids." Unsaturated fatty acids having a given number of carbon atoms have a lower melting temperature than saturated fatty acids having an equivalent number of carbon atoms. Thus, linoleic acid (unsaturated, with 18 carbon atoms) melts at 23°F (−5°C) while stearic acid (saturated, with 18 carbon atoms) melts at 157.3°F (69.6°C). When a lipid is made up of many unsaturated fatty acids, it will be liquid at room temperature. On the other hand, lipids, such as suet or beef fats, containing comparatively few fatty acids with unsaturated groups, are solid at room temperature. Fatty acids which contain none or only 1 unsaturated group are not especially subject to oxidation, but those with more than 1 unsaturated group oxidize readily. However, oxidation proceeds much faster in fatty acids containing 1 or more sites in the carbon chain in which there is first a group of 2 carbons not fully saturated with hydrogen followed by a carbon saturated with hydrogen (an isolated methylene group) followed by another group of 2 carbons not fully saturated with hydrogen. In linoleic acid (see preceding chemical formula), the isolated methylene group is identified by the dotted enclosure. Notice the pair of incompletely saturated carbons on each side of the isolated methylene group. The greater the number of sequences of this particular configuration (arrangement of atoms) in the fatty acid molecule, the faster the rate of oxidation. Thus, when linoleic acid has an isolated methylene group (a sequence of 2 carbons not saturated with hydrogen, 1 carbon saturated with hydrogen and 2 carbons not saturated with hydrogen), it is said to oxidize 10 times as fast as when it does not have an isolated methylene group.

Oxygen, or a source of it, must, of course, be present for oxidation of fats to take place. However, a large amount of oxygen is not required for this reaction, and in foods, it is very difficult and commercially impractical to package them (under a vacuum or in an inert gas, such as nitrogen) so that sufficient oxygen is absent to prevent changes in fats. The reason for this is that in foods oxygen dissolves to some extent in the water present and also becomes trapped or occluded in the tissues.

Certain sources of energy accelerate the oxidation of fats. One of these is heat, and generally, the higher the temperature at which a fat is held, the faster the rate of oxidation. Light of certain wavelengths, especially wavelengths in the ultraviolet or near ultraviolet regions, accelerates the oxidation of fats. High-energy radiations, such as cathode, beta and gamma rays, also greatly accelerate the rate at which fats oxidize.

Metals and their compounds accelerate the oxidation of fats. The probable formation of metal soaps with small amounts of free fatty acids present in fats may be the active agents in accelerating oxidation when pure metals are involved. Much less than a part per million to only a few parts per million of metals is required to accelerate the oxidation of fats. It is considered that the following is the order of decreasing activity as pro-oxidants for fats—copper, manganese, iron, nickel, zinc and aluminum.

When fats oxidize, short chain carbon compounds containing one or more groups in which 2 adjacent carbons are not saturated with hydrogen are formed. These compounds have odors and flavors that are generally undesirable and unacceptable. This change in fats and in the fats present in foods is called rancidity. (This should be distinguished from hydrolytic rancidity, in which lipase hydrolyzes fat, splitting off short chain fatty acids, such as butyric acid.) Butyric acid has an undesirable taste and odor.

Oxidative-rancidity odors are generally sharp and acrid and usually may be described as linseed-oil-like, tallowy, fishy or perfume-like. The more solid fats, as beef and mutton fat, are more apt to be tallowy when rancid, while pork fat, vegetable oils (soybean oil, cottonseed oil, corn oil, etc.) and fish and whale oils usually have the linseed oil flavor and odor when rancid. Under some conditions, vegetable or marine oils may be fishy or have a perfume-like odor when rancid. This appears to be a preliminary stage which eventually develops to the linseed oil-like odor and flavor.

Oxidative rancidity may be a cause of deterioration in many forms of fish, meat and poultry; there are instances in which oxidation of fats may be responsible for undesirable changes in shellfish and even vegetables.

When food products are dried, even those which contain small amounts of fat, such as vegetables, shrimp and some fish, their fats may soon oxidize, since whatever fat or oil is present is now exposed to oxygen over a large area. The colors of certain vegetables (e.g., carrots), spices (e.g., red pepper and paprika), and crustaceans (e.g., shrimp) are due to chemical compounds known as carotenoids. These compounds contain many groups in which adjacent carbon compounds are not fully saturated with hydrogen. If the small amounts of fat present undergo the first stages of oxidation (peroxide formation), a fading of color may occur in these compounds, as from deep yellow or red to light yellow or the color may be bleached out entirely. Off-flavors and odors may or may not be associated with such changes in these foods, but the color change represents a deterioration of the product.

NONENZYMATIC BROWNING

There are several types of nonenzymatic browning. One type is caused by a chemical reaction known as the *Maillard reaction*. This chemical change is initiated by a combination of an amino acid and a sugar. The amino acid may be present in the food as a separate entity (free amino acid) or it may be an amino acid which is present in the food as part of a protein. The sugar must be of a reactive type, containing a reactive portion known as a carbonyl group.

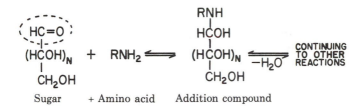

Sugar   + Amino acid   Addition compound

The carbonyl group in the sugar is identified by the broken-line circle. A carbonyl group also occurs in other chemical compounds, the aldehydes and ketones. Once the reaction is initiated, it proceeds through a long series of chemical changes to the point where complex compounds (the formulae of which are not exactly known) are formed. These compounds are flavorful and brown or black in color and cause a change in both the color and flavor of the food.

When sugars are heated at high temperatures, they turn brown and then black. This reaction involves the dehydration or removal of water from the sugar resulting, through a series of reactions, in the formation of furfurals, ring compounds of 4 carbons with 1 to 2 side groups. These ring compounds further combine to form complex chemical compounds which are brown or black in color and which have an odor and flavor entirely different from that of the sugar. This is called caramelization, although the flavor produced may be different from that which we know as caramel candy where the components of milk are involved in the typical flavor of the confectionery.

Caramelization proceeds very fast at high temperatures such as are attained when sugars are heated directly. However, at lower temperatures, as those encountered in the normal handling of many foods [usually above 50°F (10°C)] when the right conditions are present, caramelization will proceed fast enough to cause a deterioration of foods.

A third type of browning may be caused by the oxidation of ascorbic acid (vitamin C). Once oxidized, the same type of compounds may be formed as in the case of caramelization of sugars.

Browning may take place in foods of all moisture contents. Hence, in canning, at least part of the difference in flavor between that of the fresh food and that of the canned product is due to reactions that are manifest by browning, and some types of food cannot be canned successfully because of browning (e.g., scallops and cauliflower). However, browning proceeds fastest in foods of low to medium moisture content. Thus, some dried foods are especially subject to changes caused by nonenzymatic browning.

As has been pointed out, the color difference (and in some cases flavor) between fresh and heat-processed foods is probably due to browning reactions. The browning reactions can be minimized if canned foods are given a high temperature-short time process (HTST process) as compared with regular processing temperatures and times. However browning reactions may proceed in HTST-processed foods during storage if they are held at temperatures above 50°F (10°C). Apparently, browning occurs slowly, if at all, when the moisture content is less than 2%, and only by freeze-drying can the moisture in foods be reduced to this low level. Therefore, these products are freeze-dried to a low moisture content and packaged so that moisture is not resorbed from the atmosphere. For this reason, fruit juices are generally not dried except by freeze-drying. The moisture content must be held at low level throughout storage if the freeze-dried product is to be stable. Dried egg products whites and whole egg mixtures (usually spray dried to a moisture content of about 5%) are quite subject to nonenzymatic browning reaction resulting in a low acceptance of these products. For some time now methods have been available to remove the sugar (glucose) from these products prior to drying, resulting in much improved products due to the prevention of adverse nonenzymatic reactions.

Sugar can be removed from foods by subjecting them to fermentation by bacteria or yeast before drying. The microorganisms consume the sugar during the fermentation process, thereby eliminating one of the necessary components required in the Maillard reaction. Another process for preventing the Maillard reaction involves the addition of two enzymes (*glucose oxidase* and *catalase*). The *oxidase* converts glucose to gluconic acid which does not combine with amino groups, but one of the products of the conversion is hydrogen peroxide, an undesirable compound, and the role of the *catalase* is to break down the peroxide.

$$R \cdot CHO + O_2 + H_2O \xrightarrow[\text{oxidase}]{\text{Glucose}} R \cdot COOH + H_2O_2$$

Glucose   Gluconic   Hydrogen
                                                              acid   peroxide

$$2H_2O_2 \xrightarrow{\text{Catalase}} 2H_2O + O_2$$

Hydrogen
peroxide

While many processed foods may be adversely affected by nonenzymatic browning, there are other foods which are dependent on this
type of reaction for their typical odor, flavor and color. Maple syrup
develops its typical color and flavor because of the Maillard reaction
which takes place between sugar and an amino acid normally present in
the sap of the maple tree. The typical color and taste of prunes is due to
nonenzymatic browning which occurs when prune plums are dried and
stored. The flavor and color of coffee are due to nonenzymatic browning
which occurs in the components of the coffee beans when they are roasted
(comparatively high temperatures are used for roasting). In the cooking
of many foods, browning flavors and odors are produced. Roasted, broiled
or fried meats, or broiled or fried fish are examples of this type of change.

THE STRECKER DEGRADATION

The Strecker degradation is a reaction which takes place between an
amino acid and certain fragments of sugar or compounds produced by
bacteria which contain reactive groups known as dicarbonyls, as follows.

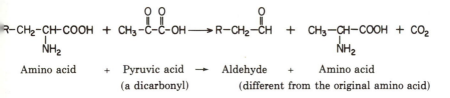

Amino acid   +   Pyruvic acid   →   Aldehyde   +   Amino acid
                     (a dicarbonyl)   (different from the original amino acid)

The reaction results in the formation of an aldehyde, a different amino
acid, and the splitting off of carbon dioxide. The new compound
containing the aldehyde group is flavorful, which may or may not be
desirable.

The Strecker degradation takes place when milk is heated to high
temperatures for long periods or during storage after milk is heated to

high temperatures for short periods and then stored at room temperature. This is probably the reason that fluid milk cannot be canned or heat-processed by ordinary methods. If fluid milk is given a high temperature-short time type of heating processing (HTST)—it will not develop the off-flavor, provided the canned produce is stored at temperatures below 50°F (10°C). Canned fluid milk, given a HTST treatment and held at room temperature, soon develops a caramelized flavor. In the manufacture of caramel confectioneries, condensed milk, cream, or some combination of the two, together with corn sugar, cane sugar, or both, are heated to high temperatures. The caramel flavor which develops is probably due to the Strecker degradation. This type of change in foods, therefore, is sometimes desirable and sometimes not desirable.

THE AGGREGATION OF PROTEINS

The aggregation of proteins causes deterioration of some foods. This change results in a bonding of protein chains to form a more closely knit network, and it would appear that in this process some of the water loosely held by the protein is squeezed out, causing drip from frozen foods upon defrosting. Protein aggregation occurs mainly in low-fat fish, such as cod, during frozen storage and is correlated with the liberation of free fatty acids from phospholipids by the phospholipase enzymes in the muscle. Recent studies tend to prove that the protein aggregation and toughening which occur in some fish during frozen storage are due to enzymes which cause a breakdown of trimethyl amine oxide (present in the flesh of some fish) to dimethyl amine and formaldehyde.

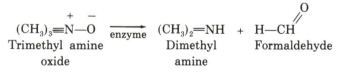

$$\underset{\substack{\text{Trimethyl amine} \\ \text{oxide}}}{(CH_3)_3 \overset{+}{\equiv} N \overset{-}{-} O} \xrightarrow{\text{enzyme}} \underset{\substack{\text{Dimethyl} \\ \text{amine}}}{(CH_3)_2 = NH} + \underset{\text{Formaldehyde}}{H - CH \overset{O}{\nearrow}}$$

It is well known that formaldehyde causes an aggregation or denaturation of proteins.

It has been theorized that in the presence of the free fatty acids, there is a reaction between free fatty acids and proteins to form a crosslinked network within the muscle, resulting in a close association of protein fibers.

Another theory of what happens during protein aggregation is that the splitting off of fatty acids destabilizes the protein molecules, causing them to form a closely bonded mass, since in normal muscle, the phospholipid itself, from which the fatty acids are split off, is bonded or conjugated with the protein.

Protein aggregation does not take place in high-fat fish, such as salmon. It is theorized that the reason is that the free fatty acids split off by phospholipases dissolve in the fat present in the tissues of high-fat fish, and due to dilution, become unavailable for bonding with the protein. There is also the possibility that the proteins in high-fat species are different from those containing little fat, hence are not destabilized by a splitting of fatty acids from phospholipids.

When protein aggregation occurs in frozen products, the tissues become tough and dry and lose succulence when cooked. This change is temperature dependent and does not occur to any extent in marine products held at temperatures below $-22°F$ $(-30°C)$. At $-40°F$ $(-40°C)$, cod held for 1 year may not be distinguished from the fresh product. The reason why the protein aggregation change does not occur at very low temperatures may be that water is not available to provide for the hydrolysis or splitting off of free fatty acids from phospholipids as brought about by the phospholipases, or that enzymes which decompose trimethyl amine oxide are not active.

Some toughening of certain meats may occur during frozen storage, but this has not been well documented and the reason for such a change has not been studied to any extent.

Section III

Food Processing Methods

10

Heating

The preservation of foods by heating started in the early 1800s, when a Frenchman named Nicholas Appert developed a method of preserving foods by sealing them in glass jars with a paraffined cork and heating the filled jars in boiling water. Napoleon had offered a prize for a method of preserving foods, and Appert received the prize in 1809.

In 1810, an Englishman named Peter Durand developed the metal cannister as a container for foods which were to be heat processed. This later became known as the can or tin can. At that time, cans were cut and soldered by hand and, after filling, were sealed by hand with solder.

In this country, canning was first started in Baltimore, but the first canning company to become really established was started in Boston by William Underwood in 1819. The Underwood Company still operates canneries in Massachusetts, in other parts of the United States and in other countries.

The process of heating foods in boiling water was, of course, not adequate to prevent spoilage, in many cases, since there are some bacteria which form spores that are not destroyed by heating in boiling water unless boiling is continued for at least 6 hr. However, it was not economical to boil foods for such long periods because of the amount of time involved and because of adverse effects of long heating on the quality of the food. It was necessary, therefore, to find methods of heating at higher temperatures. Since water boils at higher temperatures when it contains salts, the first attempt at using higher temperatures for heating canned foods consisted of heating these products in boiling salt solutions, for instance, in a solution of calcium chloride. The higher temperatures attained in this way shortened the time needed to destroy the bacterial spores. (As seen in Chapter 6, if spores are allowed to survive a preservation process, they will grow out under suitable conditions and cause spoilage of the foods and may even cause food poisoning. However, the salt water was extremely corrosive to the metal containers in which the foods were packed, so the process was eventually abandoned.

The next development was to process canned foods in steam under pressure. Steam under pressure has a higher temperature than steam that is not under pressure, and the higher the pressure, the higher the temperature of the steam (see Table 10.1). This method allowed for the heating of canned foods under conditions which would destroy bacterial spores without causing excessive deterioration of the foods. This meant, of course, that time was a critical element in the heating effects on food.

At first, steam pressures were obtained by placing some water in the bottom of a heavy metal container called a retort, filling it with cans of food, sealing it and heating the retort from the outside with flames from

TABLE 10.1. BOILING POINTS OF WATER AT 0–20 PSIG (0–1408 G/CM²) (AT SEA LEVEL)

| Pressure (at Sea Level) | | Temperature at Which Water Boils | |
(psig)	(g/cm²)	(°F)	(°C)
0	0	212.0	100.0
1	70.4	215.4	101.9
2	140.8	218.5	103.6
3	211.2	221.5	105.3
4	281.6	224.4	106.9
5	352.0	227.1	108.4
6	422.4	229.6	109.8
7	492.8	232.3	111.2
8	563.2	234.7	112.6
9	633.6	237.0	113.8
10	704.0	239.4	115.2
11	774.4	241.5	116.4
12	844.8	243.7	117.6
13	915.2	245.8	118.8
14	985.6	247.8	119.9
15	1056.0	249.8	121.0
16	1126.4	251.6	122.0
17	1196.8	253.4	123.0
18	1267.2	255.4	124.1
19	1337.6	257.0	125.0
20	1408.0	258.8	126.0

urning gas. Such a procedure was difficult to control, and on occasion, xcessive pressures were encountered which burst the retort and resulted n injuries. Eventually, methods of developing the steam outside of the etort and controlling the steam pressure in the retort by valves were volved. This process was essentially the same as that in use today.

As previously stated, the first metal containers for canned foods were nade by hand and, after filling, had to be hand-sealed by soldering. This nethod of sealing was still being used by some small food canning plants n the early 1900s. During this era, however, a method of machine nanufacture of cans and covers was developed. The cover contained a im to which a plastic gasket could be added, and the rim could be sealed ightly by machine by first crimping it over and under the flanged top of he can body and then pressing the two together by a second roller peration. The plastic gasket made the can air-tight by filling the tiny oids produced between the rim of the can cover and the flange of the can s they were brought together. This is essentially the method used today or hermetically (air-tight) sealed metal containers. Normally, cans are lescribed in terms of the dimensions of the diameter (D) and the height H) (see Fig. 10.1), and each of the dimensions is given in numbers aving 3 digits each. The first digit is in inches (1 in. = 2.54 cm), and the ext 2 digits are in 1/16 of an inch (0.16 cm). Thus, when a can is lescribed as a 202 × 214 can, this means that its diameter is 2 2/16 in. 5.40 cm) and its height is 2 14/16 in. (7.30 cm). Not all cans are round, owever, and as in the case of sardine cans, 3 dimensions are given, the

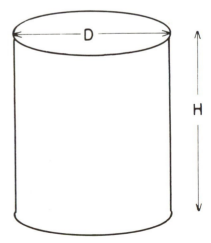

FIG. 10.1. D—DIAMETER AND H—HEIGHT OF A METAL CONTAINER

length, the width and the height. Various lacquers have been developed for lining cans, especially for preventing food discoloration that could occur as a result of interactions between the food and the can. Other innovations such as quick-opening cans have appeared on the market in recent years.

PRETREATMENT OF FOODS

It is generally necessary for foods to undergo a treatment prior to the canning process, but these pretreatments differ depending on the foods, and no attempt will be made to cover all of them here. Some pretreatments are applied to many different foods. One of these, usually applied to vegetables, is blanching. Vegetables are first washed, usually in water and detergent, then rinsed. They are then passed over belts, where any remaining foreign matter, such as weeds, stalks, etc., can be removed by hand. Blanching consists of heating in steam (no pressure) or hot water [usually about 210°F (98.9°C)] until the temperature of the food is brought up to about 180°–190°F (82.2°–87.8°C) in all parts (see Fig. 10.2), then cooling in water. Blanching shrinks the product, providing for a better fill of the container and removing gases, thus allowing better vacuum after sealing. The blanching process also destroys enzymes in the food which otherwise might react during the initial heating in the retort and cause discoloration or off-flavors in the product. When very high temperature, very short time methods of heat processing are used, enzymes might not be inactivated, which would cause development of off-flavors in the food. However, this can be prevented by blanching. Finally, blanching tends to fix the natural color of vegetables and it provides a clearer brine in the canned products.

VACUUM IN CANS

The Need for Vacuum

Foods are packed under vacuum for several reasons. If canned foods were not under vacuum, the cans would swell should they be stored at higher temperatures or lower pressures than those at which they were packed. Thus, cans packed at sea level at ordinary temperatures would swell due to expansion of gas within the can if the cans were shipped to Denver, Colorado, which because of its high altitude (about 1 mi [1.61

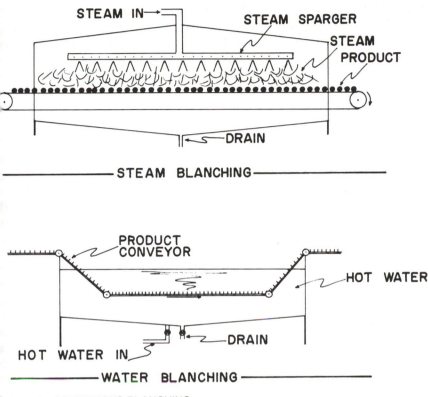

FIG. 10.2. CONTINUOUS BLANCHING

m] above sea level), is under reduced pressure, or if the cans were shipped to tropical areas or held in unusually hot places. When cans are in a swelled state, they are normally suspected of containing food that has spoiled, because when bacteria grow, many of them produce gases. Thus, a perfectly good canned food might be discarded because it could be suspected of being spoiled. Another reason for canning foods under vacuum is to remove oxygen which, of course, is in air. During heat processing, oxygen reacts with the food, causing undesirable changes in color and flavor of the product, as described in Chapter 9.

Finally, if at least some of the air is not removed from the container prior to sealing, then the product may have to be cooled under pressure (usually air pressure). Otherwise, the can will buckle (a permanent distortion along the cover seam), and such cans will be discarded. The reason for the buckling can be explained. During the heating process, the retort is under pressure, e.g., 20 psi (1.41 kg/cm²). As the temperature inside the can reaches the high temperature in the retort, moisture in the food or brine vaporizes, exerting pressure against the inside of the can. In addition, the residual air in the can tends to expand as the internal temperature increases, and the expanding air also exerts a pressure against the inside of the can. By the end of the process, the can is under equilibrated pressures from both the inside and the outside (see Fig. 10.3). In this situation, the can is not unduly stressed, because the internal pressure is counterbalanced by the outside pressure and vice versa.

At the end of the process, the external pressure, relative to the can, may be immediately released by shutting off the steam supply to the retort and opening the retort valve to the outside, allowing all its internal steam to escape. However, the internal pressure in the can cannot be released immediately because the can cannot be vented to the outside. Thus, the wall and lids of the can are under internal pressure, causing distortions, possibly even damage, to the seals (formed by the lids and the can body). This problem can be eliminated by replacing the hot steam in the retort with air under pressure. In this way, the heat source is removed without altering the pressure equilibrium. With the removal of the heat source, the contents of the can will cool down, condensing water vapor and cooling the residual air, returning it to its original volume, thus reducing the internal pressure until it no longer pushes against the inside of the

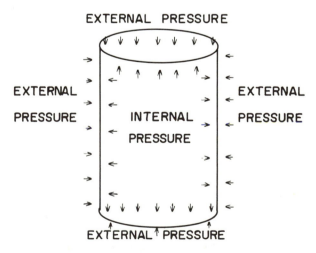

FIG. 10.3. PRESSURE EQUILIBRIUM DURING THERMAL PROCESSING

can. Removing as much air as possible from the container prior to processing tends to minimize buckling. However, even with vacuum packing, cans of large diameter must be cooled under pressure to prevent distortion of the container.

Some foods, such as whole kernel corn, may be packed under a very high vacuum in order to provide for good heat penetration, when only a small amount of liquid (canner's brine) is used with the foods. When this is done, a special (beaded) can with protruding ridges around the central part of the can body to strengthen the can may have to be used (see Fig. 10.4). Otherwise, the container body may panel or become flattened due to the difference in pressure between the inside of the can and atmospheric pressure on the outside.

Obtaining the Vacuum

Vacuum in canned foods may be obtained in several ways. One of the most common methods is to add hot food to the container. In this way, the residual air is removed, resulting in a partial vacuum. A second method is to add cold food to the container and to preheat it by passing it through a steam box uncovered, or only partially sealed, prior to sealing. Using either of these techniques, the heat causes the product and the air in the headspace to expand, pushing air out of the container. In addition, the water vapor in the headspace displaces air, and trapped air in the food is driven out. In this condition, the can is sealed, and when the product is cooled it will be under vacuum, since much of the air has been removed from the container.

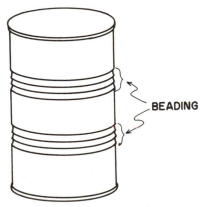

FIG. 10.4. BEADED CAN

A vacuum in canned foods can also be obtained by subjecting the container to a mechanical vacuum just prior to sealing it in the chamber A good vacuum can be obtained in this manner, but there are some limitations. For instance, if the product is packed in liquid, such as vegetables packed in canner's brine or fruit packed in syrup, when the vacuum is applied, much of the liquid may be flashed out. This is caused by the dissolved and occluded (trapped bubbles) air in the liquid which comes out as a gas when a sudden vacuum of high intensity is applied The sudden release of air causes some of the liquid to spatter out of the can. To avoid this, liquids must be subjected to vacuum treatment prior to filling into the can in order to remove dissolved and occluded air.

The third type of vacuum used in canning foods is called a steam jet vacuum. Just before the cover is placed on the can to be sealed, a jet of steam is forced over the contents of the can. This does not provide a high vacuum and only removes air from the headspace of the food in the can It is used mainly for materials packed without liquid.

LIQUIDS IN CANS

Vegetables and fruits are usually packed in liquid. Canner's brine, a weak solution of sugar and salt, is ordinarily used for vegetables, and sugar solutions which may be as concentrated as 55% sugar or as dilute as 25% sugar are used for fruits. These liquids afford some protection against heat damage because they permit convection heating, which occurs at a faster rate than does conduction heating. With solid packs such as tuna fish, corned beef hash, and even concentrated soups, such as pea or mushroom, heating takes place in the container through conduction. In convection heating (occurs when liquids are present), a hot layer of liquid rises along the sides of the can body (heat rises) travels over the top to the center and flows down the central axis as more hot liquid moves up the sides. This mixing serves to speed up the transfer of heat. Thus, in Fig. 10.5, the heat transfer pattern in convection can be seen to be more widespread.

In the conduction heating of solid packs, heat penetrates from all sides of the container, but since solid foods are poor conductors of heat, heat travels toward the center of the container only relatively slowly.

In convection heating, the slowest heating point in the container is along the central axis 3/4 to 1 1/2 in. (1.91 to 3.81 cm) from the bottom (can in the upright position). In conduction heating, the slowest heating point in the container is along the central axis at the geometric center of the container.

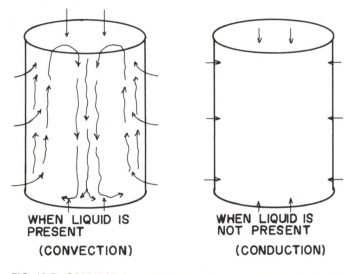

WHEN LIQUID IS
PRESENT
(CONVECTION)

WHEN LIQUID IS
NOT PRESENT
(CONDUCTION)

FIG. 10.5. CONVECTION AND CONDUCTION HEATING PATTERNS
IN CANS OF FOODS

FILLING THE CANS

Foods must, of course, be filled into cans before the cans can be sealed, and this is usually done by machine. For instance, with peas, a central hopper, which is kept filled with peas, is located above a rotating plate with openings through which the peas can fall. When the plate rotates to the point at which the plate opening is just below the hopper, the opening is just above a bottom container (below the plate) which when filled will hold a certain volume or weight of peas. The peas thus fall down and fill the bottom container which then rotates to a point where the peas are released into a can. With vegetables, after the food is added to the container, canner's brine is added automatically to fill the can, filling all voids and covering the vegetables. This liquid is usually added hot, and since such foods are not packed under vacuum, the hot liquid provides whatever vacuum will be present in the container after processing and cooling. Canner's brine is held in a reservoir on the filling machine and is heated in this reservoir. As will be explained later, the temperature of the brine should be held at 170°F (76.7°C) or higher.

Some products, such as whole tomatoes, must be placed in the can by hand. This is done with the use of a table containing openings along its edges. The peeled tomatoes are placed in the center of the table, and workers seated around the table pull the tomatoes over to the openings and allow them to fall through into a can located below the opening

Other foods, such as asparagus spears, must be entirely hand-packed The spears must be arranged within a metal band; the band is pulled tight in order to allow the product to be inserted into the can which has been laid on its side and is then released once the end of the group of spears has been inserted well within the can opening.

Semiliquid packs which heat by conduction, such as concentrated pea soup or mushroom soup, may be filled hot and can be filled volumet rically by automatic means. Solid packs like tuna fish and corned bee may be filled by hand. A small amount of hot liquid (brine or oil) may be added to provide some vacuum, or the product may be passed through a steam (exhaust) box to be heated and provided with some vacuum before the can is covered and sealed.

SEALING THE CANS

Cans are sealed automatically by machine. In sealing, the cover falls onto the top of the can automatically, the base plate of the sealer (upon which the can rests) raises the can with cover up tightly against the chuck. The edges of the can cover and the flanged body top are subjected to the action of two different rotating rollers. The first roller crimps the cover and body flange so that the edge of the cover is bent around and under the edge of the body flange; the second roller flattens and presses the top seam together so that it forms a tight seal with the help of the plastic gasket located in the outer rim of the cover (see Fig. 10.6).

In the canning operation, it is imperative that the sealing machine provide a tight seal. This can be checked by removing the cover to observe the configurations of the cover hook and body hook (see Fig. 10.7). The dimensions of the seal components may then be measured. It is known that such dimensions should fall within certain limits Tools for exposing the seam components and special micrometers for measuring their dimensions are available. A simple way to determine that the base plate, first operation (first roller) and second operation (second roller) of the sealing machine have been adjusted properly is to fill a can with boiling water, seal, cool, then measure the vacuum in the can by means of a gauge which reads in inches (1 in. = 2.54 cm) of vacuum. The vacuum gauge has a sharp shaft which is pushed through the cover of the can, and a rubber gasket prevents air from entering the can. If a vacuum of 10 in. (25.4 cm) or more is obtained, it can be assumed that the can is effectively sealed.

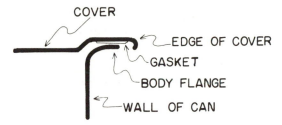

I. Prior to sealing operation

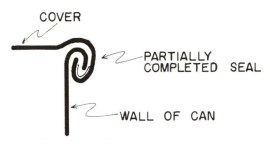

2. After first roller operation

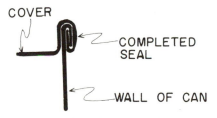

3. After second roller operation

FIG. 10.6. CROSS SECTIONS OF COVER AND BODY OF CAN AT THREE STAGES OF SEALING

The sealing operation should be checked at the start of the day's operation and periodically throughout the processing period. The importance of this is illustrated by the fact that a few years ago, the Food and Drug Administration picked up over a million cans of one food product that had not been properly sealed. The product had to be reprocessed. Such inadequacies in operational procedures can be responsible for the bankruptcy of food processing companies because of food poisoning lawsuits, leading to FDA stoppage of production, etc., that may result when cans are improperly sealed.

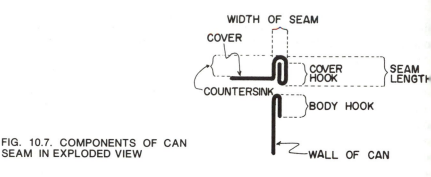

FIG. 10.7. COMPONENTS OF CAN
SEAM IN EXPLODED VIEW

THE HEAT PROCESS

The next operation in food canning is heat processing. It is probably the erroneous opinion of many people that heating times and temperatures for canned foods are applied more or less arbitrarily. Heating times and temperatures for canned foods are based upon the destruction of the spores of *Clostridium botulinum*, since this organism has been known to survive an inadequate process and thereafter grow out producing toxins which can cause sickness and, in some instances, death. To obviate this possibility, a system was set up to ensure the destruction of any spores of *Clostridium botulinum* which might be present in the food.

The first step in developing the new process was to find the most resistant spores of *Clostridium botulinum* which turned out to be those of certain type A strains. (There are 7 known types of *Clostridium botulinum*, each having a number of strains. More about this organism is found in Chapter 6.) The reason that the minimum process was necessarily based on *Clostridium botulinum* is that the toxin produced by the various strains of this particular organism is the most powerful toxin known to man. In one experiment performed by the authors, the toxin produced by one strain of type B was so powerful that when a bite-size piece of beef containing it was diluted about 10^8 (100,000,000) times, all the mice into which it was injected died. It is considered from these results that if a human had eaten a small amount of that meat, he would surely have died. Next, the heating times at various temperatures (for instance at 220°, 230°, 240°, and 250°F [104.4°, 110°, 115.6°, and 121.1°C]) that would destroy 60 billion spores of the organism were determined. A curve of thermal death times was then constructed so that the time at any temperature required to destroy 60 billion spores could be btained from the curve. The curve was generated by plotting the ogarithm of the time in minutes (y axis) versus the temperature °F) (x axis).

Next, thermocouples were placed in cans of food at the slowest heating oint (geometric center for conduction heating foods, 3/4–1 1/2 in. 1.91–3.81 cm) above the bottom of the center line for convection heating oods), and as the cans were heated in the retort (retort at a particular emperature), the temperature of the slowest heating point in the roduct was recorded at intervals of every few minutes after the steam as turned on. With the data from these 2 experiments, thermal death imes for spores of *Clostridium botulinum* and temperature of the roduct at the slowest heating point as it was heated in the retort could e tabulated, as shown in Table 10.2. Thus, the lethality rate could be alculated.

Next, a curve was drawn on rectangular coordinates in which the ethality rate was plotted on the ordinate, y axis, against time of heating n minutes on the abscissa, x axis (see Fig. 10.8). It should be noted that he amount of lethality accumulating before the temperature reached 12°F (100°C) was negligible, since more than 400 min at this temerature are required to destroy 60 billion spores of *Clostridium otulinum*.

Once the curve was drawn, the processing time required to destroy 60 illion spores of *Clostridium botulinum* could be obtained. For instance, f 1 in. (2.54 cm) of distance on the y axis were equivalent to 1/100 of a ethal effect (lethality rate), and 1 in. (2.54 cm) on the x axis were equal o 10 min of time, then 1 in.^{2} (6.45 cm^2) under the curve would provide /100 × 10 or 1/10 of a lethal effect, and it would require 10 in.^{2} (64.5 cm^2) nder the curve to obtain one lethal effect or to destroy 60 billion spores f *Clostridium botulinum*. The time under the curve, after the heat was urned on, corresponding to 10 in.^{2} (64.5 cm^2) would then be determined, nd this would be the processing time, steam on to steam off.

There are other methods for determining the processing time for anned foods, but these need not be discussed here.

ABLE 10.2. CALCULATION OF LETHALITY RATES

Time (t) After Steam Turned on (min)	Temperature (T) of Slowest Heating Point at Time t (°F)	Thermal Death Time (TDT) of Spores at Temperature T (min)	Lethality Rate (1/TDT)
t_1	T_1	TDT_1	$1/TDT_1$
t_2	T_2	TDT_2	$1/TDT_2$
t_3	T_3	TDT_3	$1/TDT_3$

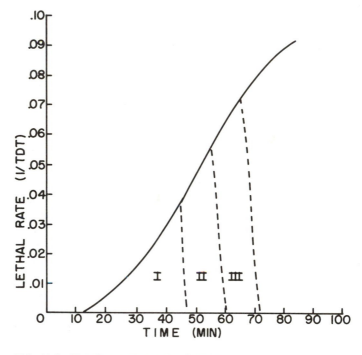

FIG. 10.8. TYPICAL LETHALITY CURVE

Actually, few canned products are subjected to a heating effect as lo
as that required to destroy *Clostridium botulinum*. The reason is tha
there are many bacterial spores that are more heat-resistant than thos
of *Clostridium botulinum,* and if a minimum botulinum kill were use
(heating sufficient to destroy 60 billion spores of *Clostridium botulinum*
the more resistant spores might survive, grow out and cause spoilage c
the food. A minimum *botulinum* kill is equivalent to the heating of a
parts of the food (this means the slowest heating part) to the equivaler
of 2.5 min at 250°F (121.1°C). Many foods are heated to the equivaler
of 7 min at 250°F (121.1°C) at the slowest heating point.

THE CONVENTIONAL HEAT PROCESSING CHAMBER (RETORT)

The ordinary retort (see Fig. 10.9) is the 3-crate retort in which 1 cra
of cans is placed on top of another in the retort. When cans are placed i
these crates, they may be allowed to fall into the crate haphazardly

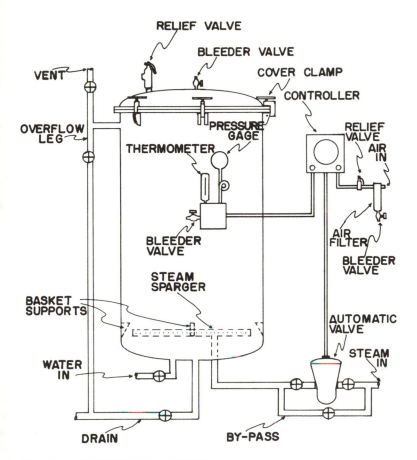

FIG. 10.9. CONVENTIONAL RETORT

ney may be placed in the crate on end in an orderly manner. If the latter ystem is used, perforated metal separators must be placed between the yers of cans. This provides for an adequate circulation of steam around ie cans which is necessary for optimum heat transfer to the product.

All retorts should be fitted with automatic steam valves which can be t to allow steam to flow into them to raise the processing temperature nly to the desired point. Retorts should also have a temperature sensor nside of the retort) attached to a recorder on the outside and a chart, iowing temperature and time of processing for each batch of product. rocessing charts should be identified with the code on the cans of each atch and should be kept on file. If the product is to be cooled in the tort after processing, an air pressure system with automatic inlet valve which can be set for a definite air pressure should be a part of th installation. In such installations, there should be an automatic pressur outlet valve attached. Otherwise, when cans with high vacuum ar cooled, the outside pressure may become high enough to cause panelin, of the cans.

All retorts must be fitted with fast-opening valves to allow venting o the retorts. When the steam is first turned on, the vent valve should b opened wide and left open until the temperature is raised to 220° (104.4°C). This removes from the retort air which would otherwise caus cold spots around some of the cans and thus prevent adequate heating The reason for this is that air is a poor heat conductor. Also, durin retorting, a small bleeder petcock at the top of the retort should be kep in the wide open position. This removes air which may come in with th steam and also provides for good circulation of the steam.

Often, especially in small canning operations, the steam pressure, an hence the temperature, are regulated by hand-operated valves, but thi should not be allowed. All retorts should be fitted with automatically operated valves. When valves are hand operated, considerable huma: error can occur.

COOLING HEAT-PROCESSED FOODS

Cans of food which have been heat processed may be cooled in th retort by allowing water to flow in after the steam has been turned off, i which case they may be cooled under a pressure of air or steam exerte over the water level in the retort. In other operations, the retort may b blown down and the crates of cans removed and moved slowly through cooling canal. In either case, the cooling water should be potabl (drinkable) and should not have a high bacterial count. The reason fc this is that as the vacuum is forming in the can due to the cooling, th gasket in some cans may be soft enough to permit microscopic amoun of the cooling water to be sucked into the cans. If the cooling water is hig in bacterial count, enough bacteria, which may cause spoilage or eve disease, may thus enter the can and contaminate the product. A illustration of what can happen in such circumstances is the typhoi outbreak in Scotland a few years ago. This was caused by corned be produced in Argentina which had been cooled in polluted water contai ing disease causing bacteria. Thus, it is desirable to chlorinate coolin water so that it contains a residual of 5 ppm available chlorine. This lev of chlorine is enough to keep the water relatively free from bacteria.

During cooling, the average temperature of the product in the ca should be brought to 95°–110°F (35°–43.3°C) as quickly as possible. Tl eason for this is twofold. Firstly, if this temperature is maintained, the urface of the can will be warm enough to evaporate moisture remaining rom the cold water bath. Should this water not be evaporated, it would ause rusting of the outside of the can, possibly spoiling the label and naking the container unsightly, so that it would have to be rejected. Secondly, if the temperature is not lowered to the above-stated range, it nay be sufficiently high to promote the growth of residual thermophilic acteria (see Chapter 7) which would cause spoilage of the food within he can.

Thermophilic bacteria form spores, and the spores of many types of hermophilic bacteria are unusually heat resistant. Therefore, in canning perations, an attempt is made to keep as many thermophilic bacteria ut of the product as possible. This can be accomplished by preventing he buildup of these bacteria at various points in the canning operation n the plant. The commercial process used today will reduce the number f thermophiles in foods to a minimum so that only a few thermophilic pores will be present in the product. The few remaining spores will cause o problems, provided the product is held at temperatures below 110°F 43.3°C). Therefore, in cooling freshly processed cans, the average product temperature is brought to 110°F (43.3°C) or below to prevent the rowth of thermophiles. This is necessary, since thermophilic bacteria equire only a few hours of growth to cause spoilage.

After or before cans are heat-processed, they must be washed to emove grease and food particles which get onto the outside of the can luring the various procedures involved in the operation. This is done by assing the can through alkaline or detergent solutions, then through inse water. If done after heat processing, the temperature of the rinsing olutions should be high enough to provide for evaporation of the water, hus preventing corrosion of the outside of the container.

After cooling, cans of food are usually packed into cases or stored in arge masses in a warehouse in conditions in which they cool very slowly o room temperature. To prevent growth of thermophiles in the product rior to canning, canner's brine in the reservoir on the filling machine hould be held at 170°F (76.7°C) or higher, a temperature at which no rowth could take place.

RECENT INNOVATIONS IN HEAT PROCESSING

There are a number of types of processing retorts for canned foods vhich have come into use in recent years and which are quite different rom the conventional retort.

Continuous Agitating Retort

One of these is the continuous agitating retort. With this system, cans enter the retort continuously on a conveyor through a special inlet which prevents loss of steam. In the retort, the cans are conveyed back and forth for whatever period of time is necessary for sterilization. Also, while being conveyed, they are rotated around their long axes. Except for solid packed foods, this action causes some agitation of the product within the can, speeding up heat penetration and shortening the processing time. Cans exit from the continuous agitating retort through a special valve and enter the cooling system which is set up much in the manner of the retort except that it is filled with cooling water.

In another type, the agitort, the cans or containers of cans are attached to a wheel which rotates during processing. In this system, the cans are rotated end over end, so the air in the headspace will travel along the sides of the can to mix the food, and back again (see Fig. 10.10). The agitation of the contents hastens heat transfer even in semi-solid packs. In this system, the cans are cooled within the agitort.

The Hydrostatic Cooker

Hydrostatic cookers (see Fig. 10.11) are now used in Europe and to some extent in this country. In this system, there is a central chamber with a narrow entrance on one side and an exit chamber on the other side. The system is partially filled with water, and when steam is turned on in the central chamber, the water is forced up to higher levels in the entrance and exit legs (chambers).

Cans enter through the warm water of the entrance chamber on a conveyor, gradually entering warmer water as they approach the steam

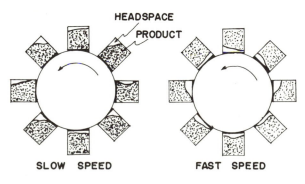

FIG. 10.10. HEADSPACE PATTERNS IN FAST SPEED AND SLOW SPEED AGITORTS

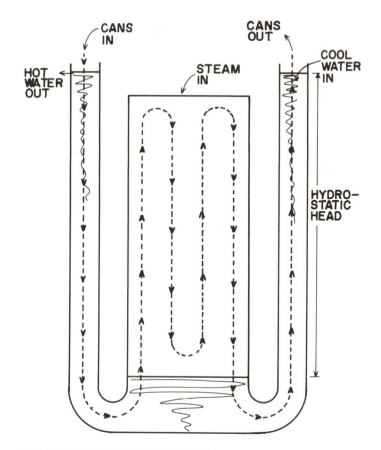

FIG. 10.11. HYDROSTATIC RETORT

hamber. They are conveyed through the steam chamber at tem-
eratures and for times which provide for commercial sterilization of the
roduct. They are then conveyed out of the steam chamber, first through
ie warm water and eventually through the cold water of the exit
hamber. The cans may be rotated around the long axis as they are
arried along. Because the pressure in the hydrostatic cooker depends on
ie water head (height of water), this type cooker is quite high (about 40
 [12.2 m]).

The agitort and the hydrostatic cooker are somewhat faster than other
eat processing methods, both because of the agitation which speeds up
eat penetration and because in such conditions somewhat higher
emperatures (up to 270°F [132.2°C]) may be used without causing
xcessive heat damage to the product. This allows for the canning of

some semi-liquid foods, such as mushroom soup or cream-style corn, in large cans. Since the labor involved and the cost of the container per unit weight of food are smaller as the can size is increased, it is economical to can foods in larger containers. This is desirable for institutions feeding large numbers of people, because it costs them less to buy and handle large cans.

HTST Process

There are systems for processing canned foods at high temperatures for short times. These are referred to as HTST processes. In such systems commercial sterilization is achieved at temperatures of 280°–300°F (137.8°–148.9°C) in 15–45 sec. Large discrete particles cannot be processed by HTST methods because they require some time for heat to penetrate their centers. HTST methods are applied only to liquids, and to foods which have been puréed (mashed bananas, concentrated pea soup, etc.).

Aseptic Fill Method

In one system, the puréed material or liquid is passed in thin layer through a heating system wherein the temperature of the product is raised quickly. The product is then pumped to a holding chamber where it is held at high temperatures for 10–20 sec, then pumped in thin layer through a cooling system wherein the temperature of the product is lowered quickly. It is next pumped to a filling system where it is forced into presterilized cans, and the cans are sealed with presterilized covers. Such systems must be presterilized with very high temperature gas or steam before the operation is started. Moreover, the filling and sealing of cans must be carried out under high temperatures, provided by steam or hot gas, to prevent the entrance of bacteria.

Cooking Under Pressure

Another HTST method of sterilizing canned foods is carried out in pressure chambers. As the pressure of the air in a chamber is raised above atmospheric pressure, the temperature at which water boils is raised; and if suitable pressures are used, temperatures of 280° (137.8°C) or higher may be attained. In these conditions, foods which contain discrete particles may be heated in open steam kettles in the pressure chamber for sufficient time to allow sterilization at the center of the food particles, then filled into cans and sealed. The cans are then inverted, allowed to stand for a few minutes, then cooled. It is no necessary to presterilize the cans in this case since the high temperature of the product, when filled, will destroy whatever bacteria may be adhering to the inside. The containers are inverted to ensure contact between the bacteria adhering to the inside of the cover and the very hot product which would, of course, destroy the bacteria. Personnel working in chambers under pressure may need to be pressurized slowly in locks when entering and depressurized slowly in locks when leaving. Otherwise, they may be subject to the bends, a condition caused by nitrogen dissolving in the blood when humans are subjected to high pressures, and coming out as gas bubbles when the pressure is suddenly released.

A system employing both cooking under pressure and aseptic filling is now being used to can such products as chicken salad.

Containers

Cans.—For general commercial applications, the speed with which cans may be handled during the process and the structural protection they lend to the contents may make them still the most desirable overall container for heat sterilized foods.

Flexible Pouches.—The use of flexible pouches as containers for heat sterilized foods offers several advantages over metal cans and glass jars. (1) Flexible pouches are lighter and require less handling both when filled and empty, and they are cheaper to ship. (2) Their shape adjusts to space limitations, so they require less space during handling and shipping. This is especially true when they are empty, as they lie perfectly flat, requiring much less space than either metal cans or glass jars. (3) When pouches are filled, they have at least one relatively small dimension—the thickness, which is about 1 in. (2.54 cm). Neither cans nor jars have a dimension this small; therefore, heat penetration into the central part of the pouch will be more rapid as it has a shorter distance to travel. Whenever there is a time benefit, there is a benefit in quality, too, since the food undergoes less thermal degradation. (4) The thickness of pouch materials is less than that of metal cans and much less than that of glass jars, and there is evidence that heat penetrates pouch materials more rapidly because of this. (5) Pouch materials are not subject to corrosion, as are metal cans (the introduction of aluminum as a material for cans partially nullifies this advantage), or to breakage, as are glass jars. (6) Pouches are more easily opened and do not require a special device like a can opener. (7) Pouches do not rely on tin which is used for coating cans. (However, a recent innovation in can manufacture requires no tin. In place of the tin, the steel is coated with a special chromium oxide.)

Unfortunately, there are also disadvantages to the using of pouches fc containing sterilized foods. (1) Pouches cannot be filled as rapidly a metal cans or glass jars with existing equipment. (2) The pouch awkward to handle when the size is increased, and the filling and sealin of large pouches is difficult. (3) The pouch does not give structur protection to fragile contents. (4) Although pouch materials are relativ ly strong, they do not resist tearing or cutting action as well as either th metal can or the glass jar.

The materials used in pouches are generally composite laminates an quite strong. They include an outer layer of polyester, which has near the tensile strength of steel and resists wear and tear, an inner layer nylon-11, also a tough material that resists wear and tear and also sea well and lends other desirable characteristics to the pouch material, an aluminum foil as a middle layer for eliminating light from the conten as well as for making the pouch material impermeable to gases. Th layers are cemented together with an adhesive.

Glass Containers.—As a container material, glass has several de sirable properties. It permits visibility of the product, imparts no tas and is noncorrosive. The use of glass containers precedes by centuries th use of metal cans and, of course, flexible pouches. The advances i closures of glass containers are probably more dramatic than th advances in the design of the containers and, in fact, have forced th changes in container design. Cork closures have been perhaps the mo important. Cork is light, compressible, hermetically sealable, ine pensive, plentiful and stable. Corks have long been in use for closin glass bottles, and still are. Ordinary cork closures (stoppers), howeve are unable to withstand the buildup of internal pressure, except whe the pressure is not too great and when they are used with bottles havi small openings, such as in wine and champagne bottles. The diamete of the openings of glass jars used for preserving fruit and vegetables a so large that a fitted cork would be easily pushed out (remember that f any given pressure, the total force increases as the area of the opening the glass container increases). Even for small openings, it was necessa to develop holddown devices, such as clamps and wire fasteners, whe high pressures were expected, as in the bottling of beer. The crimpe metal crown with cork, and later, plastic, liner was, and still is, a effective closure for beer and soda pop bottles.

Variations of screw-top, wide-mouth jars evolved to the modern Maso jar (named after its inventor). Tight fitting, screw-type closures find wide variety of uses in modern glass containers for commercial pasteurized and heat-processed foods as well as in home-canning opera

:ions. In home-canning, a Putnam-type closure (wire clamp with an ?ccentric lever that forces a glass lid tightly against a rubber gasket that 'its over the lip of the jar) is widely used. The Putnam-type container is]uicker to close and to open and generally requires less strength for ✧sage.

However, it was not until more advanced jar closures were developed ;hat canning in glass became significant. These closures include the ?hoenix cap (see Fig. 10.12); the Sure Seal cap and the Vacuum Side 3eal cap, types found in many tumbler-sized tapered jars like those used 'or cheese spreads and jellies; and the Amerseal cap, a modified screw ype cap as found in capping jars for apple sauce, jellies and wide-mouth 'ruit juice bottles (see Fig. 10.13).

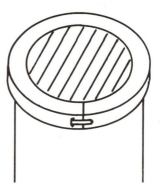

FIG. 10.12. PHOENIX 2-PIECE CAP

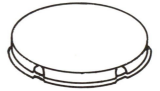

FIG. 10.13. AMERSEAL CAP

:ANNING OF ACID FOODS

Acid foods are considered to be those which have a pH of 4.5 or lower. ✧cid foods which are canned need not be heated at high temperatures to ✧ttain commercial sterilization. The reason for this is that bacteria, ✧cluding those which form heat resistant spores, are more easily ✧estroyed by heat when present in acid solutions. Moreover, spore- ✧rming bacteria generally will not grow in foods having pH values of 4.5

or less. There are some exceptions to this; for instance, *Bacillus thermo-acidurans* may grow in tomato juice (maximum pH 4.5) and cause spoilage.

Acid foods are ordinarily processed by heating the cans in boiling water until all parts of the product have reached 180°–210°F (82.2°–98.9°C), and then are cooled. An exception is tomato juice which is now often processed by flash heating to 250°F (121.1°C), holding at this temperature for 0.7 min, cooling to 200°–210°F (93.3°–98.9°C), filling into presterilized cans, sealing, and inverting the can so that the sterilizing effect of the heat [200°–210°F (93.3°–98.9°C)] at that pH will act upon the can cover.

Those foods which have a pH of 4.5 or less are apples and apple juice, apricots, blackberries, blueberries, boysenberries, cherries, cherry juice, all citrus fruits and their juices, currants, gooseberries, loganberries, papaya juice, peaches, pears, pickles, pineapple in various forms and pineapple juice, plums, prune juice, raspberries, rhubarb, sauerkraut and sauerkraut juice, strawberries, tomatoes, tomato juice and young-berries.

WAREHOUSE STORAGE OF CANNED FOODS

When canned foods have been heat-processed and cooled and the cans have been cleaned and dried, they are either stored in warehouses in bulk until labeled, cased and shipped out, or they are labeled, cased and stored in a warehouse until shipped out. Lithographed cans may be used, in which case labeling is not necessary.

Warehouses should be maintained so that the temperature does not rise much above 85°F (29.4°C) or fall below 50°F (10°C). Very high temperatures may promote the growth of thermophilic bacterial spores present in small numbers in the food. Very low temperatures may lower the temperatures of the cans to the point that in sudden hot spells the cans will sweat (condense moisture), eventually causing external corrosion of the cans.

11

Drying

The preservation of foods by drying is probably the oldest food preservation process practiced by man. It is believed that many foods, especially grains and fruits of high sugar content, were preserved by primitive man by allowing them to dry in the sun. Spices and fish, cut into thin strips, were also preserved in this manner.

There are a number of different methods of drying foods for preservation. The most important are sun drying, tunnel or cabinet drying, drum drying, spray drying and freeze-drying.

PRETREATMENT

Foods to be dried must be washed, and some peeled and cut. Others may be precooked. Cut fruits are subject to darkening through enzyme action, and must be either blanched, treated with salts or exposed to the fumes of burning sulfur (to provide sulfur dioxide). Certain vegetables may be pretreated in the same manner. Sulfuring may also be required to limit nonenzymatic browning (the Maillard reaction). Browning refers to the development of brown color. Various dried egg products (egg white, dried egg yolk and dried whole egg products) are also subject to browning and are susceptible to the development of off-flavors. In this case, the reaction involves a combination of a small amount of glucose, which is naturally present, with the proteins. Because of this, dried egg products, especially egg white, are either allowed to undergo a natural fermentation (this involves the growth of bacteria) or are treated with enzymes (glucose oxidase and peroxidase). Enzyme treatment converts the glucose to a compound which does not react with proteins. The natural fermentation method of removing sugars is not desirable from the standpoint of sanitation. However, by holding the product at 130°F (54.4°C) for some hours after drying, any disease-causing bacteria which might have survived or even increased in the product during the natural fermentation process should be eliminated.

Since prunes are naturally coated with a thin layer of wax, drying is greatly speeded up by predipping the fruit in dilute lye solution, then in hot water, prior to drying.

METHODS OF DRYING

Sun Drying

Sun or natural drying is still used in hot climates for the production of dried fruits or nuts. This may be done in direct sunlight or in shaded areas where the drying is accomplished by the hot dry air. It should be apparent that sun-dried fruit is produced only in areas where the climate provides periods of relatively high temperatures, relatively low humidities and little or no rainfall. Prunes, grapes, apricots, peaches and pears are dried in this manner. Some of these fruits are also dried in tunnel or cabinet dehydrators.

In sun drying, small fruits are prepared and spread on trays to dry in the sun for several days, then stacked to complete the drying cycle in shaded areas. Larger fruits, such as apricots, peaches and pears, are halved and pitted, and apples are peeled, cored and sliced prior to drying. Such fruits are sulfured to prevent enzymatic browning. Sun drying times vary between 4 and 25 days, depending upon the size of the product, the type of pretreatment, etc. During sun drying, precaution must be taken to prevent contamination from wind-blown dust and dirt. Moisture contents in sun dried products vary between 10 and 35%, depending upon the tendency of the dried product to absorb moisture. After drying, some fruit may require moisture-vapor proof containers.

Hot Air Drying

When mechanical dehydrators are used, the product is placed on metal mesh belts in a tunnel, or in a cabinet on trays where controlled, elevated temperatures are used. Heated air is circulated by blowers and the air temperature, relative humidity and air velocity are controlled. Hot air driers of this type are classified as parallel flow, counter flow, direct flow or cross flow, depending upon the direction in which the product moves in relation to the direction of flow of the heated air. In bin, loft and fluidized-bed driers the heated air is blown upwards through the product.

The hot air used in tunnel or cabinet driers may or may not be recirculated. If it is recirculated, the relative humidity must be carefully regulated since during each passage over the food, the air takes up moisture, raising its relative humidity. (Relative humidity is the percentage of moisture in the air at a particular temperature based on the total amount of moisture which air can hold at that temperature.) Actually, the relative humidity may be controlled by the amount of air recirculated.

The air in dehydrators is heated either by steam tubes or coils, or by being mixed directly with the combustion gases of gas or oil. Electric resistance heaters are used in rare instances. In all cases, except natural draft driers, the hot air or gas is circulated by blowers or fans of different designs and is discharged from the drier through a ventilator which may be equipped with a fan to increase its capacity, thus increasing the amount of air that can be circulated through the drier (see Fig. 11.1).

The time required for the drying of a particular product depends upon the characteristics of the raw material (moisture content, composition,

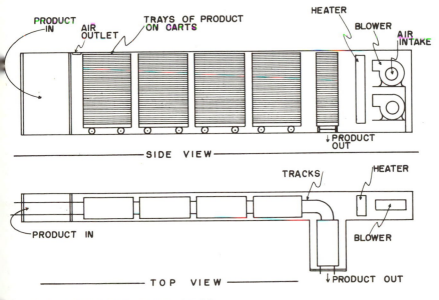

FIG. 11.1. CONTINUOUS TUNNEL DRIER

shape and size), the temperature and humidity of the air in the drier, the rate of air circulation in the drier, etc.

Initially, drying of foods occurs through evaporation of moisture from the food surface. Later, in the drying cycle, drying involves the diffusion of water, water vapor, or both, to the surface of the food.

In the initial stages of drying, air velocities are usually regulated at about 1000 ft/min (304.8 m/min), but in the later stages, the air velocity is usually lowered to about 500 ft/min (152.4 m/min), since this rate will remove all moisture available at surfaces at this point of the drying cycle. High initial rates of drying are said to prevent adherence of the dried particles to drying trays and belts, thus facilitating unloading. Water-repellent plastics, such as polyethylene or Teflon, may be used for coating trays and belts to prevent sticking of the dried food.

Due to evaporation of moisture which lowers the temperature, product temperature is below that of the air in the drier during the initial stages of drying and up to a point where approximately 1/2 of the moisture has been evaporated. The product temperature then starts to rise, and at the end of the drying cycle approaches that of the air.

Foods undergo some form of breakdown, or loss in quality, when they are exposed to heat. The amount of damage they undergo increases as the temperature to which they are exposed increases, and it also increases as the time of exposure is increased. Therefore, it is important to control both time and temperature during the drying cycle. In the beginning of the drying cycle, higher temperatures can be used for two reasons. (1) The amount of surface water in the foods is highest in the beginning of the drying cycle, so most of the energy of the hot air is expended in vaporizing surface water, and (2) the tendency of the food to be heated by the hot air is partially counterbalanced by the cooling effect of the evaporation of the surface water.

In the later stages of drying, the amount of surface water in the product is relatively low; therefore, it is prudent to lower the temperature of the air to a point where the energy is just sufficient to vaporize the surface water. If the air temperature is not lowered, the excess energy will go into raising the temperature of the product, which by this stage is not being cooled sufficiently by evaporative cooling. For the drying of vegetables, the initial temperature of the air is 180°–200°F (82.2°–93.3°C). In the later stages of drying, the temperature is reduced to 130°–160°F (54.4°–71.1°C).

Fluidized-bed Drying

In fluidized-bed drying (a special type of hot air drying), the product is fed in at one end to lie on a porous plate and is agitated and moved along towards the exit at the other end by hot air which is blown up through the product. The air which has passed through the product, picking up moisture, exits through an outlet at the top of the drier.

Drum Drying

Milk, fruit, and vegetable juices and purées, cereals, etc., may be dried with drum driers. These products are allowed to flow onto the surface of two heated stainless steel drums rotating in opposite directions with little clearance between them. The product dries on the drums and is scraped off by stationary blades fixed along the surface of the drum (see Fig. 11.2). Refrigeration may be used to lower the temperature of the dried product quickly. Drum drying can also be carried out under vacuum, in which case the drying is accomplished at lower temperatures and the product is protected from oxidation. In the ordinary drum drying, the process is controlled by varying the moisture content of the raw material (preconcentration), the temperature of drum surfaces, the space between the drums, the speed of rotation of the drums and the amount of vacuum applied.

Spray Drying

Milk, eggs, soluble or instant coffee, syrups and other liquid or semi-liquid foods are spray dried (see Fig. 11.3). The liquid material is sprayed into the top of a chamber simultaneously with hot air, which is also

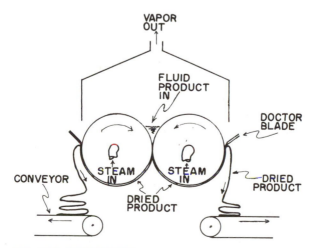

FIG. 11.2. DRUM DRIER

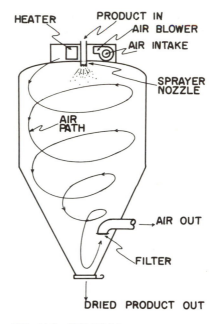

FIG. 11.3. SPRAY DRIER

blown in at the top. The cool, moist air exits near the bottom, and the dried particles fall to the bottom and are collected by gravity flow, or by the aid of scrapers which may also be used to remove dried material from the walls or bottom. Cyclones (conical-shaped collectors) may be used to collect particles escaping with the exit air.

Particle size is an important factor in spray drying, and the liquid is, therefore, dispersed into the drying chamber through a pressure nozzle or by centrifugal force generated by a disc rotating at high speed. Both methods atomize the liquid producing droplets to a size which will be dried by the heated air through which they must fall on their way to the bottom of the drying chamber. The material may be preconcentrated in some instances, or aids, such as gums, pectin or milk solids may be added prior to spray drying. In any case, it is necessary to lower the moisture content and temperature of the product to the point where particles will neither stick together nor stick to the wall of the spray drier.

In spray drying, the temperature of the food reaches a maximum of about 165°F (73.9°C) and only remains there for a short period of time. However, in spray drying the moisture content can be lowered only to about 5%. Thus, spray drying does not cause much heat damage to foods, but because of their relatively high moisture contents, spray dried foods may undergo some spoilage during storage. The shearing action of the

atomization step of spray drying may affect the functional characteristics of some proteins, such as those of egg products (whipping qualities,
etc.).

Spray drying tends to produce fine powders which neither wet nor
disperse readily when reconstituted. The material thus tends to clump
and form a mass through which water does not penetrate. In order to
avoid this, a process known as agglomeration may be used. In this case,
the dried material is slightly rehumidified; for instance, dried milk
powder may be heated with steam under controlled conditions which
surface-moistens the product. By other methods, the preconcentrated
product is injected into the drying chamber with a steam injector. The
latter procedure causes the dried product to form in the shape of beads
or bubbles which break into flakes that disperse readily in water.
Vacuum puff drying may also be used to produce a product which
disperses in water without forming clumps.

Freeze-drying

The freeze-drying of foods is carried out by first freezing the product
and then subjecting it to a very high vacuum wherein temperatures are
high enough to assist in the evaporation of moisture but low enough to
prevent melting of the ice in the product (see Fig. 11.4). In this method,
the water, existing as ice in the food, is evaporated directly as a vapor
without passing through the liquid phase (sublimation). The vapor is
condensed outside the evaporation chamber. The resulting product has a
honeycombed structure containing much surface area, and because it
maintains its original shape, its specific gravity is reduced considerably,
as is the moisture content.

In freeze-drying, the maximum surface temperatures used depend
upon the composition of the food. Some foods, such as vegetables and
mushrooms, can withstand 180°F (82.2°C); others, such as fatty fish,
require temperatures as low as 100°F (37.8°C). The vacuum applied
must be very high (less than 0.02 in. [0.05 cm] mercury), and most foods
require even higher vacuums for good results.

As in the case of foods dried by other methods, it may be necessary to
blanch or to treat the foods prior to freezing them, or to treat them with
chemicals to provide a source of sulfurous acid to inactivate or inhibit
enzymes. This is especially the case with vegetables (garlic and onions
excepted) and mushrooms. Meats are sliced or cut into small enough
portions to permit reasonably fast drying. Chicken is usually precooked,
boned and diced prior to freezing and drying.

While it has been stated that faster freezing rates with the resulting
uniformly distributed small ice crystals are preferable for foods which are

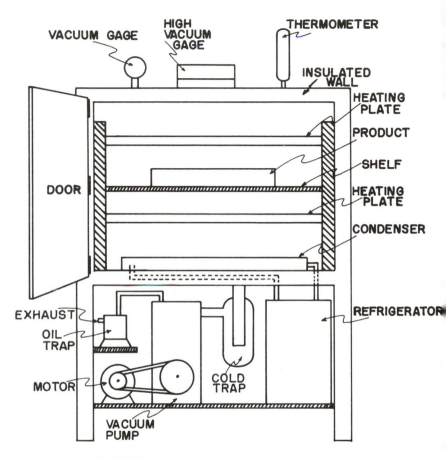

FIG. 11.4. FREEZE DRIER

to be freeze-dried, some investigators have found that slower rates of freezing result in better rehydration (reabsorption of water) of the dried food, hence better quality. The temperature to which the frozen food is brought prior to the drying process will depend on the product itself and should be low enough to rule out significant amounts of melted material. For most foods this temperature is around −7°F (−21.7°C), but for some fruit juices may be as low as −26° to −30°F (−32.2° to −34.4°C).

During freeze-drying, by most commercial methods, the highest food temperature is that of the dried surface layers. The temperature of the ice in the food is determined by the degree of vacuum present in the evaporation chamber. The higher the vacuum, the lower the ice temperature.

Freeze-dried foods are dehydrated to lower moisture contents than those dried by other methods. Usually, the moisture content of the freeze-dried food is below 3%. Since large surface areas are present in freeze-dried foods, it may be necessary to use an inert gas, such as nitrogen, to break the vacuum and so prevent chemical changes, such as oxidation of fats, which may occur shortly after contact with air is made.

Puff Drying

Puff drying may be used for some foods which are temperature sensitive, such as fruit or vegetable juice concentrates. With this method, the product may be heated in an oven and suddenly subjected to a high vacuum. In other applications, the product is evacuated to remove air, then heated with steam, then puffed by reapplying the vacuum. Puffing may also be attained by raising the temperature of the food under conditions which raise the temperature of the water in the food above 212°F (100°C) to provide pressure. The product expands or puffs when the external pressure is released. This is done with some cereals, providing a porous, puffed structure.

RECONSTITUTION OF DRIED FOODS

Dried foods ordinarily must be reconstituted (water must be added back to them) before they can be eaten. During drying and during storage, changes take place in the foods which affect the rate and extent to which water will be taken up by the dried product. Changes in protein and other components cause dried foods to have a somewhat lower rehydrated moisture content than that of the raw material from which the dried product was made. There is also a redistribution of soluble components during rehydration.

It has been observed that some dried foods rehydrate better when the water is held at low temperatures (below 40°F [4.4°C]). Other foods are said to rehydrate better when the water used is at higher temperatures, with some foods rehydrating satisfactorily in boiling water.

PACKAGING OF DRIED FOODS

In the packaging of dried foods, some products require only minimal specifications (cereals, some vegetables, etc.). Others require packages which are essentially moisture-vapor proof. Hygroscopic materials, especially some dried fruit juices, which readily take up moisture, must be packaged to prevent moisture from entering.

Freeze-dried foods have a very low moisture content. If their low moisture content is not retained, they lose their desirable characteristics because of nonenzymatic browning (see Chapter 9). Moreover, since the water has been removed, such components as fats are exposed to the oxygen of the air over a large surface area, subjecting them to an accelerated rate of oxidation which eventually leads to rancidification and off-flavor development.

Certain freeze-dried foods, such as fatty meats (especially pork), lobster meat, crab meat, shrimp, etc., require protection against both oxidation and moisture absorption. They must, therefore, be packaged under vacuum or in an inert atmosphere in packages which are impervious to both moisture and oxygen, such as hermetically sealed metal containers and flexible pouches made of laminated aluminum foil and plastic. Fruit juices, dehydrated whole milk and certain freeze-dried egg products must also be protected against both oxygen and moisture.

Some products, such as certain freeze-dried vegetables and mushrooms, need only be protected against moisture.

THE EFFECT OF DRYING ON MICROORGANISMS

The main purpose of drying foods is to lower their moisture content to a particular level that will exclude the growth of microorganisms (bacteria, molds and yeasts). For any given moisture content, one food may support the growth of microorganisms while another will not. Whether or not microorganisms will grow in a dried food depends upon what is called the water activity of the food. Water activity is defined as the equilibrium relative humidity/100, and it depends upon the equilibrium relative humidity which is that relative humidity at which the food neither picks up nor loses water. That is, if placed at this relative humidity, as much water would be evaporating from the product as would be taken up by it. The equilibrium relative humidity, of course, depends upon the composition of the food. The lower the water activity of a food, the less probable that microorganisms will grow. Generally, molds will grow at lower water activities than yeasts and yeasts will grow at lower water activities than bacteria (see page 194). For this reason, molds are more apt to grow in dried foods than are yeasts or bacteria. In drying foods, therefore, the moisture content is lowered to the point where microorganisms will not grow and it is kept that way through packaging which excludes moisture.

Water activity can also be lowered by soluble components, such as sugar or salt. Thus, certain syrups and salted, partially-dried foods (e.g., fish) are relatively stable as far as the growth of microorganisms is concerned, although there may be conditions in which they become subject to the growth of yeasts or molds.

DETERIORATION IN DRIED FOODS

Oxidative Spoilage of Dried Foods

Regarding the chemical changes which may take place in dried foods, oxidation of fats is one of the chief causes of deterioration. This is especially the case with fish, shrimp, crab meat, lobster and other seafoods, and also a factor of some concern with meats such as pork. The pigment which provides the red color of cooked crustacean foods may also be changed or entirely bleached through the oxidation. Packaging under vacuum or in an inert atmosphere, such as nitrogen, and in such a manner as to exclude oxygen, may be used to protect dried foods against oxidation of fats. There may be instances in which antioxidants can also be added to the food to inhibit oxidation of fats. Antioxidants are chemicals which tend to interfere with the type of oxidation to which unsaturated (adjacent carbons of the fatty acids not fully taken up with hydrogen atoms) fats are subject. For instance, the tocopherols (vitamin E) are antioxidants. Very small amounts of chemical antioxidants are required to provide a considerable protection against the oxidation of fats. For instance, only about 0.02% of a chemical antioxidant may be added to fats or oils which are subject to oxidation. However, antioxidants are fat-soluble and are not generally soluble in water. For many foods, therefore, in which the fat is solid or generally distributed throughout the food, there is no known method of getting antioxidants into the fat itself. Hence, there is no good method of preventing rancidification of the fat of some foods through the use of antioxidants. The packaging of dried products in such a manner as to exclude light is another procedure which may be used to assist in the prevention of fat oxidation since light energy accelerates fat oxidation and rancidification.

Nonenzymatic Browning in Dried Foods

Nonenzymatic browning is another cause of deterioration in dried foods. This may be due to caramelization (a dehydration of sugars) or to the combination of certain sugars and proteins, either process leading eventually to the formation of brown or black colored compounds which not only cause off-flavors but also discolor the food. The best protection against such changes is to lower the moisture content to the point where the rate of nonenzymatic browning is greatly reduced. In some foods this may mean lowering the moisture to less than 2%. Sulfurous acid or a source of this compound may also be used to inhibit nonenzymatic browning. Browning may also change proteins, lowering their nutrient quality and affecting their rehydration properties.

Enzymatic Changes in Dried Foods

Enzymatic changes may take place in dried foods during drying, storage or rehydration. Generally, such changes are prevented by pre-blanching of the food, or by using sulfurous acid to inhibit enzyme action. Lowering and retaining the moisture content to 2% or less will inhibit enzyme changes during storage, but this process will not prevent such changes from taking place during rehydration of the food.

12

Refrigeration at Temperatures Above Freezing

Unlike drying or freezing, both methods of preserving foods in certain areas of the world for centuries, the refrigeration of foods at temperatures above freezing is of comparatively recent origin, but the history of this method of food processing is well documented. Ice and snow had been used by the Romans and the early French for the preparation of iced drinks, but the application of refrigeration at temperatures above freezing as a means of extending the storage life of foods was started in the United States in the middle 1800s. It is now the most popular method of food preservation, and it is estimated that more than 85% of all of our foods are refrigerated (temperatures above freezing) at one point or another in the chain of food handling from harvesting to consumption.

At the start, ice, harvested from lakes and ponds that had frozen over in the winter, was put in ice-warehouses and used during warm weather to keep foods cool in ice boxes, the forerunners of the modern refrigerator.

The next step was to develop mechanical refrigeration systems which would produce ice to replace natural ice, and, eventually, mechanical refrigeration systems were used to cool rooms, trucks, boxes, etc., where food could be held without the use of ice. The use of ice was at its peak in the United States as late as the early 1930s.

The first household refrigerators, of course, were ice boxes in which ice was placed in a chamber at the top in order to cool a lower chamber where food was kept. Water from melting ice was allowed to run down to a container beneath the ice box and this had to be emptied periodically.

It is interesting to note that in using ice boxes some housewives would wrap the ice in newspaper in order to conserve it. However, even though the ice lasted longer it did so at the expense of the degree of refrigeration attained.

At this point, the refrigeration capacity of ice deserves mention. The cooling capacity of ice is such that if one should supercool 1 lb (0.45 kg) of it to 20°F (−6.7°C), then this ice would have the capacity to lower the temperature of 1 lb (0.45 kg) of food about 6°F (−14.4°C) before any of the ice would melt. At that point, the temperature of the ice would have risen to 32°F (0°C). However, when heat is added to ice that is at 32°F (0°C), the ice temperature no longer can be raised. Instead, the ice begins to melt, but the temperature remains at 32°F (0°C). It so happens that when 1 lb (0.45 kg) of ice melts, it lowers the temperature of 24 lb (10.9 kg) of food about 6°F (−14.4°C). It can be seen from this that the cooling effect of ice is greatest at the point when it is melting.

Since little was known about the theory and principles of refrigeration in the early days, much research was required to develop effective systems for circulating cold air in cold storage rooms. As a matter of fact, in fishing boats, fish were first refrigerated by a pile of ice in one corner of the pen instead of being applied directly to the fish as it is today.

EFFECTS OF REFRIGERATION ON EATING HABITS

The use of refrigeration has greatly affected our eating habits. Prior to the use of refrigeration, the only fresh fish available was that consumed in or near fishing ports. The major fish of commerce in those days was cod. The reason for this was that the cod could be preserved by salting and drying and sold to other countries or in areas of this country away from the coast. Halibut, now considered to be an excellent food fish, was thought of as a trash fish or a pest which oftentimes ate the cod from the hooks by which they had been caught.

Before refrigeration came into commercial use, many parts of the country, especially the northern areas, were unable to consume many varieties of fruits and vegetables except during the summer months when these items could be produced on local farms. Thus, the chief vegetables used were relatively stable, like cabbages, turnips, potatoes, carrots and beets.

With the advent of refrigeration, however, it became possible to ship any perishable food—fruit, celery, fresh beans, corn on the cob, tomatoes, etc., from warm climates where they were grown to colder climates where these products were not in the growing season. Now, of course, refrigerated shipments are made around the world.

MECHANICAL REFRIGERATION

While ice is still used for refrigerating fish and some shellfish, and in some cases, fresh produce during shipment and during holding or display at the retail level, most foods are refrigerated by mechanical systems.

A mechanical refrigeration system consists of an insulated area or room (the refrigerator) and a continuous, closed system consisting of a refrigerant, expansion pipes or radiator-type evaporator located in the refrigerator, a pump or compressor and a condenser (see Fig. 12.1). The compressor and condenser are located outside the refrigerator. The refrigerant, usually ammonia, or one of the freons, flows into the expansion pipes as a liquid. Here, it evaporates to a gas, and in changing from the liquid to the gaseous phase, it absorbs heat. The gas is pulled into the compressor by the suction action of the pump and is then compressed into a smaller volume. The latter action causes the gas to heat up. This heat must be taken out and this is done by passing the condensed gas through a system of pipes or radiators usually cooled by water, sometimes by forced air. Cooling the condensed gas liquefies it, whereupon it is returned to the evaporator in the refrigerator. The

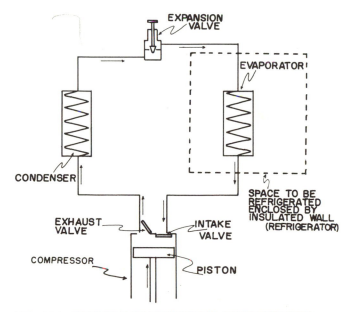

FIG. 12.1. PRINCIPLE OF MECHANICAL REFRIGERATION

conversion of the gas to a liquid also produces heat which is transferred to the water or air of the condenser. Special valves at both ends of the evaporator allow the required flow of liquid refrigerant in, and of gas out, of the expansion system in the refrigerator.

There are a number of ways in which refrigeration may be applied to the insulated area which is to be cooled. Expansion pipes where the refrigerant is evaporated may be located along the walls of the refrigerator. In this case, natural circulation of air (the cold air being heavier) may be depended upon to refrigerate areas within the room away from the expansion pipes, or some type of forced air circulation may be used. In some instances, radiation-type evaporation units are used. A fan which blows air through the radiator fins provides circulation of cold air throughout the refrigerator. An indirect method of cooling refrigerators depends upon the use of refrigerated brine in an outside container, cooled (usually below the freezing point of water), then sprayed into a chamber within the refrigerator and returned to the outside cooler. The cold brine spray refrigerates and humidifies the air within the chamber and causes it to circulate outside of the chamber throughout the refrigerator. Refrigerated brine circulating through pipes may also be used.

Refrigerators should be constructed so that walls, ceilings and floors are sufficiently insulated to prevent excessive heat flow from the outside into the refrigerated area. The inner surfaces of the ceiling and walls should be of washable material, preferably of glazed tile. The floor should be made from impervious material (cement, unglazed tile or epoxy) and should be able to withstand heavy loads and whatever trucking to which it may be subjected without breaking or cracking. The floor should either be sloped toward the door or to a central drain so that wash water used to clean the ceiling, walls and floor can be removed from the refrigerator. The drain should not empty into the waste sewerage system but into a sink or sump which then empties into the waste sewerage system. In this manner, back-up of waste material into the refrigerator may be avoided.

REFRIGERATION PRACTICES

Raw materials, such as fish, meats, poultry, vegetables and fruits should be held in a different refrigerator from cooked foods, or the cooked foods should be placed in impervious containers before they are placed in the same refrigerator with the raw foods. This precaution is necessary to prevent bacteria on raw materials from contaminating the cooked foods.

The available refrigeration capacity for all refrigerators used to hold foods at temperatures above freezing should be sufficient to take care of ɔeak loads. That is, when comparatively large amounts of material are to ɔe placed in the refrigerator within a short period of time, the refrigɛration capacity must be adequate to cool all parts of all products down ɔ the desired holding temperature within 1–3 hr. Often this is not ɔossible and some alternative must be used. The usual alternative is to ɔrecool the product to or near to the desirable holding temperature ɔefore it is placed in the refrigerator. The product can be precooled in an ɪnsulated area where refrigerated air can be blown through the product, ɔr in a chamber where the product can be placed under vacuum. (The ɔasis of vacuum cooling is accomplished by evaporation of water from ɹhe product under low pressure.) Heat exchangers and thin film ɛvaporators may be used for liquid or semi-liquid materials.

Regardless of whether the product is precooled or cooled in the ɹefrigerator, the overall refrigeration capacity must be sufficient to take ɹare of peak loads.

Products should not be placed in the refrigerator in large bulk form ɪnless they have been precooled. Individual units may be placed on trays ɑnd the trays placed on racks to facilitate cooling of the product within ɹhe refrigerator. Once cooled, the individual units may be stacked closer ɹogether to conserve space. Liquids which are to be held in large ɋuantities must be precooled before placing in the refrigerator, or they ɑan be poured into small containers to facilitate cooling.

Large commercial refrigerators should not open directly to the outside ɔut rather to an ante room which in turn opens to the outside, to reduce ɹhe amount of heat entering the refrigerator from the outside. The doors ɰhould be of the swinging, self-closing type.

Fish, meats and poultry should be held at temperatures above, but as ɳear 32°F (0°C), as possible, to maintain good edible quality and prevent ɰpoilage because of excessive enzyme action or excessive bacterial ɹrowth. Since these products generally freeze below 31°F (−0.56°C), a ɡood temperature range for them is 31° to 35°F (−0.56° to 1.67°C). Such temperature range is attainable in the well-regulated refrigerator.

13

Freezing

HISTORY OF FREEZING

The preserving of foods by freezing has a history that dates from antiquity, this type of preservation being used by such ethnic groups as Eskimos, and Indians of certain cold areas. Fish caught in the winter months in cold climates were frozen and held frozen in the cold ambient air. Red meats were also frozen and held in natural, freezing, ambient conditions.

According to some records, the earliest use of artificial freezing was in the mid 1800s, when fish were frozen in pans surrounded by ice and salt. In the late 1800s, fish, meats and poultry were frozen by ammonia refrigeration equipment, with fish being the most important in terms of volume. The commercial freezing of fruits and vegetables started in the early 1900s, the former preceding the latter.

By 1970, U.S. production of frozen foods reached a value of nearly $8,000,000,000 (about 15,000,000,000 lb [6,750,000,000 kg]) (see Fig. 13.1).

However, so-called "quick-freezing" as we know it today was started in the United States in the early 1920s. There is no exact definition for quick freezing, but one definition states that the fall in temperature from 32° to 25°F (0° to − 3.9°C) must occur in 30 min or less. Clarence Birdseye is credited with starting the quick-freezing of foods, and he must be given credit for promoting the development of frozen foods.

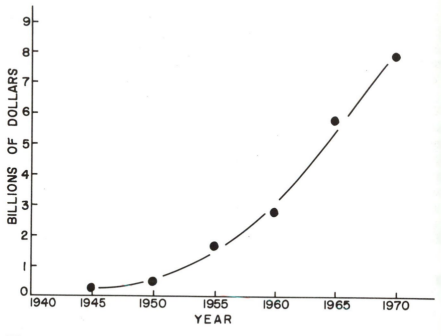

FIG. 13.1. U.S. PRODUCTION OF FROZEN FOODS

THE PRESERVATIVE EFFECT OF FREEZING

Foods generally contain large amounts of water. For example, meats
are about 3/4 water. Bacteria and other microorganisms require water, as
well as nutrients, in foods to live and carry out their vital physiological
processes, such as metabolism and reproduction. When the temperature
of a food is continually lowered, water in the food starts to freeze soon
after the temperature goes below 32°F (0°C). Of course, the freezing
temperature of food is somewhat below that of pure water, mainly
because the water in foods is associated with dissolved substances that
tend to lower the freezing point of the system.

The freezing of foods is relatively complex. As water molecules in the
food begin to form ice (crystallize), there is a tendency for randomly
distributed water to form the orderly network pattern of ice crystals and,
as the water freezes, molecular freedom of movement becomes restricted.

When foods are allowed to freeze slowly, water molecules, even though
they are slow moving, have time to migrate to seed-crystals, resulting in
the formation of large ice crystals. When foods are made to freeze

apidly, the sluggish water molecules do not have enough time to migrate to ice crystals at any distance and instead are "frozen in their tracks," so to speak, forming relatively small ice crystals made up of local water molecules. In most cases, small ice crystals are more desirable than large ice crystals, and rapid freezing methods are employed.

Even when food is rigid from exposure to temperatures of less than 28°F (−2.2°C), some of the water remains in liquid form, and it is not until the temperature of the food is lowered to about −76°F (−60°C) that all detectable water is converted to ice. Figure 13.2 shows the temperatures at which various percentages of the water are frozen. Thus, the preservative effect of freezing on foods is due mainly to its making water unavailable for use by bacteria and other microorganisms, and its slowing of the rate of chemical reactions, literally immobilizing the system. Various methods are used for the freezing process as far as freezing preservation is concerned.

FREEZING METHODS

Air-blast Freezing

Air-blast freezing is one of the most commonly used procedures for freezing foods. The foods are packaged and placed on racks, and the

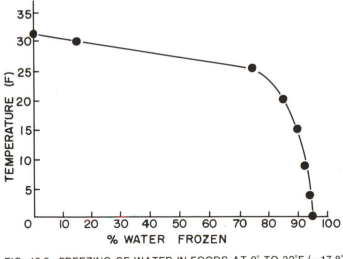

FIG. 13.2. FREEZING OF WATER IN FOODS AT 0° TO 32°F (−17.8° TO 0°C)
See metric conversion tables in Appendix.

racks are wheeled into insulated tunnels (see Fig. 13.3) where air at −20°
to −40°F (−28.9° to −40°C) is blown over the product at a speed o
500–1500 ft/min (152.4–457.2 m/min). When the temperature of the
product reaches 0°F (−17.8°C) in all parts, the packages are put into
cases and the cases are placed in storage at 0°F (−17.8°C) or below. Air
blast freezing may also be applied to packaged products placed on a belt
and carried through cold air tunnels.

A modification of air-blast freezing is used to obtain free flowing frozen
products, such as peas. In this case, the product is frozen, prior to
packaging, on a belt operating in a refrigerated, insulated tunnel. Air at
−20°F (−28.9°C) or below is blown over the belt as it moves along. The
frozen product empties from the belt into a hopper from which it is
promptly removed, then packaged, cased and stored. One drawback of
the cold air-blast method is that moisture is lost to the cold air since the
product is not packaged. Peas, for example, lose about 5% moisture when
frozen by this method.

Plate Freezing

In plate freezing (see Fig. 13.4), layers of the packaged product are
sandwiched between metal plates. The refrigerant is allowed to expand
within the plates to provide temperatures of −28°F (−33.3°C) or below
and the plates are brought closer together mechanically so that full
contact is made with the packaged product. In this manner, the
temperature of all parts of the product is brought to 0°F (−17.8°C) or
below within a period of 1.5–4 hr (depending upon the thickness of the

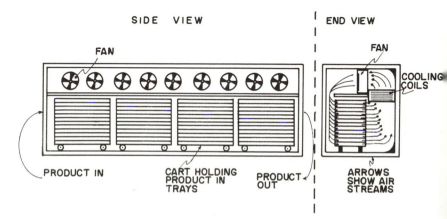

FIG. 13.3. TUNNEL BLAST FREEZER

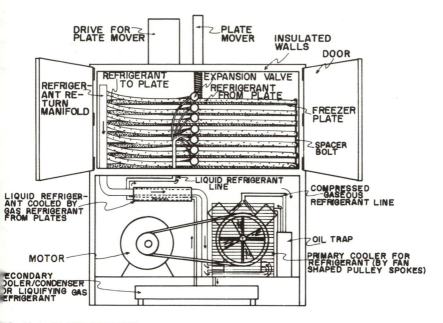

FIG. 13.4. PLATE FREEZER

roduct). The packages are then removed, put into cases, and stored.

Continuous operating plate freezers are now in use. In one such system, he freezer is loaded at the front and unloaded at the rear after ompletion of the freezing cycle. This is done automatically and con- nually. In another continuous system, the packages are fed auto- atically on belts which place them in front of eight levels of refrigerated lates. The packages are forced into the spaces between the plates and he plates closed to provide contact. As freezing proceeds, the packages e advanced so that with each opening of the plates the packages are dvanced by one row with a new set of packages entering the front row. y the time the packages reach the far side of the plates, the foods are ompletely frozen, and they are pushed out of the freezer or unloaded to cased and stored.

iquid Freezers

A recent development in the freezing of foods is the use of liquid trogen or freon. Liquid nitrogen has a temperature of −320°F 195.6°C), and liquid freon has a temperature of −21°F (−29.4°C). The dividual food portion in this case is placed on a moving stainless steel mesh belt in an insulated tunnel where it is sprayed with liquid refrigerant (see Fig. 13.5). Excess refrigerant is recovered, filtered and recycled. The food leaves the freezer in the frozen state and is thereafter packaged, cased and stored. This method provides very fast freezing and is being used mostly for certain marine products, such as various forms of frozen shrimp.

Slow Freezing

Some foods, such as whole fish, fruit in barrels (used for the manufacture of jams and jellies), etc., are frozen in bulk by placing them in a cold room on racks (individually or in pans) or standing them on cold room floors. The temperature of such rooms may be as low as −10° to −30°F (−23.3° to −34.4°C), and some air circulation may be used within the storage room. In such instances, freezing proceeds at a slow rate. Fish frozen in this manner are immersed in cold water or sprayed with cold water (in pans) to form a glaze (coating of ice) which helps to protect it against dehydration during frozen storage. The glaze is built up in layers and, during long storage periods, must be replaced as it is lost by sublimation.

GENERAL CONSIDERATIONS OF FREEZING PRESERVATION OF FOODS

There are three methods of freezing foods: fast freezing, sharp freezing and slow freezing. There is no good definition for differentiating among the various freezing rates. It is obvious that with some methods, as for food frozen in bulk, the product is frozen slowly.

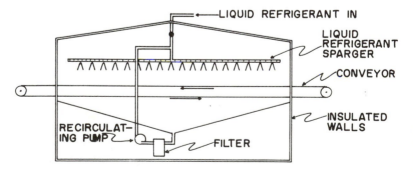

FIG. 13.5. CONTINUOUS LIQUID-REFRIGERANT FREEZER

It is generally agreed that the quality of foods that are frozen quickly is better than the quality of foods that are frozen slowly, and that the lower the temperature to which the food is brought the better the retention of the characteristics of the fresh product. The reasons are: (1) rapid freezing results in the formation of a large number of very small ice crystals which are evenly distributed, and this causes less damage to the tissues of the food; (2) soluble components move about within the food to a lesser degree when the product is quick-frozen since the time required for solidification of the food is shortened; (3) the rates of chemical and biochemical changes are reduced or prevented by decreasing the temperatures rapidly to a certain point.

While the freezing rate is important to the quality of the frozen food, it should be noted that the temperature at which a frozen product is held after freezing is more important than the temperature to which it is brought during freezing. It is obvious that if a food is frozen to 0°F (−17.8°C), then stored at 10°F (−12.2°C), the same changes will take place and at the same rate as if the food were brought only to 10°F (−12.2°C) originally. In fact, additional damage is incurred when the product goes initially from 10° to 0°F (−12.2° to −17.8°C) and back to 10°F (−12.2°C), since any freezing cycle has an adverse effect on the quality of foods. Because frozen foods are stored for much longer periods than those required for freezing, storage changes are of much greater significance to the quality of the product than are the initial changes due to freezing. It has been shown by government research workers that changes which cause deterioration of many frozen foods occur approximately twice as fast at 5°F (−15°C) as at 0°F (−17.8°C), twice as fast at 10°F (−12.2°C) as at 5°F (−15°C), and so on. Actually, frozen storage warehouses usually attempt to maintain temperatures of 0°F (−17.8°C) or lower, and if storage temperatures of −30° to −40°F (−34.4° to −40°C) could be maintained, chemical and biochemical changes in foods during storage would be negligible over storage periods of one year or more. Unfortunately, it is presently considered to be uneconomical to maintain the temperature of storage warehouses at −30° to −40°F (−34.4° to −40°C).

Whereas changes often take place in frozen foods during storage after freezing, the major changes take place during distribution after freezing. Frozen foods may be carried to retail outlets in trucks which are not refrigerated or in which the refrigeration is not adequate to hold the temperature at 0°F (−17.8°C). Frozen foods may be delivered to the retail outlet and left on the unloading platform at ambient (outside air) temperatures for several hours, sometimes even subjected to the direct rays of the sun. At the retail outlet, products are sometimes placed out of refrigeration, either in an open-top display case above the load line, or out of the case entirely. Open-shelf type frozen food display cases now used in many retail stores are usually incapable of maintaining temperatures of 0°F (−17.8°C), especially around the products displayed at the front of the shelf, since refrigerated air is blown from the back of the shelf to the front, and cold air, being heavier than warm air, tends to flow out and down to the floor area in front of the case. As a matter of fact, in stores using this type of display case, the cold air can be felt as one walks through the aisle where such storage cases are located. Since high storage temperatures greatly accelerate deteriorative changes in frozen foods, loss of their quality occurs more often during distribution and display at retail stores than during freezing and subsequent storage in frozen food warehouses.

PREPARATION OF FOODS FOR FREEZING

Many foods preserved by freezing require some type of pretreatment, but only a selection will be described here. Citrus juices are oftentimes concentrated to about 1/5 of their original volumes, then diluted back to 1/4 of the original volumes with fresh juice prior to freezing. This dilution with fresh juice is done to add back flavor components since most of the flavor compounds are removed during vacuum concentration of the juice. When this product is to be reconstituted it requires the addition of water equal to three times its volume.

Except for a few, such as onions, vegetables must be blanched before they are frozen and stored. Vegetables are blanched by heating in steam or hot water [about 210°F (98.9°C)], while being carried along a perforated metal belt or screw conveyor, until the temperature of all parts reaches 185°–200°F (85°–93.3°C) (see Fig. 10.2), then cooling with sprays of water or in water flumes. Vegetables are blanched prior to freezing because enzymes are present, and if not inactivated by the heat, they would cause certain spoilage reactions to take place in the food during frozen storage. These reactions would produce off-flavors (hay-like flavors) in the vegetable.

Blanching times vary with the type of food, the type of heat used (steam or hot water) and the bulk of material being heated. The heating must reach a high enough temperature in the product and remain at that temperature long enough to inactivate the particular enzymes which cause off-flavors. Very high temperatures, used for sterilizing certain types of canned foods, applied for a few seconds, will not destroy all the enzymes; the time is too short. Blanching times used for frozen vegetables may be relatively short [about 1 min in water at 210°−212°F (98.9°−100°C) for peas] or relatively long (about 9 min in steam for corn on the cob).

Besides destroying enzymes which would cause off-flavors, blanching has a cleaning effect and destroys contaminating bacteria. Moreover, blanching fixes the color of green vegetables, removing air from beneath the surface tissue layers and giving the product a bright green appearance. Although the bright green color of some frozen vegetables appears artificial, it is actually due to the removal of air during blanching, and possibly to some effect of the heating on the green chlorophyll of the vegetables. Chlorophyll is the compound that is responsible for the green color in plants in their natural state.

PACKAGING

Packaging is an important factor in the freezing of foods. It is not possible to maintain high humidities in frozen storage rooms because moisture present in the atmosphere of freezer storage rooms tends to crystallize out as ice on the expansion pipes or other apparatus used to cool the rooms. Packages used for frozen foods, therefore, must be reasonably moisture-vapor proof. Otherwise, the product will dehydrate, causing undesirable changes in its appearance and accelerating loss in flavor and juiciness. Toughness and other manifestations of deterioration in quality will also be accelerated. Even when food is packaged in vapor-proof material, however, there can be some moisture lost from the food. This may happen when there is too much space within the package not occupied by the food itself. This results in the formation of what is called "cavity ice" or ice within the package. In such instances, the product, which is warmer than the package, loses water to the cavity space by evaporation, since any water vapor in the cavity space is condensed on the colder package. Under these conditions, there is a transfer of water from the food to the inside of that part of the package surrounding the cavity. Cavity ice formation may cause the same undesirable changes to take place as when moisture is lost to the atmosphere outside the package.

The frozen-food package should have sufficient mechanical strength; that is, it should have adequate "burst strength" and "tear strength" at low temperatures and high "wet strength" against water exposure.

During freezing, the package will be subjected to stresses due to the expansion of the product. Because of this and possible "cavity ice" formation, the package should conform closely to the shape of the food.

A suitable flexibility is, therefore, a desirable property of the packaging material.

Packages should be liquid-tight since some frozen foods (fruits packed in syrup) may have some free liquid which could leak out, and some are thawed in the package, which could also cause leakage.

Transfer of moisture-vapor through packaging may take place through pores or cracks in the container or by diffusion of moisture through the packaging material. The loss of moisture through seals, and package imperfections such as pores, may account for the greatest loss of moisture from frozen foods.

Tin or aluminum cans, waxed paper tubs or cylindrical containers, rectangular paper cartons treated with special waxes, plastic bags of Saran or polyethylene and aluminum foil wraps or aluminum dishes are all used for the packaging of frozen foods. Many of the paper containers are used in combination with cellophane, waxed paper liners or over wraps.

It should be noted that, in many cases, frozen foods are inadequately packaged. It would, for instance, be desirable for some foods to be packaged under vacuum in material which is impermeable to both gas and moisture. This could be done in metal containers, in laminated plastic containers or in aluminum foil and plastic laminated container of certain types.

PROBLEMS WITH FREEZING OR THAWING OF BULK FOODS

Most of the changes occurring in frozen foods take place during frozen storage rather than during the freezing process itself. There are some instances, however, when foods are frozen very slowly, during which time undesirable changes occur. When strawberries in sugar are frozen in bulk form in cold rooms, the freezing time is long, and the deterioration during freezing is equivalent to that which would occur in 2 years of storage at $0°F$ ($-17.8°C$) after freezing.

The defrosting of bulk-packed foods, fruit in barrels, liquid egg frozen in 30 lb (13.6 kg) tins, etc., may be responsible for a considerable loss of quality. When frozen, bulk-packed foods are allowed to defrost at room temperature, the outside layers of the food will have been held at room temperature for a long period of time before the inner layers become defrosted. This period of holding at room temperature may result in enzyme changes or in loss of quality due to the growth of microorganisms.

QUALITY CHANGES DURING FROZEN STORAGE

A number of different changes may take place in frozen foods during storage. These changes may be physical, chemical, enzymatic, and, in rare cases, microbial. Microbial changes occur when refrigeration is inadequate.

Desiccation or drying out is a kind of change which takes place in frozen foods under conditions of poor packaging or varying storage temperatures. Such loss of moisture in poultry occurs first around the area of the feather follicles, causing a speckled or "pock-marked" appearance. Protein and fat changes may also be accelerated when moisture is lost from the surface areas of frozen foods.

Crystallization is a physical change which may occur in some types of frozen foods. Certain types of dairy products, such as ice cream and concentrated milks or creams, sometimes undergo this type of change, which is due to the crystallization of lactose or other sugars which do not readily dissolve upon defrosting, causing an undesirable texture called "sandiness." A similar change may occur in some types of sweetened citrus juices during long storage.

Loss of volatile flavor components may occur in some frozen foods, such as fruit, during frozen storage because these compounds boil or evaporate at temperatures that are even lower than the temperature at which the foods are stored. This causes a loss of the typical flavor components of the foods.

The breaking of gels or emulsions during defrosting may take place with some foods. High-moisture fruits, such as tomatoes, the liquid of which is held by pectin gels, are subject to this type of change and have, therefore, not been frozen successfully. Foods packed with a white sauce or gravy are also subject to this type of physical change and it has been found that fluctuation of storage temperatures greatly accelerates the curdling and "weeping" of the gravy or white sauce.

Protein denaturation is a general term for the physicochemical change occurring especially in flesh-type foods during frozen storage. It results in a toughening of the tissues and dryness or loss of succulence. It is considered that this type of change in stored frozen fish is due to enzymes which break off fatty acids from phospholipids in tissues or to enzymes which decompose trimethylamine oxide and produce formaldehyde. The free fatty acids combine with protein chains, which causes them to aggregate or bind themselves into a closer matrix, resulting in the squeezing out of water (drip occurring during defrosting) and a firming or toughening of the tissues. Protein changes of this type occur especially in lean fish but also, to some extent, in poultry and meats. Grinding of tissues of this kind appears to accelerate protein changes during frozen storage.

As might be expected, the oxidation of some of the components of frozen foods is a cause of some of the major changes which may take place during storage. Oxidation of ascorbic acid or vitamin C is a change of this kind. In fruits in which enzymes have not been inactivated by heating this may be accelerated by enzyme action, but it can also occur through straight atmospheric oxidation. While this change may cause no loss of edible quality, it results in a loss of the nutritional quality of the food.

The fats of foods may oxidize during frozen storage resulting in a change recognized as rancidity. The fats of meats may undergo this change but fish fats are especially subject to this kind of deterioration since the fatty acids in fish fats contain many adjacent carbons not fully saturated (combined) with hydrogen. They are, thus, very subject to reacting with oxygen, a process that eventually leads to the formation of compounds which cause the off-flavors that are recognized as rancid flavors. Fatty fish become rancid faster in frozen storage than do lean fish, but even lean fish are subject to rancidification during relatively short periods of frozen storage.

The colors of fruits and vegetables may deteriorate through oxidation during frozen storage, resulting in a less desirable appearance.

Enzymes may cause various changes in foods during frozen storage. It is not possible to freeze and store whole uncooked lobster, since, during storage, enzymes react with the proteins causing the meat to become soft and crumbly after cooking. On the other hand, if whole lobsters are cooked to inactivate the enzymes and then frozen and stored, the oil in the digestive gland (tomalley) oxidizes and becomes rancid causing off-flavors which spread into the meat itself.

In fruits, enzymes, called polyphenolases, accelerate the oxidation of certain chemicals leading to the formation of brown or black colored compounds. This is the reaction that one sees when an apple or peach is cut and allowed to stand at room temperature for a short period of time. In preparing apples for freezing, therefore, peeled apple slices are treated with a salt which liberates sulfurous acid (sulfur dioxide) and held under refrigeration for sufficient time to allow this compound to diffuse into the tissues. The sulfur dioxide inactivates or inhibits the enzymes that promote enzymatic browning. Also, in preparing sliced peaches for freezing, small amounts of ascorbic acid (vitamin C) are added to the syrup since this compound is a reducing agent (counteracts oxidation reactions).

In general, microorganisms will not grow in frozen foods unless they are held at temperatures above 15°F (−9.4°C). At this temperature the nay be enough free liquid in the food to allow the growth of some microorganisms, especially molds. However, at such temperatures, quality degradation from other factors occurs at such a rate that the food would soon be spoiled because of other changes not due to the growth of microorganisms.

Frozen foods have not often been involved in the transmission of food-borne disease although, if grossly mishandled, as by defrosting and then holding out of refrigeration for some time, they may constitute a public health hazard.

SHELF-LIFE OF FROZEN FOODS

The storage life of frozen foods, at temperatures which today are considered to be economically practicable, is not without definite limits. Some idea of the length of time that frozen foods may be expected to retain high quality may be obtained from Table 13.1.

TABLE 13.1. APPROXIMATE TIME[1] (IN MONTHS) OF HIGH-QUALITY SHELF-LIFE OF SOME FOODS

Product	0°F −17.8°C	10°F −12.2°C	20°F −6.7°C
Orange juice (blanched)	27	1	4
Peaches	12	<2	0.2
Strawberries	12	2.4	10 days
Cauliflower	12	2.4	10 days
Green beans	11−12	3	1
Green peas	11−12	3	1
Spinach	6−7	<3	0.75
Raw chicken	27	15.5	<8
Fried chicken	<3	<1	<0.6
Turkey pies or dinners	<30	9.5	2.25
Beef (raw)	13−14	5	<2
Pork (raw)	10	<4	<1.5
Lean fish (raw)	3	<2.25	<1.5
Fat fish (raw)	2	1.5	0.8

It should be noted that the above storage life times refer to those periods when a quality difference between the frozen and the fresh product can first be detected by an expert panel and do not refer to spoilage or rejection times.

THAWING

While freezing is one of the most effective means of preserving foods over long periods, the need to thaw them prior to reprocessing in food plants or for domestic use represents one of the undesirable aspects of freezing preservation. Thawing is time-consuming and, in some cases, is associated with loss of product quality. Evidence shows that it normally takes food longer to thaw than to freeze under similar heat transfer conditions. In other words, it takes longer for the temperature of the food in 2 to go from $-10°$ to $60°F$ ($-23.3°C$ to $15.6°C$) than it takes the temperature of the food in 1 to go from $60°$ to $-10°F$ ($15.6°C$ to $-23.3°C$) (see Fig. 13.6).

This is because the thermal conductivity of ice is about four times greater than that of water, and the rate at which the temperature of ice changes is about nine times more rapid than that of water. Thus, in thawing, the time during which the product temperature permits bacterial growth is longer than during freezing. A more important consideration, however, is that during freezing, ice occurs first on the surface where most of the bacterial contamination occurs (except in materials such as hamburger, stews, etc., where the contamination can occur throughout). Thus, during freezing, the immobilization of surface microorganisms occurs early in the process before much deterioration can occur, whereas in the thawing process, the surface is thawed first and surface microorganisms are provided with good growing conditions for nearly the entire duration of the thawing period.

There is some evidence to indicate that, as far as the quality of the product not associated with bacterial decomposition is concerned, the faster a frozen food is defrosted, the better the result. Foods packaged in small units defrost in a few hours during holding at room temperature and during this time are not subject to an undesirable amount of decomposition due to bacterial growth. However, foods frozen in bulk, such as barrels of fruit, 30 lb (13.6 kg) containers of egg mixtures, fruit, large fish, or blocks of meat used for sausage products, may present a defrosting problem. Because bulk-frozen foods take long to defrost and because it is well known that the rate at which the food defrosts is dependent on the temperature to which it is exposed, there may be a tendency to defrost the food at relatively warm temperatures. When this is done, the outside portion of the food is subject to bacterial decomposition or the growth of yeasts and molds before the inner portions defrost.

Some methods of alleviating the problems associated with the defrosting of bulk-frozen foods have been developed.

Refrigerator defrosting (holding at temperatures of $35°-40°F$ [$1.7°-4.4°C$]) is probably the best method of defrosting bulk-frozen foods

FIG. 13.6. FREEZING AND THAWING FOOD WITH A START-ING TEMPERATURE DIFFEREN-TIAL OF 70°F
See metric conversion tables in Appendix.

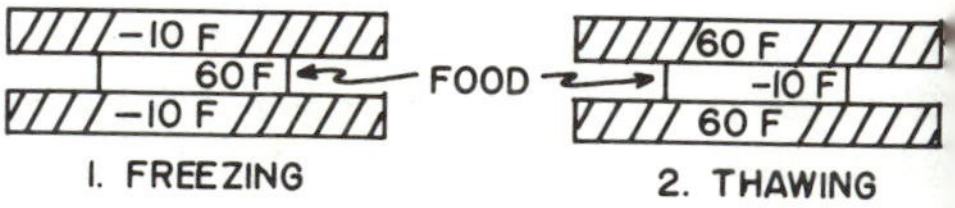

1. FREEZING   2. THAWING

when no fast method is available. This would apply to large whole fish, fruit in barrels and apples in 30 lb (13.6 kg) containers, since bacterial or mold growth would be limited under these conditions. However, in industrial processing, where bulk-frozen products are thawed as an intermediate step in the manufacture of the company's line of products, the refrigeration space required may be so large as to discourage this practice.

In defrosting eggs in 30 lb (13.6 kg) containers, it may be possible to use a machine which breaks or grinds the product up into a kind of slush which can be used in that form in preparing baked foods.

Large blocks of meat used for manufacturing sausage products can be ground in the frozen state and used as such, although this may cause problems in further handling of the meat emulsion.

By the use of microwave energy, food can be thawed rapidly and with virtually no quality loss. That is because microwaves, by their unique character, cause a temperature rise throughout the product nearly simultaneously. When heating is not uniform, overheating or hot spots can occur, but this can be overcome, as we shall see later. The microwave beam can penetrate foods with an alternating current. In alternating current, the charge alternates between positive and negative. Because water molecules are polar, that is, they have positive and negative ends, they are put into a twisting motion due to the alternating current which attracts first the positive end of each molecule then the negative end at a rate of millions of times per second. The twisting action of the water molecules creates considerable friction which generates heat. Ice is not affected by microwaves, but neighboring unfrozen water molecules when foods are frozen, most, *but not all*, of the water is frozen) generate the initial heat that melts adjacent ice to release more water which accelerates the heating.

Since the heat generated in foods by microwaves is quite rapid (about 10 times more rapid than by baking), when uneven heating in a frozen product does occur, the temperature differences within a food can become great. This, however, happens only under certain conditions, and it can be dealt with quite easily. For this condition, and also when one wants to ensure very uniform temperature control, one solution is to apply the microwave energy in intermittent bursts. By this technique, the absorbed thermal energy generated during a burst of microwaves is allowed to be distributed by conduction during the intervals between the bursts, thereby permitting the temperature of the food to increase more uniformly, albeit more slowly. Modern developments, such as wave guides, have improved the distribution of microwave energy.

The particular advantage of using microwave energy for thawing food is that deterioration by microorganisms is not a factor. The feasibility and benefits of microwave thawing of frozen meats and fish have been adequately demonstrated, especially for thawing frozen shrimp blocks.

14

Addition of Chemicals

The practice of preserving food by the addition of chemicals is quite old, and ordinary table salt (sodium chloride) has been used as a preservative for centuries. It might be surprising to think of a naturally occurring substance as a chemical preservative, but it will be seen that many chemical substances used in the preservation of foods occur naturally. When they are used with the proper intent, they can be used to preserve foods that cannot be easily preserved by other means. They should not be used as a substitute for sanitation and proper handling procedures. Sometimes chemicals are used together with other processes, such as holding at refrigerator temperatures above freezing.

To preserve foods, it is necessary either to destroy all of the spoilage microorganisms that contaminate it or to bring about conditions that prevent the microbes from carrying out their ordinary life processes. Although preservation is aimed mainly at bacterial spoilage, it must be remembered that there are other types of spoilage factors, such as oxidation.

SODIUM CHLORIDE

When sufficient salt is added to food, it makes water unavailable to microorganisms. Since microorganisms require water to survive, they cannot exist when their water requirement is diminished by the addition of salt. There are other means by which we can reduce the amount of water available to microorganisms, in other words lower the water activity (see Chapter 7).

Microorganisms will not grow in foods below certain levels of water activity since water must be available for growth. Generally, bacteria need high water activities for growth—many species requiring water activities of 0.99–0.96. There are, however, some species among the halophilic (salt-loving) bacteria which will grow at a water activity of 0.75, which is essentially that of a solution saturated with sodium chloride.

Yeasts grow at lower water activities than most species of bacteria many types growing at values of 0.90 and some at values as low as 0.81 Some molds may grow at water activities as low as 0.62.

Foods are preserved by drying because drying lowers the water activity of the particular products. Also, high sugar content may be used in some instances to lower water activities to the point where microorganisms will not grow.

While the preservative effect of sodium chloride is due primarily to the lowering of the water activity of foods, the chloride ion in ordinary salt has some inhibitory effect on the growth of microorganisms.

Some precautions must be observed in the salting preservation of flesh-type foods, such as fish or meats. When these products are salted several days will be required before enough salt has diffused into all parts of the product to inhibit the growth of microorganisms. If, therefore precautions are not observed, the growth of spoilage or even disease causing bacteria may occur in some parts of the food before enough salt has diffused into the product to inhibit growth. The usual procedure is to hold products under refrigeration during salting until there has been an adequate "take up" of salt throughout the food. Fish and meats should never be held at temperatures above 60°F (15.6°C) during salting Preferably, holding temperatures during such procedures should be at 40°F (4.4°C) or slightly below.

Salted, undried meats, such as corned beef, should be held at 40°F (4.4°C) or below at all times after curing since there are some micro-organisms which may grow in the salt contents present in such products Chipped beef, which is dried as well as salted, has a low enough moisture content to prevent the growth of all microorganisms and may be held at room temperatures.

Salt cod, which has a moisture content of 40% or higher, should be held at temperatures of 40°F (4.4°C) or slightly below since it is subject to spoilage through bacterial growth. On the other hand, well dried salt cod and certain types of salted and smoked herring (which have dried during smoking) may be held at room temperatures without spoilage.

ACIDIFICATION

Acidification is a method of food preservation. As has been pointed out, all microorganisms have a pH at which they grow best (see Chapter 7), and a range of pH above or below which they will not grow. Generally, it is not possible to preserve all foods by adding acid to the point where no microorganisms will grow. Most foods would be too acid to be palatable. Enough acid may be used to inhibit the growth of microorganisms provided that such treatment is combined with some other method of preservation. Certain dairy products, such as soured cream, and fermented vegetables, such as sauerkraut, are preserved through lactic acid produced by the growth of bacteria together with the holding of these products at refrigerator temperatures above freezing. When sauerkraut is canned, it is given a heat process sufficient to destroy all spoilage and disease microorganisms.

Pickles are preserved by the addition of some salt, some acid and a heat process sufficient to raise the temperature of all parts of the food to or near 212°F (100°C).

Pickled herring are preserved by the addition of some salt, some acetic acid (vinegar) and the holding at refrigerator temperatures above freezing. In this case, the nonacid part of the acetic acid molecule has an inhibiting effect on the growth of microorganisms.

FATTY ACIDS

The salts of certain fatty acids have an inhibitory effect on the growth of microorganisms. Thus, sodium diacetate (a mixture of sodium acetate and acetic acid) and sodium or calcium propionate

$$CH_3-CH_2-\overset{\displaystyle O}{\overset{\displaystyle \|}{C}}-ONa$$
sodium propionate

are added to bread and other bakery products to prevent mold growth, as well as the development of a slimy condition known as "ropiness," due to the growth of certain aerobic, spore-forming bacteria (see Chapter 7 for definition of spore-forming bacteria). Caprylic acid, $CH_3-CH_2-CH_2-CH_2-CH_2-CH_2-CH_2-COOH$ or its salts or the salts of other fatty acids may be used in cheeses to prevent the growth of mold.

As pointed out previously, it is the nonacid part of the molecule of fatty acids or their salts which inhibits the growth of microorganisms. It is believed that the effect of these compounds is the destruction of the cell membrane of microorganisms.

SULFUR DIOXIDE

Sulfur dioxide (SO_2) is used in some foods to inhibit the growth of microorganisms. Sulfur dioxide may be used as such, or as a source of this compound: for instance, sodium bisulfite ($NaHSO_3$) may be added to foods. Sulfur dioxide inhibits a rather narrow range of microorganisms and is usually applied together with another chemical inhibitor to prevent the growth of undesirable yeasts or bacteria in fruit juices, which are stored prior to fermentation, to produce wine or vinegar.

The inhibiting effect of sulfur dioxide may be due to the prevention of the utilization of certain carbohydrates as a source of energy or to the tying up of certain compounds concerned with the metabolism of some microorganisms.

SORBIC ACID

Sorbic acid, CH_3—CH=CH—CH=CH—COOH, inhibits the growth of both yeasts and molds. This compound is most effective at pH 5.0 or below. This compound can be metabolized by humans, as are fatty acids, hence is generally recognized as safe. Sorbic acid is used in certain bakery products (not yeast-leavened products, since it inhibits yeast growth), in cheeses, and in some fruit drinks, especially for the purpose of preventing molding. It is believed to inhibit the metabolic enzymes of certain microorganisms which are required by these species for growth and multiplication.

SODIUM NITRITE

Sodium nitrite, $NaNO_2$, is added to some food products to inhibit bacterial growth and to enhance color. It is added to most cured meats, including hams, bacon, cooked sausage (such as frankfurters, bologna, salami, etc.), and to some kinds of corned beef. Nitrite provides for the red or pink color of the cured and cooked sausages and of the other cured products after cooking. The nitrite combines with the red coloring material of meat, the myoglobin, and prevents its oxidation. If the meat were not treated with nitrite, it would discolor during cooking or during storage. When red meat is heated, as in cooking, the color turns from red to gray or brown due to the conversion of myoglobin to the oxidized form, metmyoglobin. On extremely long or extremely high heating or on exposure to light and air (oxygen), even the nitrited myoglobin may be oxidized to metmyoglobin, with the result that the red or pink color is lost.

In addition to stabilizing the color of cured or cured and cooked meats, the industry claims that nitrite acts as a preservative in that it tends to prevent the growth of spores of *Clostridium botulinum* which may be present. *Clostridium botulinum,* of course, is a disease-causing organism.

Nitrite is also used in some fish products, such as smoked whitefish and chubs, for the specific purpose of preventing the outgrowth of *Clostridium botulinum.*

Sodium or potassium nitrite may not be used in meats or on fish which are to be sold as fresh. In cured products, it is allowed in concentrations not to exceed 0.02% (200 parts of nitrite per million parts of the food). There is some question as to whether or not nitrites should be allowed in foods in any concentration. It has been found that nitrite-cured products, especially those which are cooked at high temperatures, such as bacon, may develop nitrosamines, compounds formed by the reaction of nitrites with amines, and nitrosamines are known to be extremely carcinogenic or cancer-promoting.

DIETHYL PYROCARBONATE

In recent years, a compound called diethyl pyrocarbonate

$$CH_3\text{---}CH_2\text{---}O\text{---}\overset{\overset{\text{O}}{\|}}{C}\text{---}O\text{---}\overset{\overset{\text{O}}{\|}}{C}\text{---}O\text{---}CH_2\text{---}CH_3$$

has come into use in Europe and, to some extent, in this country. This compound is used primarily in fruit juices or fruit juice drinks to prevent the outgrowth of yeasts which, when present, could cause fermentation. It has been shown that diethyl pyrocarbonate is effective only when microorganisms are present in low concentrations.

In water (which would be present in fruit juices), diethyl pyrocarbonate quickly decomposes to ethyl carbonate and thence to ethyl alcohol (drinking alcohol) and carbon dioxide. It is not clear whether the microorganisms are destroyed by the ethyl carbonate or by small residuals of the diethyl pyrocarbonate that may not decompose. Either the original compound or the intermediate decomposition product must be the active agent which affects microorganisms, since in the concentrations that this compound is allowed in foods, sufficient ethyl alcohol could not be formed to provide any inhibiting or destructive effect.

OXIDIZING AGENTS

Oxidizing agents, such as chlorine, iodine and hydrogen peroxide, are not ordinarily used in foods, but they are used to sanitize food-processing equipment and apparatus and even the walls and floors of areas where foods are processed. Thus, there is no doubt that small residuals, especially of chlorine or iodine, can get into foods.

Hydrogen peroxide may be used to destroy the natural bacterial flora of milk, prior to inoculating with cultures of known bacterial species, for producing specific dairy products. In such cases, all of the residual hydrogen peroxide must be removed by treating with the enzyme catalase.

$$2H_2O_2 \xrightarrow{\text{catalase}} 2H_2O + O_2$$

Hydrogen peroxide   Water   Oxygen

This treatment with catalase must be carried out prior to the inoculation of milk with cultures of desirable bacteria, otherwise the hydrogen peroxide will destroy the added bacterial culture, the growth of which is the objective of culturing milk.

Chlorine may be used in water for cleanup as a solution of the liquefied gas, as a solution of sodium or calcium hypochlorite or as a solution of some other compound containing chlorine, such as chloramine-T. Often all of the water used in a plant in which foods are processed will be chlorinated to the extent that there are a few parts of chlorine per million parts of water. This is done as an added precaution to prevent the growth of food-spoilage or disease-causing bacteria in various parts of the plant. For disinfecting equipment and apparatus, higher concentrations of chlorine are used, usually 50–200 ppm. The bactericidal action of chlorine is attributed to its reaction with cell protoplasm to produce lethal chloramines.

Iodine is used for the disinfection of equipment, etc., in a form known as an iodophor. In this case, elemental iodine (crystalline iodine) is added to a detergent in which it forms a complex which, when dissolved in water, reacts to allow part of the iodine to dissolve and become available as a disinfectant. Actually, in use in food plants, the iodophor s merely dissolved in the water used for sanitizing equipment, etc.

Oxidizing agents are believed to destroy and inhibit the growth of microorganisms by destroying certain parts of the enzymes essential to the metabolic processes of these organisms.

BENZOATES

Benzoic acid

or its sodium salt is allowed in food in concentrations up to 0.1%. Para-hydroxybenzoic acid

or its esters, for instance, propyl para-hydroxybenzoic acid, may also be used. The benzoates are most effective in acid foods in which the pH is as low as 4.0 or below. The parabenzoates are said to be more effective than the benzoates over a wider range of pH and against wider groups of microorganisms than are the benzoates.

The benzoates and parabenzoates have been used as preservatives mainly in fruit juices, syrups (especially chocolate syrup), candied fruit peel, pie fillings, pickled vegetables, relishes, horseradish, some cheese, etc.

The probable reason that the benzoates and the related parabenzoates have been allowed as additives to food is that benzoic acid is present in cranberries as a natural component in concentrations that are higher than 0.1%.

Investigations have indicated that benzoates prevent the utilization of energy-rich compounds by microorganisms. It has also been found that when bacteria form spores in the presence of benzoate, the spore may take up water and germinate to the point of bursting and shedding the spore wall, but enlargement and outgrowth into the vegetative cell with subsequent cell division and multiplication does not occur.

ANTIBIOTICS

For some years now there has been known a group of chemical compounds called antibiotics. These are compounds produced by one species or strain of a microorganism but effective in preventing the growth of other microorganisms. No detailed discussion of the formulas and chemistry of antibiotics will be given here since generally they vary considerably and are quite complex. Antibiotics are used primarily to destroy disease-causing bacteria and to prevent disease in man and animals. These diseases are primarily caused by bacteria, and it should be noted that antibiotics are not effective against viruses and therefore are useless for treating or preventing viral diseases.

The reason for discussing antibiotics at this point is that the tetracyclines, especially aureomycin (chlortetracycline) and terramycin (oxytetracycline), were once allowed in foods (meats, poultry and some fish) in this country and are still used in some countries.

The tetracyclines are wide spectrum antibiotics, that is, they are effective against a rather wide range of bacterial species. On the other hand, some antibiotics are relatively specific, and penicillin, for instance, is most effective against cocci, the spherical-shaped bacteria.

The tetracyclines were eventually banned from human foods on the basis that they are not completely destroyed during the cooking of food and the small residuals remaining after cooking might permit the buildup of a bacterial flora in humans (types of bacteria present in the intestines and in other parts of the body) that would then be resistant to treatment with antibiotics. This could well happen, since one of the shortcomings of using antibiotics to extend the storage life of refrigerated foods was found to be that, in food plants, a resistant bacterial flora was eventually built up. Obviously, when the foods became contaminated with the resistant bacteria, the antibiotics were not effective and were then of no use.

Antibiotics are still used in animal feeds, and the purpose here is not only to prevent disease but also to inhibit microorganisms in the intestine which utilize the food material therein and prevent its availability to the animal. Thus, antibiotics in animal feeds provide for a greater gain in animal weight per pound of food consumed.

The antibiotics nisin and tylosin are polypeptides, the primary units present in proteins. Nisin is not allowed in foods in the United States but it is allowed in small concentrations in canned foods in some countries on the basis that it is produced by bacteria. It is used in the manufacture of many dairy products in some countries. Nisin in canned

ods is considered to prevent the outgrowth of spoilage bacteria that
ay survive heat treatments tailored to destroy only disease-causing
acteria.

NTIOXIDANTS

One of the ways in which food spoils involves the oxidation of fats.
Vhen this occurs, foods develop off-flavors, off-odors and sometimes off-
olors. Among the measures taken to prevent oxidative degradation of
ods is the addition of certain chemical compounds which, because of
neir effect, are called antioxidants.

There is a large number of compounds having antioxidant properties
ut differing in some respects. True, or primary, antioxidants are
henolic in nature (contain a phenol ring

r have similar structures in the molecule), and they may be derived
om natural or synthetic sources. Secondary antioxidants normally are
cids that include ascorbic, citric, tartaric and phosphoric acids. Al-
ough phenolic antioxidants are effective when used alone, their effect
enhanced synergistically when acid antioxidants are added.

Tannins, which are widespread in plants, are natural antioxidants, as
e vitamin E, vitamin C, spices, sugar and others.

The use of antioxidants has greatly reduced the problem of rancidity in
ts and oils and in foods containing fats and oils or to which fats or oils
e added during processing.

THER COMPOUNDS

A number of chemical compounds may be used in the solutions
ith which fruits are washed. The compounds are used mainly on citrus
uit to prevent mold growth, but some may be used in the washing
lutions applied to apples, pears and quinces, as well as to citrus fruits.
cluded among these compounds are hexamine (hexamethylenetetra-
ine), sodium chlorophenate, 2-4D (dichlorophenoxyacetic acid), 2-4-5T
richlorophenoxyacetic acid), diphenyl and boric acid. Some of these

compounds are applied to oiled papers used for wrapping citrus fruits.

The latter group of compounds is toxic to humans and is generally not allowed in foods. Used on citrus fruits, these compounds are applied to the peel, which is not generally consumed by humans. However, diphenyl is allowed in citron (candied citrus peel), which is eaten by humans, in concentrations up to 100 ppm of the candied peel. This is probably allowed on the basis that citron is eaten only rarely and then in comparatively small amounts.

Section IV

Handling and Processing of Foods

15

Meat

Of the many foods that he obtains from the land, man tends to have a preference for animal foods, mainly beef, pork, poultry and lamb, as well as their by-products, e.g., cheeses, milk and eggs. The population of the United States consumes about 40 billion pounds (18.1 MMT) of meat, of which about half comes from beef. Other sources in the order of importance are hogs, poultry and sheep.

The use of horses as a source of meat is not generally practiced. However, from time to time, these animals have been used as substitutes. This has occurred during wars when there was a shortage of meat or more recently, due to the high price and shortage of beef animals. While some believe that they could tell the difference between horse meat and beef, there is strong evidence to show that this is not easily done. The biggest difference actually occurs in the fat, with the color being the predominant factor. The fat in horse meat is quite yellow as compared to that of beef. When the fat in horse meat is removed and the fat of beef is added to the horse meat in its preparation, then the consumer is unable to tell the difference between horse meat and beef. The tastes imparted by the fats may also be significantly different.

A market also exists for domestically produced game-type meats. These include buffalo, bear, elk, kangaroo, rabbit and others. They are usually expensive and selected mainly as gourmet items. The meat is generally tough and very "gamey." In general, this type of meat is prepared in moist heat.

The proportion of meat-derived foods that man consumes is related to the general affluence of the society in which he lives. The reason for this is that the use of animal flesh as food tends to use up a much greater amount of the calories, proteins and other nutrients which might be available directly from plants. It is reported that it takes about 10 lb (4.5 kg) of plant food to produce about 1 lb (0.45 kg) of beef. Thus, in deprived societies, the population is compelled to derive most of its nutrition from plants rather than from animals.

Animal foods offer more than just palatability. Since animals are biologically similar to humans, it should be obvious that animal foods contain many of the nutrients that are required by man to carry out his own body functions. For example, in most cases, the animals are good sources of the essential amino acids, as well as the vitamins and minerals required by man.

Meat and meat products include the muscle tissues of cattle, hogs, sheep and other animals, as well as the organs of these animals. The organs which are used are the tongue, heart, brain, liver, kidneys and sometimes the lungs. Some of the by-products derived from the animals that are used in the meat industry include the intestinal walls, used as casings for making different types of sausages. Fat is used for the production of lard and tallow; the tallow is used as a source of raw material for making soap and candles. Other by-products include leather for shoes and a variety of other commodities, wool for textiles, etc., gelatin, which is used in the production of gelatin desserts and other similar food products, blood which is used in some sausages and in feeds, bone which is used mainly in fertilizers and feeds, animal scraps from the slaughtering plants which are used in the manufacture of feeds, and a variety of enzymes and other chemicals which are used in different industries, especially in the food and pharmaceutical industries.

Except for tongues, head meats are generally used in the production of sausages.

Tripe, prepared from the first and second stomachs of cattle, is cooked for about 3 hr and sold as is, or it may be cured. The sweetbreads (pancreas and thymus) are marketed either fresh or frozen.

Both gelatin and animal glue are prepared from the collagen derived mainly from the bones and the hide, but also from other parts, such as the sinews, ears, snouts and trimmings.

Oxtails, although quite tough, have a distinctly desirable flavor, and are used in making oxtail soup.

When livestock animals are grown for meat production, the males are castrated. There are two reasons for this. Upon castration, the hormone balance in the animal is affected so there is an increased amount of fat deposited throughout the musculature. Because of this, the meat tends to be more succulent and more tender. The second reason for castration is that the strong odor usually associated with viable males tends to disappear.

The quality of the meat has been related to the amount of fat which is distributed uniformly through the muscle. This effect is known as marbling. Quality is also related to the age of the animal. The younger the animal, the better the quality. Marbling tends to increase as the animal matures. It is quite possible that in an old animal which has considerable marbling, the meat may not be tender at all. In this case, marbling is not a good indicator of the texture of the meat. Generally speaking, the marbling—that is, the desirable marbling—must be paralleled with young age. That is to say, the marbleized meat of young animals will almost surely be tender, while the marbleized meat of old animals will almost surely be tough.

Generally, animals are allowed to rest for a period just before slaughter. This is done in order that they not use up all their muscle sugar (glycogen). When the animal is slaughtered with an amount of glycogen in its body, the glycogen is converted under anaerobic conditions to lactic acid. The presence of this acid has a preservative effect on the meat. On the other hand, if the animal is not rested before slaughter and it uses up its glycogen, there will not be enough lactic acid formed to have the preservative effect, and thus the meat will spoil sooner.

After slaughter, the carcass portions are aged in cold rooms at low temperatures just above freezing, usually 35°F (1.7°C). This process may take up to one month, depending on temperature, humidity and other conditions, but ordinarily, meat is held for about 12 days prior to consumption. The aging is accelerated if the beef is stored at higher temperatures, but the higher temperatures permit the growth of bacteria and surface spoilage to occur. However, since the spoilage is on the surface, this problem can be resolved by storing the meat at the higher temperatures in the presence of ultraviolet light which has a bactericidal effect. Sometimes a processor is willing to accept the microbial growth in order to hasten the tenderization. Since the bacteria and mold growth create a certain amount of spoilage on the surface of the meat, the processor must then trim the surface in order to remove the effects of microbial growth and spoilage. The main purpose of aging is, of course, to tenderize the meat. Meat cuts can be tenderized just prior to consumption by the addition of commercially-prepared enzymes. Enzymes have a proteolytic effect; that is, they tend to break down the proteins. However, this is a slow process if added only to the surface, and it is believed that injection into the meat or into the bloodstream of the living animal just before slaughter is a more effective way of producing tender beef.

With regard to texture, it can be stated rather generally that the smaller the amount of connective tissue in the meat, the more tender the meat. Thus, cuts such as the bottom round, which contain relatively more connective tissue, are tough, and such cuts as the rib or the loin, which contain less connective tissue, are more tender.

When meat is heated in the presence of water, as in boiling or in the making of stew, the connective tissue is changed to a sort of tender gelatin and it becomes more palatable. On the other hand, when the meat is heated without water, such as in an oven in dry heat, the connective tissue tends to become tougher. The amount of connective tissue present in the meat depends on the location of the meat in the animal. The tenderloin, for example, has very little connective tissue, especially in the younger animals.

Although boiling tends to make meat less tough, one of the disadvantages is that taste components are extracted from the meat, and, therefore, boiled meat is less tasty. Of course, the water in which the meat has been boiled acquires the meat flavor components and can be concentrated and sold as a meat extract, which can be used as a base for soups and stews. Also, little browning (which enhances meat flavor) occurs during boiling.

Some points to remember in the cooking of meat are: the higher the temperature during cooking, the more the shrinkage; the lower the temperature during cooking, the higher the quality and the more uniformity of doneness throughout the meat. A certain amount of aging is beneficial from a quality standpoint. In fact, some chefs will hold the meat at room temperature for a considerable period before they finally put it into the oven. In the case of roasts, it has been found that an oven temperature of 375°F (190.6°C) should be maximum and 325°F (162.8°C) about minimum.

Of the different types of protein of which meat is composed, the contractile proteins are of some importance. They include myosin, actin and tropomyosin, plus some other minor ones. When these proteins are extracted from the cells in the meat, either with salt solution or with mechanical agitation, such as chopping or grinding, they form a sticky substance which has binding qualities. When a mass of comminuted meat is heated, the contractile proteins coagulate, and they hold the small pieces of meat together. This is the way that so many formed products, such as sausages, meat loaves, etc., are produced.

Unfrozen meat should be stored in a temperature range of 28° to 38°F (-2.2° to 3.3°C). The relative humidity should be in the range of 85-90%. The high humidity will prevent excessive drying and shrinkage. In addition, high humidity tends to preserve the white color of the fat in the meat. The storage room should have a reasonable circulation of air to ensure more uniform distribution of the humidity and heat removal. Meats stored for long periods are generally frozen and stored at temperatures in the range of -10° to 0°F (-23.3° to -17.8°C).

Since many diseases can be transmitted from animals to humans, all meats shipped interstate in the United States are subject to inspections for wholesomeness under the authority of the Meat Inspection Act. "Wholesome" meat is safe to eat and is without adulteration. In addition to the wholesomeness inspection, meat is graded for quality. Although the lean is more valuable than fat, the best cuts of beef have some fat, and the fat is well distributed, giving the marbling effect cited earlier. The prime grade is of the highest quality, but in general, this is limited in amount and sold only to restaurants, hotels and the better eating places.

It should be noted that in federally-regulated plants, the inspection for wholesomeness is mandatory. The grading of the meat is optional.

BEEF

The most important beef animal breeds in the United States are the Angus, Hereford, Galloway and Shorthorn. Less important are the Santa Gertrudis, the Chambray, a cross between a Charolaise bull (French) and a Brahma cow, and the Brangus, a cross between an Angus bull and a Brahma cow. The latter breeds were developed to produce cattle resistant to screw worm (an insect infesting cattle), Brahma cattle being resistant to this parasite.

Much of the cattle population of the United States is raised on the Western range, which includes the Great Plains, the Rocky Mountains and the intermountain and Pacific Coast regions. The raising of calves produced in these areas may be finished elsewhere. These feeding areas include the beet pulp feeding area of Colorado, the cottonseed cake section of Texas, the corn belt area (especially Iowa) and certain southern states, especially Florida, and California.

Discarded dairy animals supply a considerable portion of the cattle slaughtered in this country. Veal calves are produced mainly in dairying areas (especially Wisconsin), and slaughter calves (3–12 months old) are also produced in southwestern areas.

Farmers and ranchers may sell cattle to terminal markets (who sell to various buyers), to local markets, auctioneers, or to packers. Usually, cattle are shipped by truck to stockyards near or at the point of slaughter.

Steers (males castrated prior to sexual maturity) and heifers (females beyond the veal and calf age which have never calved) make up the major portion of beef animals. Cows (females which have calved), bulls (mature males) and stags (castrated after reaching sexual maturity) make up the remainder of the beef animals.

At the point of sale, cattle may be graded in the live state. Grading is based on conformation (shape, build and breed of the animal), finish (quantity and distribution of the fat) and quality (quality of the hair and hide, moderate bone mass, etc.)

The term "conformation" is used to describe the structural characteristics of the beef animal. For the best conformation, the animal should be short and compact. A rather blocky, or large-bodied, short-legged appearance is good.

Basically, the best structural characteristics or conformation is valuable to the retail butcher because apparently he is able to make more profitable cuts from the animals having the top conformation, and because there is a relationship between good conformation and good quality.

In the United States and many other countries, animals must be rendered insensible to pain before slaughter (see Fig. 15.1 for general slaughtering scheme). With cattle, this may be done by hitting them on the head with a maul or the bolt from a specially devised gun. Due to religious laws, cattle for kosher meats are slaughtered by first cutting the throat with a special knife, in one motion, severing the jugular vein, and then stunning. In ordinary slaughter, after the animal is stunned, it is shackled at the hind legs and raised from the floor in an upside-down position, after which it is bled by severing its jugular vein.

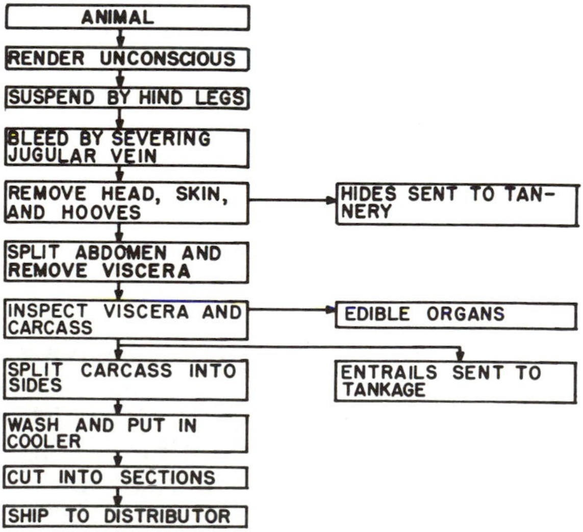

FIG. 15.1. GENERAL SCHEME FOR SLAUGHTERING ANIMALS

After bleeding, the head is skinned and severed at the neck. The body is lowered to the floor or to a cradle, the skin is cut ventrally and pulled back from the belly portion and the lower legs are removed. Hooks are inserted in the cords of the hind leg section, and the body is raised to the half-hoist position. Most of the skin is removed in this position after which the carcass is raised to full hoist and skinning is completed. Also, in this position, the sternum (breastbone) is opened with a hand saw, the belly is split, the bung and esophagus are tied off and the viscera are removed. The viscera **are** inspected by veterinarians, and the carcass is split into halves with a saw. The hide is inspected, salted and sent to the tannery.

The split carcass is washed, and, if it is not to be boned out soon after cooling, it is covered with a cloth to shape the surface fat. Sides handled in this manner are sent to a cooler where 24–48 hr will be required to bring the temperature of all parts to about 35°F (1.7°C).

Dressing, including skinning, may be carried out completely while the animal is hanging from a rail. Mechanical instruments, such as hide pullers, may be used to increase the efficiency of rail dressing.

In the past, when slaughtered beef was not to be used in local areas, it was shipped out as sides or, more often, as quarters. Today, a considerable amount of beef is cut into various primal cuts, such as rounds, ribs, loins, etc. (see Fig. 15.2). The various sections, usually without bones, are packaged, usually under vacuum, in a film that is impervious to moisture and oxygen. The packages are boxed and shipped to distributors and retailers. Some retail chains prepare their own primal cuts from sides or quarters, packaging them in film under vacuum and shipping them in master cartons or in wire or plastic boxes to their retail stores. Since the meat has been boned and trimmed of excess fat, meat department personnel at the stores have little to do but cut the meat into individual portions, package the various cuts and place them on display.

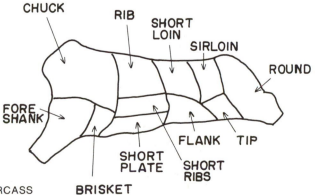

FIG. 15.2. CATTLE CARCASS

Primal cuts packaged in moisture and oxygen impermeable film, unde
vacuum, and held at temperatures between 32° and 40°F (0° and 4.4°C)
are said to have a storage life of at least 21 days.

Inspectors from the USDA evaluate beef carcasses and specify variou
grades. These are "prime" (the best beef from steers and heifers
"choice" (the next best from heifers, steers, and young cows), "good
(third best), "standard," "commercial," "utility," "cutter" and "can
ner." The last grades are less desirable as beef in the order listed. Th
grades normally used for fresh beef are mainly "prime," "choice" an
"good." A veal carcass is shown in Fig. 15.3.

PORK

There are many breeds of hogs which, in this country, have becom
pretty well mixed in recent years in order to produce meat-type anima
(more lean and less fat) rather than lard-type animals. Part of the reaso
for getting away from the high-fat type of hog has been that other oils an
fats have largely replaced lard, which previously was the chief fat use
for cooking. Some breeds of hogs, such as the Poland-China, Berkshir
Chester White and Hampshire are not mixed.

The north central states, especially Iowa, produce most of the hog
raised in the United States. Hog production is said to be increasing in th
southern states.

Hogs are fed soybean meal, tankage (heated slaughterhouse wastes
meat scraps, fish meal or milk, or combinations of these as a source c
protein. Carbohydrate foods consist mainly of corn, but other grains ar
used. Minerals may be added as supplements. Antibiotics may be adde
as feed supplements to lessen disease and to provide for a bett
utilization of food. In this instance, the antibiotics serve to limit th
growth of bacteria in the intestines which themselves utilize food, thu

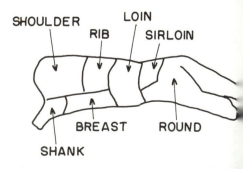

FIG. 15.3. VEAL CARCASS

using some portion of the feed to be unavailable to the animal. Some
tamins, especially B-12, may be added to hog feed, since they may
rve as growth factors.

Slaughter hogs are classified as barrows (males castrated prior to
xual maturity), gilts (young females which have not produced young
gs), sows (mature females which have produced young or which have
ached an advanced stage of pregnancy), stags (males castrated after
aching maturity) and boars (uncastrated males).

Hogs are graded according to conformation, finish and quality as U.S.
o. 1, U.S. No. 2, U.S. No. 3, medium, and culls, this grading being
sted in the order of desirability.

Prior to slaughter, hogs are rendered insensible by electric stunning or
exposure to an atmosphere of carbon dioxide. The hog is then
ackled by the hind legs and raised to a conveyor after which a knife is
serted into the neck to sever the jugular vein. After bleeding for about
min, the hog, still on the conveyor, is passed through a water bath
ater temperature about 140°F [60°C]) where it is immersed for 5 or 6
in. This facilitates removal of hair in a machine which has mechanical-
-driven beaters. The loose hair is washed away with a water spray.

The carcass is then suspended by the hind legs by inserting the ends of
gambrel under the achilles tendon in each hind leg and attaching the
mbrel to a moving conveyor. Remaining hair is scraped or singed off.
During dressing, the head is partially severed, the carcass is opened
d the viscera are removed. The edible organs are inspected by
terinarians and removed to refrigerated areas to be sold later as such or
ed for further processing. The carcass is then split and removed to the
oler where the temperature of all parts should be brought to 35°F
.7°C) within a period of 18 to 24 hr.

Hogs are generally not shipped as sides. Usually, they are cut up into
ms, loins, shoulders, bellies and back fat before shipment to dis-
ibutors or to processing plants which manufacture various pork prod-
ts. See Fig. 15.4 for hog carcass.

IEEP

The bulk of the sheep produced in the United States is said to be made
of crossbreeds developed for the production of both meat and wool.
re breeds include "Cheviots," "Lincolns," "Oxfords," "Shropshires,"
Rambouillets" and "Merinos."

Sheep are raised on ranges in arid and semi-arid regions and on farms.
farms, flocks of sheep are smaller than those raised on ranges. Where

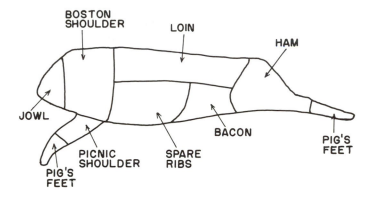

FIG. 15.4. HOG CARCASS

conditions allow, lambs may be fattened on the range, but during wint storms, the feed must be supplemented with protein concentrates, cor hay, barley and oats.

Fattening of lambs may take place in the corn belt states, especial during the fall and winter months. This may be done in corn field stubble fields or on pastures. Some fattening occurs while the anima are confined to barns. Most lamb fattening is carried on from Septemb through May.

Sheep are classified as ewes (females), wethers (males castrated pri to reaching maturity), rams (uncastrated mature males) and laml about 3–12 months in age which do not show their first permanent teet Lambs are identified by the break joint—the temporary cartilage in th leg bone just above the hoof. As the animal grows older, the break joi loses its redness and sawtoothed appearance, and the cartilage replaced by bone. Spring lambs are those from new crops markete before July 1 and weighing 70–90 lb (31.8–40.9 kg). As with beef catt and hogs, grading of slaughter lambs is based on conformation, fini and quality.

Sheep are sold directly to the ultimate buyer or indirectly through commission agent. Since most sheep are produced west of the Mississip and are consumed in the East, they must be transported for comparativ ly long distances for slaughter. Lambs and sheep are shipped both truck and by rail, in double-decked cars.

Lambs are slaughtered by inserting a knife below the jaw to sever t jugular vein while the lamb is shackled by the hind legs to a conveyc The hind legs are skinned and the joints of the forelegs are broken ju above the feet. The whole pelt is then removed after which the carcass

pened and the viscera removed. After removal of the head, the carcass
s washed and removed to the cooler. Inspection of the viscera, carcass
and edible organs is done by veterinarians. As in the case of beef and
pork, diseased sheep are discarded and used for tankage. In coolers, the
emperature of all parts of the carcass should be lowered to 35°F (1.7°C)
within 24 hr.

The lamb carcass is not split prior to shipment to the distributor or
retailer. Sheep carcasses are classified into two categories—mutton and
amb (the latter is under 12–14 months of age). Lamb and mutton
carcasses are graded as prime, choice, good, utility and cull. A sheep
carcass is shown in Fig. 15.5.

CURED MEAT PRODUCTS

Much pork and some beef are cured or processed in some manner.
Veal, lamb and mutton are ordinarily not cured.) During the cutting up
of these animals, there are always some portions which are trimmed off
from the carcass. Much of the trimmings from both beef and pork
carcasses is used for the production of fresh or cooked sausage. Some of
the fat which is trimmed off, especially from beef, is sold as such to be
used for rendering. Fat rendered from trimmings of this kind is mostly
used for the manufacture of soap. Bones may be used for the manufacure of bone meal and, in some instances, for the production of gelatin.
Pork skin is used almost entirely for the manufacture of gelatin.

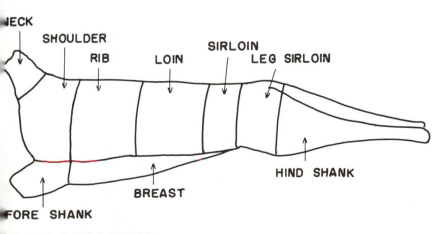

FIG. 15.5. SHEEP CARCASS

Curing agents for meat products may include only salt (sodium chloride), such as that used for the manufacture of New England style corned beef. However, more often, curing agents include salt, sodium nitrate, sodium nitrite and table sugar (sucrose) or glucose. Sometimes reducing agents, such as vitamin C or the related iso-ascorbic acid, are used to facilitate the coloring effect of nitrite, and in many instances spices, and materials such as monosodium glutamate, are used as flavor enhancers. Nitrite also inhibits the growth of *Clostridium botulinum*, a producer of a powerful toxin, and imparts a characteristic flavor.

When freshly cut, the color of meat is purplish red due to its pigment (myoglobin). After cutting the meat, the myoglobin soon becomes loosely bound to oxygen to form oxymyoglobin (bright red), and, if exposed to oxygen for some time, the bond becomes more permanent and the meat takes on a gray or brown color as metmyoglobin is formed. If nitrite is allowed to react with myoglobin, oxymyoglobin or even metmyoglobin under good reducing conditions, it replaces the oxygen and forms a compound known as nitrosomyoglobin. When meat containing nitrosomyoglobin is cooked to coagulate the protein, this compound is called nitrosohemochrome. The latter two compounds are pink or red in color and they are somewhat more permanent than oxymyoglobin and less subject to oxidation to form the brown or gray metmyoglobin.

In previous years, curing of large portions of meat, such as hams or bacon, was carried out by placing these meats in contact with a solution or dry mixture of sodium chloride (table salt) and sodium nitrate and allowing the salts and their products to diffuse into the flesh. This required much time, as much as 60–90 days, and, in the meantime, the meat might spoil because of bacterial decomposition before the salts had penetrated sufficiently to inhibit bacterial growth. Moreover, only nitrate in the form of sodium nitrate ($NaNO_3$) was used, and in order to obtain a red or pink color, nitrite (NO_2) had to be present, since the nitrate itself will not combine with meat pigment in this manner. In the long cures, nitrite was formed from the nitrate by the reducing action of bacteria which grew in the pickle which was added or formed. Today both nitrite (NO_2) and nitrate (NO_3) are used in curing solutions which are injected into the meat by one of several methods. The injection methods greatly shorten the curing time, since the distances over which the salt must diffuse, to reach all parts, are very much shortened. These new curing methods have resulted in the elimination of much of the spoilage of cured meats and in products having a more uniform cure, hence better quality.

In the curing of hams, a brine [55–80% saturated with salt, containing 20–50 lb (9.1–22.7 kg) of sugar, about 1.5 lb (0.68 kg) of sodium nitrite

nd 1 lb (0.45 kg) of sodium nitrate per 100 gal. (378.5 liters) of pickle]
s used. Small amounts of ascorbic acid or iso-ascorbic acid may also be
added. Ascorbic acid is added to provide reducing conditions in the
meat, which facilitates the formation of nitrosohemochrome. Phosphate
(for the retention of water) is allowed in cured meats in amounts not to
xceed 0.5%. This pickle is pumped into the ham through the exterior
liac artery. The hams are placed in tierces (casks) and may or may not
•e covered with pickle. Curing in this manner may require 5–10 days,
depending upon the size of the product. The curing should be carried out
t a temperature of 40°F (4.4°C) or below.

In smoking, moistened hardwood sawdust is usually burned in a
mechanically controlled smoke generator (located outside the chamber
where the product is hung) for purposes of treating with smoke and heat.
Oak and hickory sawdust or logs (logs may be used in a special smoke
enerator) are mostly used for the production of smoke. The temperature
f the smokehouse can be regulated by thermostatically controlled
heating units within the smokehouse.

Hams to be canned may be dry rubbed with a mixture of 70–80 lb
31.8–36.3 kg) salt, 25 lb (11.4 kg) sugar, and 1 lb (0.45 kg) sodium nitrite.
n this case, they will be held in the cure mix for about 14 days.

After curing, the hams are washed, hung to dry (with or without
overing with a stockingette), then smoked. The time and temperature of
moking must be sufficient to raise the temperature of all parts of the
iam to at least 137°F (58.3°C). This is a USDA regulation which has
een introduced to make certain that a roundworm, *Trichinella spiralis*,
s destroyed. This parasite infests hogs, bears and certain other animals
nd may infest the human. Cured hams to be boiled or canned may be
moked, but generally are not. Hams for canning are cured, skinned,
oned and placed in the can (some gelatin and various spices or flavoring
gents may be added), and the cans are sealed under a slight vacuum
nd then heated in an agitated water bath at 165°–180°F (73.9°–82.2°C)
ntil an internal temperature of 150°–165°F (65.6°–73.9°C) is reached.
3oiled hams are placed in metal molds before cooking. Neither boiled nor
anned hams are commercially sterile and must be refrigerated in order
o prevent spoilage.

In preparing bacon, the belly portion of the hog is stitch-pumped with
he curing solution. This is done by machines which, through needlelike
rojections, inject the curing solution simultaneously and automatically
ito the meat in a uniform pattern. The amount of pickle used for bacon
5 5–10% of the green (uncured) weight. It is composed of a 65–75%
aturated salt solution containing 20–100 lb (9.1–45.4 kg) of sugar, 1–1.5
o (0.45–0.68 kg) of sodium nitrite, and ¾–1 lb (0.34–0.45 kg) of sodium

nitrate per 100 gal. (378.5 liters) of pickle. The pumped bellies may b
placed in cover pickle (the pickle covers the product) or dry rubbed wit
curing salts during curing. Curing times are short, and in some cases, th
product is smoked almost immediately after pumping with pickl
During smoking (a period of 12–15 hr), the product is heated to a
internal temperature of 120°–125°F (48.9°–51.7°C). This is done t
develop the red color in the lean portion of the product. If the ten
perature were raised to 137°F (58.3°C) in all parts, as required for ham;
much fat would be rendered out and lost. Apparently, regulator
agencies consider that bacon will be cooked prior to consumption so as t
raise the temperature of all parts to 137°F (58.3°C) or higher.

After smoking and cooling, bacon slabs are squared up (uneve
ends cut off), the product is sliced by machine, and packaged (usuall
under vacuum) so that a representative portion of the product can k
viewed (through the plastic film) without opening the package.

Canadian bacon is prepared in a similar manner to that of regul;
bacon, except that pork loins, trimmed of fat, are used instead of bellie
Also, curing times are somewhat longer than for regular bacon an
during smoking, the internal temperature must be raised to 137°
(58.3°C) or higher.

Pork shoulders and shoulder picnics are cured and smoked much in tl
same manner as are hams.

Special pork products, such as capocollo, are prepared from bonele;
pork butts (part of the shoulder). They are dry cured for at least 25 da;
at temperatures not lower than 36°F (2.2°C). Special spices are used ;
the curing mixture with table salt, nitrite and nitrate as the ma;
ingredients [not more than 1 oz (28.3 g) of nitrite for each 100 lb (45.
kg) of meat]. After curing, the product is smoked for not less than 30 k
at a temperature not lower than 80°F (26.7°C), and then held in a dryin
room for at least 20 days at a temperature not lower than 45°F (7.2°C
These times and temperatures are specified to make certain tl
Trichinella spiralis has been destroyed, since this product is eat
without cooking.

Some beef products are cured, or cured and dried. Among these a
corned beef, pastrami, and chipped or dried beef.

Corned beef is usually produced from the brisket (lower forwa
portion of cattle near the foreleg). The fresh briskets are pumped wi
pickle and placed in cover pickle for 7–14 days at a temperature of 40°
(4.4°C) or below. In most cases, nitrite is used as an ingredient of tl
pickle, but in New England style corned beef, only salt is used.
producing pastrami, table salt, nitrite (in limited amounts as stat
above), and nitrate together with spices, are used in curing solutions

Corned beef and pastrami are often sold as the cured product to be cooked in the home, restaurant or delicatessen. However, they may be cooked and sold in the sliced (packaged) or unsliced form.

Corned beef (containing nitrite) is also canned as corned beef hash, a mixture of cooked corned beef, cooked potatoes and flavoring agents, such as onions. Some corned beef is cooked and canned as such.

Chipped beef is produced from beef hams or other portions which make up certain parts of the round section of the hind quarter. Usually cutter and canner grade cattle are used for this product. The beef hams are placed in tierces and covered with a pickle of high salt content containing not more (usually less) than 2 lb (907.2 g) of nitrite per 100 gal. (378.5 liters) of pickle. They are held in this manner for 50–65 days at a temperature below 40°F (4.4°C) (but above 34°F [1.1°C]). After curing, the product is hung in a room of low relative humidity and allowed to dry to the point where the salt content is 10–14%. It is then sliced in thin sheets and placed in glass jars which are capped under vacuum. This product is not heat processed. The high salt content and the nitrite are sufficient to prevent the development of food spoilage or disease-causing bacteria. The vacuum packing is necessary to retain the pink color of chipped beef, and it prevents the development of mold growth.

Regarding the stability of cured meats, it should be pointed out that impurities in the salt used in the curing process may considerably affect the development of rancidity. For example, salt may contain traces of copper, iron and other metals which are natural catalysts that initiate and accelerate rancidity.

SAUSAGE PRODUCTS

The meat ingredients used in a variety of sausage products may come, in part, from trimmings resulting from the cutting of beef quarters into rounds, loins, etc., and from the cutting of hogs into loins, hams, shoulders, etc. However, the chief source of meat for sausage comes from low-grade cattle and hogs. Bull meat is a prime ingredient for certain types of cooked sausage.

Fresh pork sausage is not cooked or smoked. The meat is ground and seasoned (usually only salt, sugar, sage and pepper are added for seasoning, but ginger may also be used). Some ice, up to 3% by weight based on the meat ingredient, may be added. The product is mixed and stuffed into natural casings. In federally inspected plants, the meat ingredient is limited to contain not more than 50% of fat. The natural

casings used for fresh pork sausage are made from the outer covering o
the small intestine of sheep. Larger plastic or natural casings may b
used for sausage.

Fresh pork sausage should be held at temperatures very near 32°
(0°C) after manufacture, since it is quite subject to bacterial spoilage, t
oxidation of the fat (rancidification) and to loss of the typical pink colo
Usually, when bacterial spoilage occurs, other types of spoilage do not

Frankfurters (see Fig. 15.6), the most popular type of sausage, ar
smoked and cooked. The meat ingredients used for frankfurters are be
(40–60%), and the rest, pork. Bull meat, lower valued portions of cattl
(boned out chucks and plates), and hearts and beef trimmings are th
usual beef ingredients. Trimmed pork fat and other pork trimming
back fat and hearts are the chief pork ingredients. Up to 20% of fill
meats, such as tongues, snouts, lips and other parts may be used. Som
frankfurters contain only beef as the meat ingredient.

Federal regulations allow the addition of up to 3 1/2% of cerea
vegetable starch, vegetable flour, soya flour, nonfat dried milk, drie
milk or any combination of these, as well as up to 15% chicken, to cooke
sausage products, such as frankfurters, but these ingredients must b

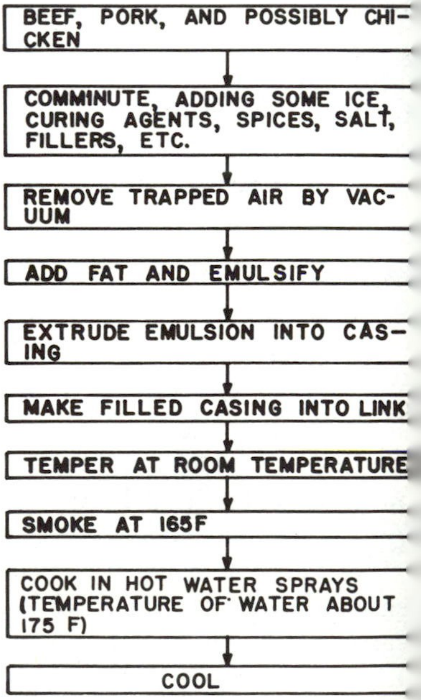

FIG. 15.6. SEQUENCE FOR PRODUCTION
OF FRANKFURTERS
See metric conversion tables in Appendix.

declared on the label.

Spices, such as pepper, nutmeg or mace, mustard, garlic and coriander are commonly used to provide flavor in frankfurters. Curing salts must be added to cooked sausage to provide for the development of color and to inhibit spoilage and the growth of *Clostridium botulinum*. Usually, 3 lb (1.36 kg) salt, 1/2 lb (226.8 g) dextrose (corn sugar), 1/4 oz (7 g) sodium nitrite, and 2 oz (56 g) sodium nitrate are used as curing agents per 100 lb (45.36 kg) meat.

To produce frankfurters, the meat is passed through a meat grinder and then finely comminuted in a silent cutter (a large metal bowl in which knives rotate at right angles to the plane of the bowl, which also rotates). While in the silent cutter, ice is added to prevent the temperature of the meat from rising above 60°F (15.6°C). It should be noted that the finished product may contain up to 10% added water (federal regulations), the purpose of which is to facilitate cutting and to provide more succulence to the finished product. Curing agents, spices, and fillers (if used), are added while the meat is in the silent cutter.

When ascorbic or iso-ascorbic acid is used to assist in color development, it is added about 1 min before the end of chopping in the silent cutter. Colloid mills or similar equipment may be used to comminute the meat instead of the silent cutter. From the silent cutter, the meat emulsion is put into metal carts, which may be placed in a vacuum chamber to eliminate trapped air bubbles. The air may also be removed in a mixer under vacuum.

The meat emulsion consists of a continuous phase, which is the water, and a dispersed or discontinuous phase, which is the fat. The fat is dispersed more or less uniformly in the water. The emulsifier, in this case, or the agent which maintains this uniform dispersion of the fat, is actually the fraction of proteins which has been solubilized, especially in the presence of salt. In preparing the emulsion, one must heed the temperature rise during preparation, and the rate at which the fat is added as well as the speed with which the emulsion is mixed. The temperature should not be too high. If it exceeds 68°F (20.0°C), or thereabouts, the emulsion may break down. The optimum rate at which the fat should be added depends a great deal on the rate at which the fat can be emulsified.

If the fat is added more slowly than the emulsification rate, then the mixing time will be unnecessarily extended, creating an increase in temperature and a decrease in the amount of fat that can be emulsified. If the rate at which the fat is added greatly exceeds the rate at which it can be emulsified, then it is quite possible that little or no emulsion will be formed.

The viscosity of the mix, the size of the fat droplets, the type of fat, and a considerable number of other factors affect the emulsion characteristics.

The meat emulsion is next placed in a stuffing machine where a powered piston extrudes the product into a stuffing horn and then into artificial (cellophane or other plastic) or natural casing (sheep or other animal intestine). The stuffed casing comes out as a long cylinder, which is then made into link form, generally by machine.

The linked sausage is hung on racks at room temperature (if ascorbic acid has not been added) to provide for the development of color; this is called tempering. The linked frankfurters are then smoked and cooked.

During smoking and heating, the frankfurters are hung in air-conditioned or natural draft smokehouses. During the first phase in the smokehouse, a temperature of 130°–140°F (54.4°–60°C) is used for 10–20 min. Smoke is then introduced, and the temperature is slowly raised to 165°F (73.9°C) and held until the internal temperature of the product reaches 140°–165°F (60°–73.9°C). After smoking, the product is cooked with hot water (170°–180°F [76.7°–82.2°C]) sprays. This may be done in the smokehouse or in a special chamber known as a Jordan cooker. Immediately after cooking, the internal temperature of frankfurters should never be less than 150°F (65.6°C), and preferably should be at 155°F (68.3°C) or slightly higher. After cooking, the frankfurters are hung in a cooler.

Color is sometimes added to the water used to cook frankfurters, and colored cellulose casings are sometimes used to enhance the appearance of the product.

Frankfurters stuffed into natural casings are not skinned after manufacturing, but those stuffed into cellulose casings are removed from the casing before they are sold. Skinning of the cooled frankfurters may be done by hand but is usually done by machine. The links of sausage are passed through a bath of warm water, then into the skinning machine, where the casing is slit with a blade, then rolled off or blown off.

After skinning, the frankfurters are packaged and immediately returned to the cooler, which should be held at a temperature near 32°F (0°C), until shipped out.

Frankfurters, like other perishable foods, eventually spoil when held for long periods at temperatures near but above freezing. When spoiled frankfurters either become slimy, due to the growth of bacteria or yeasts, or they turn green on parts of the surface. Greening is due to hydrogen peroxide produced by the growth of bacteria of the Lactobacteriaceae group. The hydrogen peroxide reacts with the nitrosohemochromagen (color complex of nitrite and myoglobin) and oxidizes it to form a green color.

At times, green rings on the inside of frankfurters, near the surface, may be encountered. This is not due to bacterial growth in the product after manufacturing. It is believed to be due to high bacterial counts in the fresh meat used for producing frankfurters. The bacteria are killed during smoking and cooking, but their end products are not destroyed and cause green rings to form during, or shortly after, manufacturing.

Continuous frankfurter manufacturing operations are now used in many instances. In this type of production, automatic stuffers extrude the emulsion into molds, which pass through an automated tempering, smoking, cooking and cooling unit. The product is packaged immediately after exit from the processing unit.

Bologna is another type of smoked and cooked sausage. The ingredients and manufacturing procedures for manufacturing bologna are similar to those used in making frankfurters, but the size of the casing, hence the size of the bologna, is much larger in diameter, and the length of the bologna link is proportionately longer than the frankfurter link. When not sufficiently cooked, bologna may develop green cores after processing.

There are many other types of meat sausage produced. Among the smoked, uncooked sausages are country-style pork sausage and country-style sausage which contains beef, as well as pork. These products contain moderate amounts of curing agents. Cooked sausages are made both with and without curing agents added, and are stuffed into natural cellulose or Saran casings. In the manufacture of Braunschweiger or liver sausage, pork liver and jowls, with or without beef, are used.

Types of sausage vary, depending on the style of cutting or grinding of the meat, the size and shape of the product, the seasonings used, the degree of smoking, and the cooking procedure. A few of the different types include: kielbasy, knockwurst, cooked salami, mettwurst and Polish sausage.

There are various loaf-type specialty products, such as chicken loaf, luncheon meat, meat loaf and head cheese, and these may be processed in casings or in metal containers. In the latter instance, they are usually placed in casings after cooking. This type of product may be cooked in hot water or baked in an oven. The products may be deep-fat fried, after baking, in order to brown the surface.

Some types of sausage are deliberately handled to permit a bacterial fermentation to occur in the product during processing. During the bacterial fermentation, lactic acid is produced, lowering the pH of the product to around 5.0 (could be slightly higher or slightly lower). The lactic acid assists in the preservation of the product and contributes to the particular flavor of this type of sausage.

Fermented sausages are either of the semi-dried or dried type. Most fermented sausages are smoked, and the semi-dried type is heated to an internal temperature of at least 137°F (58.3°C). Many dried types of fermented sausage are not heated to temperatures above 90°F (32.2°C).

Curing salts are among the ingredients used in the production of fermented sausages, but nitrate sometimes replaces nitrite. During processing, bacteria remove oxygen from the nitrate, reducing it to nitrite. As stated earlier, nitrite provides the typical red or pink color in the meat.

After mixing with the curing agents, the meat for semi-dried sausage is held in pans for 48–72 hr in a refrigerated room 38°–40°F (3.3°–4.4°C). The meat is then remixed, stuffed into casings, and held at 50°–60°F (10°–15.6°C) for 12–48 hr, then smoked. Smoking is accomplished in stages: first, at a temperature of 80°–90°F (26.7°–32.2°C) for 12–16 hr, then, at 100°F (37.8°C) for 24 hr, and finally, at 137°F (58.3°C) for 4– hr. During the processing of semi-dried sausage, conditions at first favor the growth of bacteria which reduce nitrate to nitrite. This is followed by the growth of lactic acid producing bacteria. To ensure the type of fermentation desired, it is possible to obtain cultures of appropriate bacteria, which may then be added as an ingredient of the product. Typical semi-dried sausages are thüringer, cervelat, Lebanon bologna, and pork roll.

Dried sausages, which are not heated to temperatures above 90° (32.2°C), do not undergo bacterial putrefaction because of their low pH (due to acid produced by bacterial growth), their low moisture content, and their high salt content. Spices and curing salts also contribute to the preservation of the product. Some dried sausages are smoked.

During the manufacture of dried sausage, the meat, curing salts and spices are mixed, and the mixture is passed through a grinder. The ground product is then placed in pans and held in a room at 38°–40° (3.3°–4.4°C) for 2–3 days. After being stuffed into casings, the sausage is held on racks in a room at 70°–75°F (21.1°–23.9°C) and at a relative humidity of 75–80% for a period of 2–10 days. Smoked varieties are then smoked at low temperatures (not above 90°F [32.2°C]). Finally, the sausage is dried by hanging in a room at 45°–55°F (7.2°–12.8°C) (relative humidity of 70–72%) where there is an adequate air movement (at least 15 air changes per hour). Drying times vary between 10 and 90 days, and the moisture loss during drying is from 20–40% of the weight of the freshly smoked product.

16

Dairy Products

Dairy products are produced from milk. In the United States, essentially all dairy products are produced from cow's milk, although minor quantities of goat's milk products may also be manufactured.

Because milk and milk products have traditionally been priced on their valued component, the cream, standards have been established mainly for this component but include provisions for other components, as well, at both the federal and state levels. The standards, found in the *Agriculture Handbook No. 51,* and published by the USDA, provide the values for the minimum requirements for fat, nonfat solids and vitamins, as well as the maximum allowable amounts of water and additives.

FLUID MILK

The Holstein breed outnumbers all others used in the United States for the production of milk. Jersey and Guernsey breeds tolerate hot weather better than Holsteins, hence may be the predominant types used for the production of milk in hot weather areas. Some Ayrshire, and Brown Swiss or Shorthorn breeds are used in certain areas.

Cow's milk contains an average of 3.8% fat (called butterfat), 3.3% protein, 4.8% lactose (a 12-carbon sugar), 0.7% ash (minerals) and 87.4% water. Milk also contains vitamins and other nutrients in small amounts, making it the most complete of foods. The young of mammalians survive on it exclusively. However, components of milk from different species vary, and occasionally the young of one species may be unable to tolerate the milk from another species, mainly because of differences in the lactose contained therein. The fat content of milk from Ayrshire and Brown Swiss, and especially from Guernsey and Jersey breeds, is slightly higher than that from Holstein cows, but the latter breed generally produces much more milk than the others.

Most milk is produced on farms which deal primarily with the raising of dairy cattle.

Microorganisms in Milk

As drawn from the cow's udder, milk seldom, if ever, is free from microorganisms; bacteria, molds and yeasts are usually present in small numbers, among which the bacteria are most significant from the standpoint of quality and the transmission of food-borne disease. The control of microbial activity in milk and milk products, especially the control of bacteria and bacterial growth, is the most important function in the handling and manufacture of dairy products.

Raw milk, when improperly handled, may undergo any of several adverse changes. It may become sour due to the growth of bacteria which produce lactic acid, or it may become foamy due to the growth of gas-producing coliform bacteria or yeasts. Raw milk may also be subject to peptonization (digestion of the casein), the formation of rope (a viscous polymer of sugars) and sweet curdling, when bacterial growth is not controlled.

In the past, a number of food-borne diseases including scarlet fever, septic sore throat, diphtheria, salmonellosis, typhoid fever, tuberculosis, and undulant fever or brucellosis have been transmitted to man through milk. Today, transmission of disease through the drinking of milk is very rarely encountered. The main reason for this is that most milk is pasteurized or heated at temperatures and for times which destroy pathogenic bacteria, should they be present. Also, dairy herds are tested for tuberculosis, and diseased animals are eliminated, as are those which show positive tests for brucellosis. Cattle may also be vaccinated to prevent infection with the organism which causes brucellosis.

Most, but not all, certified milk is pasteurized. However, certified milk must be produced from herds that have been inspected, tested and found to be free from disease, and the milk must be drawn and handled under the best conditions of sanitation. There are bacterial standards for certified milk which limit the number and types of bacteria which may be present.

Sources of Bacteria in Milk and Methods of Limiting Bacterial Contamination

Some bacteria are normally present in the udder of the cow, and these may contribute to the bacterial flora of milk. However, unless the udder is infected, it is not considered to be an important source of such microorganisms. Other sources of microbial contamination of milk are the body of the cow, milking machines and other equipment and utensils, the air in the milking barn, and the hands, nose and throat of those attending to the milking process. In the handling of milk upon delivery to the processing plant or dairy, further sources of contamination may be encountered.

In order to limit the number of bacteria present in raw milk, certain precautionary procedures are ordinarily applied. The flanks, udder and teats of the cow should be washed, treated with a sanitizing solution and dried before milk is drawn. Large dairy farms often have a special wash pen for cows to be milked. Utensils, including the milking machine, should be cleaned and disinfected either with live steam or with a solution of chlorine (about 200 ppm of available chlorine). Bulk milk tanks may be cleaned manually with detergent and water at about 130°F (54.4°C), then sanitized with chlorine solution, or cleaned mechanically with detergent and water at 150°F (65.6°C), and finally sanitized with chlorine solution. Outlet valves and the outside of the tank must be cleaned and sanitized manually. Cleaning in place (CIP) may be used to clean, sanitize and rinse the milk pipe line, the teat cup assembly, and the bulk milk tank if a vacuum or pressure system is available.

The water supply used on dairy farms should be potable and located so that there is no possibility of contamination with animal or human discharges. Where municipal sewerage systems are not available, human wastes and floor washings from milk handling areas should be disposed of in septic tanks or cesspools. Manure should not be allowed to accumulate near milk handling areas and is best disposed of by spreading in thin layers on pastures.

Flies may be controlled in milk-handling areas, at least to some extent, by flytraps which utilize entrapment liquids, by poisons, such as formaldehyde, or by electric fly killers. The control of other insects, such as cockroaches, may require the use of an approved insecticide.

The walls and floor of the milking area should be kept clean, and the room should be reasonably well ventilated and free from dust. Personnel involved in the handling of milk should not have a history of intestinal disease. Flush toilets, in rooms which do not open directly into the area where cows are milked, should be provided for personnel, and hand-washing and sanitizing facilities should be available at or near the area where milk is drawn. The hands of the milker or milking attendant should have been cleaned, sanitized and dried before milking is started.

Handling of Milk on Farms

Milking is generally carried out by machine rather than by hand. At first, a small amount of milk is drawn and examined for impurities, after which the cups of the milking machine are applied to the teats of the cow. The cups are connected to a hose which leads to a holding container, and milk is collected by suction and a rhythmic squeezing action. After milking, the cups are immersed in a non-irritating bactericidal agent before applying them to the teats of another cow. Milk may be collected in enclosed metal containers which, when milking is completed, are emptied into 10 gal. (37.8 liter) milk cans. The milk is strained through cloth or fine metal screening prior to emptying into the cans. The covers are placed on the filled cans, which are then immersed in refrigerated water to cool the milk prior to delivering it to a receiving station, country manufacturing plant or town or city fluid milk plant. A more recent method of milking and milk handling (the method now most often used) is to deliver the milk directly from the cow to a bulk cooling tank through a glass or stainless steel pipe line. The bulk milk tank is cooled by refrigerated water or refrigerant sprayed directly or expanded into a jacket which covers the outside of the tank. During cooling, the milk is slowly agitated mechanically to provide for faster heat transfer. Milk should be cooled to approximately 40°F (4.4°C) within 2 hr after it is drawn, although there appears to be some factor in freshly drawn milk which inhibits the growth of microorganisms for a period of several hours.

Milk may undergo a number of adverse changes during its handling. If held for any length of time without adequate cooling, it is subject to the various types of spoilage due to the growth of microorganisms. In addition, milk may develop various off-flavors due to the feed consumed by the cows, especially when they have fed on wild onions, french weed or ragweed. Large proportions of beets, beet tops, potatoes, cabbage or turnips in the fodder provided for cows may also cause the development of off-flavors in the milk. Lipase, an enzyme present in cow's milk, may cause a hydrolysis of fat, splitting off butyric acid, which causes an off flavor and off-odor. Milk which has been cooled, then warmed to about 85°F (29.4°C), then recooled or homogenized in the unpasteurized state is subject to this kind of off-flavor development. Off-flavors due to the oxidation of some of its components may occur in milk, especially if traces of copper are present, since copper catalyzes this type of reaction. Milk, therefore, should be kept out of contact with equipment which contains copper.

Transportation of Fluid Milk

Milk is transported from the farm to the receiving station or to the fluid milk processing plant in pickup tankers. Pickup tankers are insulated, stainless steel tanks, usually having a holding capacity of more than 5000 gal. (18,925 liters) on a trailer which is handled by a motorized vehicle. No refrigeration is provided for pickup tankers, since the insulated container prevents a significant rise in the temperature of the milk during the period required for transportation and delivery to the processing plant. The production of whole milk involves a series of steps, the important ones of which are shown in Fig. 16.1.

When receiving the milk, the operator of the tanker first tests the product, which has been stored in the bulk tank, for odor and flavor and, if not suitable, the milk is rejected. If acceptable, the volume of the product in the bulk tank is measured with a rod. It is then agitated, after which a sample is taken in a glass or plastic bottle from which the butterfat content will be determined, since the farmer is paid on the

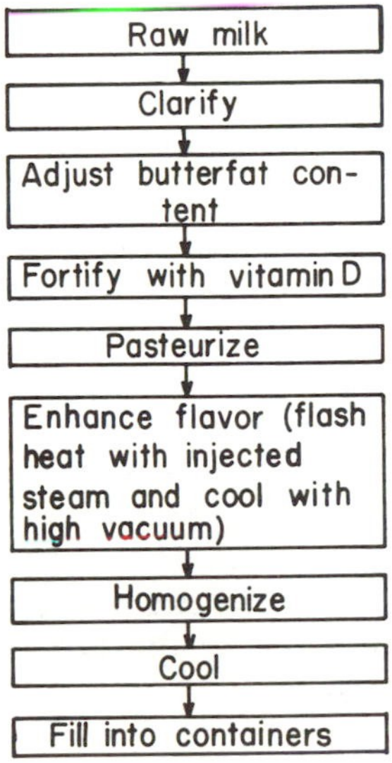

FIG. 16.1. PRODUCTION OF WHOLE MILK

basis of butterfat content. The milk is then pumped from the bulk tank into the tanker through a sanitized plastic hose after which the hose is capped. The final step is to prepare a weight ticket for the farmer and tabulate the weight, temperature, etc., of the product on a record sheet.

Pickup tankers, including auxiliary equipment such as hoses, must be cleaned and sanitized, as are dairy farm milking and milk holding equipment, after delivery of the product to the processing plant.

Processing of Fluid Milk

Milk, as delivered to the processing plant, is first clarified while cold. Clarification consists of passing the milk through a centrifuge, which is similar to a cream separator but operated at low speed. This treatment is sufficient to separate out dirt and sediment which might be present depositing them as a layer on the inner surface of the centrifuge bowl The clarifier is not operated at sufficient speed to separate the cream from the milk. After clarification, the milk is usually pumped into a storage tank equipped with an agitator. While in the storage tank, the milk is sampled, and the butterfat content is determined. It is then standardized by adding enough cream or skim milk (milk from which cream has been removed) to provide the fat content required by state regulations. The milk is then fortified with vitamin D at the rate of 400 USP units per quart (0.95 liter).

The next step in fluid milk processing is to pasteurize it. During pasteurization, milk must be heated in all parts to 145°F (62.8°C) and held at this temperature for 30 min, or it must be quickly heated to 161°F (71.7°C) and held at this temperature for 15 sec, heated to 191°F (88.3°C) and held for 1 sec or heated to 194°F (90°C) and held for 0.5 sec then cooled. Pasteurizing at temperatures above 145°F (62.8°C) is called the high-temperature short-time (HTST) method, and both heating and cooling are carried out over a short period of time. Milk may be pasteurized in insulated vats heated by coils carrying hot water, or in vats heated by hot water sprayed within a jacket surrounding the sides and bottom of the vat. With low temperature pasteurization, the milk is ordinarily agitated during heating and cooling (see Fig. 16.2). Plate heaters and coolers (see Fig. 16.3) or tubular heaters may be used to pasteurize and cool milk. When tubular heaters are used, the product travels in one direction through an inner tube while the hot water for heating or the refrigerated liquid for cooling passes in the opposite direction through an outer tube surrounding that which carries the product.

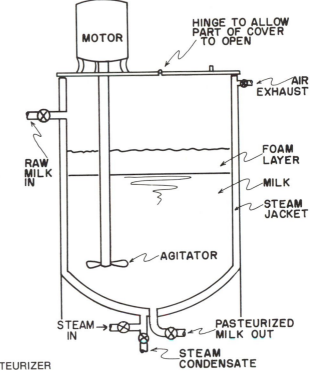

FIG. 16.2. BATCH PASTEURIZER

Today, milk is usually given what is called a flavor treatment to provide a product which is uniform in odor and taste. During flavor treatment, milk is instantly heated to about 195°F (90.6°C) with live steam (injected directly into the product) after which it is subjected to a vacuum of about 10 in. (25.4 cm) in one chamber and to a vacuum of about 22 in. (55.9 cm) in another chamber. The high vacuum treatment serves to regulate flavor, to cool the milk to about 150°F (65.6°C) and to evaporate water which may have been added through the injection of steam.

While the milk is still hot, it is usually homogenized by passing it through a small orifice which breaks up the fat globules to a small size, preventing the separation of cream from the milk.

After pasteurization, with or without flavor treatment and homogenization, milk is quickly cooled to about 35°F (1.7°C). This is

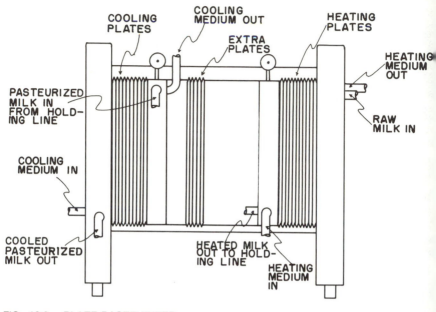

FIG. 16.3.   PLATE PASTEURIZER

done by the same general procedures used in heating, except that refrigerated water or brine, or directly expanded ammonia is used in the coils, vat jacket, outer tubes of the pasteurizer or tubes of the plates or the cooling side of the plate pasteurizer. During HTST pasteurization and during flavor treatment and homogenization, milk is passed through the heating and cooling cycles at such a rapid rate that at no time is it held for long periods at high temperatures.

After processing and cooling, milk is filled mechanically into containers made of waxed or plastic-coated cardboard of different volumes up to 2 qt (1.9 liters) and of semi-rigid plastic of 1 gal. (3.8 liters), and the containers are sealed. In this state, milk should be held as close to 32°F (0°C) as possible until consumed.

In plants producing fluid milk or milk products, all equipment including tanks or vats, pasteurizers and coolers, homogenizers, pipe lines, pumps, etc., should be of sanitary design. There should be no threaded pipes. Joints should be of the clamp type which can be easily disassembled for cleaning and sanitizing. All surfaces which contact milk or milk products should be readily accessible for cleaning and sanitizing regardless of whether they are to be cleaned in place (CIP) or disassembled. The suitability of equipment for cleaning and sanitizing, and the frequency with which this is done, are equally important.

Skim or low-fat milk may be produced by centrifuging whole milk to remove the butterfat as cream. Low-fat milk usually contains less than 2% of butterfat and is ordinarily fortified with vitamins A and D prior to pasteurizing and cooling.

When milk is centrifuged at comparatively high speed, the fat or cream (which is lighter in weight than the other ingredients) is separated from the milk at, or near, the center of the centrifuge bowl. Separation is ordinarily done prior to pasteurization, and after the milk has been warmed to 90°–110°F (32.2°–43.3°C), depending upon the point of separation, speed of rotation of the centrifuge, etc., cream may be separated as approximately 40% butterfat (heavy cream), 30% butterfat (all purpose cream) or 20% butterfat (light cream). The creams higher in butterfat may also be diluted with skim milk to provide the various fat densities or to produce a product known as "half-and-half" (about 10.5% butterfat).

Since cream tends to spoil more quickly than milk, during pasteurization it is given a more drastic heat treatment than that given to milk. When batch pasteurization is used, cream is heated to 150°–155°F (65.6°–68.3°C) and held at this temperature for 30 min prior to cooling. When the HTST (hot-temperature, short-time) method is used, cream is heated to 166°–175°F (74.4°–79.4°C) and held at this temperature for 15 sec prior to cooling. Table cream (light cream or half-and-half) is usually homogenized after pasteurization. All cream, after pasteurization [cooling to 35°F (1.7°C) and containerizing] should be held at 35°–40°F (1.7°–4.4°C) until consumed or subjected to additional processing.

In recent years, a special fortified low-fat milk has become popular. This product is manufactured by removing the fat, adding sodium caseinate (a derivative of casein, the main protein in milk) and vitamins, homogenizing the mixture, pasteurizing and cooling.

Some bacteriological standards for milk and cream as issued by the U.S. Public Health Service are listed in Table 16.1.

TABLE 16.1.   UNITED STATES PUBLIC HEALTH SERVICE BACTERIOLOGICAL STANDARDS FOR MILK AND CREAM

Product	Maximum Standard Plate Count per ml of Product		Maximum Coliform Count per ml of Product	
	Raw	Pasteurized	Raw	Pasteurized
Certified Milk	10,000	500	—	—
Grade A Milk	200,000	20,000	50	0
Grade B Milk	1,000,000	50,000	100	0–1
Grade A Cream	400,000	60,000	—	—
Grade B Cream	2,000,000	100,000	—	—

Large quantities of skim or low-fat milk are dried. This may be done by spraying atomized droplets of milk into a chamber through which heated air is circulated (spray drying, see Fig. 11.3) or milk may be dried by allowing it to flow over the surfaces of two heated metal drums which rotate toward each other (drum drying, see Fig. 11.2). With the latter method, the dried milk is scraped from the drums, as they rotate, by metal scraper blades. Dried milk (usually the spray dried type which contains about 5% of moisture) may be rehumidified to slightly higher moisture content after drying. This treatment agglomerates the fine milk particles to form clumps of milk powder. This treatment yields a product which dissolves or disperses in water much more readily than the finely powdered dried milk. It is, therefore, considered to be an instantly soluble product.

Milk is used for the manufacture of a wide variety of popular dairy products, and these are described in the following paragraphs.

OTHER DAIRY PRODUCTS

Ice Cream

Regular ice cream may contain about 10% butterfat (added as cream), milk or skimmed milk, sugars (sucrose and/or dextrose), gelatin or vegetable gums (to provide body), eggs, and flavoring such as vanilla, fruit, fruit extracts, fruit juices, cocoa, chocolate and nuts. Low-fat ice cream may be made by using less butterfat (as cream) and adding more milk solids or sodium caseinate or both. Generally, the butterfat content of ice cream is regulated to a certain minimum by state requirements and low-fat mixtures must be labeled as something other than ice cream, e.g., ice milk, etc.

There are various systems and procedures for the manufacture of ice cream. A typical method is described in Fig. 16.4. Liquid ingredients (milk, cream, concentrated skim milk, etc.) are mixed in one tank and liquid sweeteners (sugar, corn syrup) in another tank. The liquids are blended together with the dry ingredients (stabilizers, emulsifiers, whey solids, etc.) and allowed to stand ("soak") for about 20 min. The mixture is then pasteurized, homogenized, cooled to about 36°F (2.2°C) while agitating and pumped to storage tanks. From the storage tanks the mixture is metered into the freezer, and as this is done, measured amounts of flavoring materials, fruit or nuts (when used) and possibly coloring are added. The freezer consists of a tube with walls refrigerated to a temperature of 5° to 15°F (− 15° to − 9.4°C) and blades that rotate

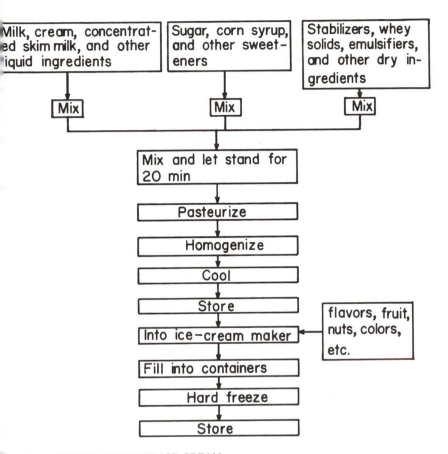

FIG. 16.4. MANUFACTURE OF ICE CREAM

at 175–225 rpm. The blades scrape the inner walls of the tube and incorporate air into the mixture as it is cooled. The blades also cause the product to move through the freezer, and at the time of exit the ice cream is plastic and has a temperature of 22° to 26°F ($-5.6°$ to $-3.3°$C). In this form it is filled into cardboard or plastic containers. The containers are capped and placed in hardening rooms which usually have a temperature of $-20°$ to $-50°$F ($-28.9°$ to $-45.6°$C). Special systems may be used for hardening ice cream. After hardening, ice cream is stored at 0°F ($-17.8°$C) or below.

Ice cream in the plastic or unfrozen state may be added to small molds which are carried on a belt through refrigerated brine. A stick is inserted at the start, if a partially frozen mix is used, or when partially frozen, if the liquid mixture is used), and the product is allowed to freeze and solidify thoroughly. The frozen portions are removed from the molds and dipped in warm, liquid chocolate. The cold ice cream hardens the chocolate, and the product is then wrapped, packaged and placed in storage at 0°F (− 17.8°C) or below. Some plastic ice cream is extruded between sheets of cookie-like pastry, prior to wrapping and hardening, to produce ice-cream sandwiches.

Since air is incorporated into ice cream as it is manufactured, the volume of the finished product is much larger than that of the liquid mixture from which it was produced. This increase in volume is called overrun. Overruns for ice cream vary between 60 and 120% and can be calculated as follows:

$$\% \text{ Overrun} = \frac{(100) \times (\text{weight per gallon of mix} - \text{weight per gallon of ice cream})}{\text{weight per gallon of ice cream}}$$

Ice Milk, Sherbets and Ices

A number of milk ices or ice milks are manufactured. These products differ from ice cream mainly in that the butterfat content is only 3–4% and the sugar content is somewhat higher than that of ice cream.

Sherbets are fruit or mint-flavored frozen desserts containing some milk solids, as in the case of milk ices. The sugar content of sherbets is somewhat higher than that of ice cream. High overruns in sherbet are not desirable, since this product tends to develop a spongy texture when overruns are higher than 25–40%.

Ices and frozen suckers contain no milk solids. The mixture consists chiefly of a solution of fruit juices and 30 to 32% sugar. Ices may be frozen in an ice-cream freezer, or the mixture may be added to molds and a stick inserted into the partially frozen material, and then completely frozen as the molds pass through a bath of refrigerated brine.

Cultured Fluid Dairy Products

In the United States, the three most utilized cultured dairy products are buttermilk, soured cream and yogurt. The main difference between buttermilk and soured cream is that buttermilk is made from fluid skim milk to which some churned cream may have been added, while soured cream is cultured from light cream which has a butterfat content of about 18%.

In the preparation of buttermilk, skim milk is pasteurized either by heating to 185°–190°F (85°–87.8°C) and holding for 30 min, or by heating to 195°F (90.6°C) and holding for 2 min. After pasteurizing, the product is cooled to a temperature of 71°–72°F (21.7°–22.2°C), then inoculated (about 0.5 to 1.0%) with a culture of *Streptococcus lactis* or *Streptococcus cremoris* (to produce lactic acid) and *Leuconostoc citrovorum* or *Leuconostoc dextranicum* (to produce flavor). The milk is incubated at the stated temperature until the acid content is about 0.8% (pH of 4.5), after which the coagulum is broken by agitation and the product is cooled to 40°–45°F (4.4°–7.2°C). After cooling, the product is allowed to stand for about 2 hr, to remove air which may have been incorporated during agitation, then filled into cartons. Buttermilk should be held at 35°–40°F (1.7°–4.4°C) until consumed. If butter granules are to be included, about 2% of the cultured milk is removed and cream is added to obtain a butterfat content of about 15% in the removed portion (sufficient to produce a butterfat content of 0.75% in the finished buttermilk). This mixture is brought to 65°F (18.3°C) and then churned by circulating through a centrifugal pump. When butter granules are formed, the churned product is cooled and added to the bulk of the cultured skim milk. In order to obtain butterfat granules in buttermilk, melted butter may also be sprayed into the cultured product to provide a butterfat content of about 1%.

Soured cream is prepared from light cream (the fat content must be 16 to 20% depending on state regulations) to which 8 to 9% of nonfat milk solids and/or 0.25 to 0.5% of stabilizers (gelatin, gums, etc.) may have been added. In some cases, small amounts of rennet (a coagulating enzyme) are also added. The cream mixture (without rennet) is first pasteurized by heating to 165°F (73.9°C) and holding for 30 min or by heating to 180°–185°F (82.2°–85°C) and holding for 25 sec. The cream or cream mixture is homogenized before cooling.. After cooling to about 72°F (22.2°C), the mixture is cultured with the same bacteria used to produce buttermilk. If rennet is to be used, it is added at the time of culturing. After incubation, to produce an acidity of 0.65 to 0.70% (pH 4.5), the product is cooled to 40°F (4.4°C), filled into cardboard containers and held at 40°F (4.4°C) for 1 to 2 days to solidify the fat. Soured cream should be held at 35°–40°F (1.7°–4.4°C) until consumed.

Yogurt is made from whole milk that has been boiled to increase the solids or from whole milk to which 1 to 5% of nonfat milk solids has been added. Stabilizers (e.g., gelatin, alginates, gums) are generally added to the milk mixture in producing yogurt. The mixture is either preheated to 150°F (65.5°C), homogenized, then heated to 185°F (85°C) and held at this temperature for 30 min for purposes of pasteurization, or it is heated to 195°F (90.6°C) and held for 25 sec in order to pasteurize it, then homogenized. In either case, the mixture is cooled to about 113°F (45°C), then inoculated with the starter culture. The starter culture consists of a mixture of *Lactobacillus bulgaricus* and *Streptococcus thermophilus* which has been added to sterile milk and allowed to grow, providing high concentrations of the respective bacteria. The inoculated milk product is then mixed and filled into cartons; the cartons are capped and packed in cases. The filled cases are then held at 106°–108°F (41.1°–42.2°C) for a period of 3 to 4 hr, then placed in a cooler where the product temperature is eventually brought to 40° to 45°F (4.4° to 7.2°C). The product is then stored at about 35°F (1.7°C). Fruit-flavored yogurt may be prepared by adding a portion of fruit or fruit mixture to the cartons before the yogurt mixture is added.

Cheeses

There are many types of cheeses of which only those most commonly used in the United States will be described.

Cottage Cheese.—In the manufacture of cottage cheese, skim milk is pasteurized at 145°F (62.8°C) for 30 min or at 161°F (71.7°C) for 16 sec, cooled to 90° to 72°F (32.2° to 22.2°C) and inoculated with the starter culture (*Streptococcus lactis* with or without a culture of *Leuconostoc citrovorum*). A small amount of rennet may also be added. The mixture is then stirred for about 10 min and incubated, during which time the curd is set. With the short-set method, the inoculated mixture is allowed to incubate for about 4 hr, whereas the long-set method requires 14 to 16 hr of incubation (more starter culture and higher incubation temperatures are used in the short-set method than in the long-set method). During the set, an acidity of 0.48 to 0.52% and a pH of 4.6 are reached. After incubation, the curd is cut by passing wire knives through it. To produce small curd cheese, the knives are set to cut the curd in 1/4–1/2 in. (0.6–1.3 cm) squares. For producing large curd cheese, the knives are set to cut the curd in 1/2–3/4 in. (1.3–1.9 cm) squares. After cutting, the curd is allowed to stand for 15 to 20 min and is then cooked. Cooking requires slowly raising the temperature of the water in the jacket of the vat to 120°–125°F (48.9°–51.6°C). After cooking, the curd is pushed to one end of the vat, and the whey is drained off. The curd is then washed by adding cold water over it and allowing it to stand for 10 min; it is then drained. The temperature of the curd is now about 85°F (29.4°C). A second and third washing and draining are used, during which the temperature of the curds is brought to 60°F (15.6°C) and 45°–40°F (7.2°–4.4°C), respectively. The product is drained for 30 min after the last washing. The curd is then salted (0.75 to 1% of salt is sprinkled over

) and mixed, after which cream may be added and mixed in (cream
containing 12 to 14% of butterfat is used) to bring the butterfat content
of the product to 4%. The cream may be added and mixed in the cheese
vat or in a blender. The cheese is then mechanically filled into cardboard
or plastic cups, capped and stored under refrigeration. Cottage cheese
should be held at 35° to 40°F (1.7° to 4.4°C) until consumed. Some
cottage cheese is pasteurized and cooled after packaging, then refrigerated.

Cheddar Cheese.—Cheddar cheese (see Fig. 16.5) is made from
whole milk. Prior to culturing, the milk may be pasteurized with heat or
treated with hydrogen peroxide (to destroy bacteria). If hydrogen peroxide is used, the milk must be treated with the enzyme catalase to
decompose residual hydrogen peroxide before the culture is added. Some

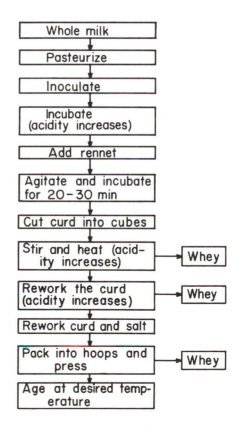

FIG. 16.5. MANUFACTURE OF
CHEDDAR CHEESE

cheddar cheese is made from raw, unpasteurized or untreated milk. If raw (unpasteurized) milk is used, the cheese must be held for at least 60 days at a temperature not lower than 35°F (1.7°C) prior to sale for consumption. This is required to destroy staphylococci, which, if present, will grow and produce a toxin [should cheese be held at 50°F (10°C) or higher] and to destroy the toxin which might be present due to the growth of these bacteria in the milk used to make the cheese. The toxin causes intestinal disturbances in humans.

In the manufacture of cheddar cheese, the temperature of the milk is brought to about 86°F (30°C), after which starter cultures consisting of *Streptococcus lactis* and *Streptococcus cremoris* are added (an amount equivalent to 0.5 to 1% of the volume of milk). The starter is mixed with the milk and allowed to incubate for 30 to 60 min. During incubation, an acid content (as lactic acid) of about 0.2% is reached. A coagulating enzyme (rennet) is then added (2 or more parts of enzyme per million parts of milk) and mixed with the milk. In order to set the curd, the mixture is allowed to stand for a period of 20 to 30 min. The curd is then cut into cubes of about 1/2–3/4 in. (1.3–1.9 cm). After stirring the cubed product for 10 to 15 min, the temperature in the jacket of the cheese vat is raised to bring the temperature of the product to 100°–106°F (37.8°–41.1°C) for a period of 30 min. The final cooking temperature is maintained for a period of 35 to 45 min, during which time the acid content of the mixture increases. After cooking, the curd is cheddared (the whey is drawn off, and the curd is heaped into a mass which mats together and is then cut into slabs which are piled one upon another). During cheddaring, acid formation continues as some of the moisture is expelled from the slabs of cheese. When the acid content has reached about 0.5%, the slabs of cheese are milled by machine or cut into portions which are about 5/8 in. (1.6 cm) square and 2 in. (5.1 cm) long. This facilitates salting. Enough salt is then sprinkled over the cheese and mixed with it to provide a salt content of about 1.5% in the finished product. The salted cheese is then placed in hoops or forms which are placed on racks to which pressure is applied. The cover of each hoop fits within the container so that as pressure is applied to the cover it presses the curd, squeezing out the whey. Pressing lasts about 24 hr or more.

After pressing, the cheese is aged. Aging time and temperature vary according to whether mild or sharp cheese is to be produced. Cheddar cheese may be aged for 12 to 18 months at 32° to 34°F (0° to 1.1°C), for 8 to 10 months at 38° to 40°F (3.3° to 4.4°C) or for 60 days at 38° to 40° (3.3° to 4.4°C).

Cheddar cheese is much lower in moisture content than is cottage cheese, which has a moisture content of about 79%. After curing, the noisture content of cheddar cheese must not exceed 39%. Also, the milk olids of cheddar cheese must contain not less than 50% butterfat.

During the aging of different kinds of cheese, various microorganisms bacteria or molds) grow and produce enzymes which modify the protein o provide the typical texture of the product. Also, microorganisms produce small amounts of chemical compounds from the components of he curd, which provide the typical flavor of the cheese. During the aging f cheddar cheese, the lactic acid streptococci continue to grow for about ? weeks after which *Lactobacillus* species grow and predominate. Eventually, micrococcal species become established as the predominant members of the bacterial flora.

In the manufacture of some types of cheese, lactic acid, or another uitable acid, may be added to the milk, instead of the cheese being ultured with bacteria, to produce the required acidity (about 0.2% of actic) to bring the pH to a level suitable for the coagulating action of the nzyme, rennet.

Process Cheese.—Various kinds of process cheese may be manufac-ured using different types of cheese (cheddar, swiss, blue, etc.), as the main ingredient. Generally, green (not fully aged) cheese is ground ogether with some aged cheese. Some water, an emulsifier (disodium)hosphate, Na_2HPO_4), salt and powdered skim milk are added, and the ngredients are heated and mixed, then extruded into molds. The molded shaped) cheese is wrapped with plastic and sometimes packaged.)ftentimes, sorbic acid in small quantities is included in the ingredients o prevent the growth of molds (fungi). The cheese may or may not be liced prior to wrapping with plastic. Process cheese may also be filled nto glass jars. Other ingredients of process cheese may include gums locust bean gum, etc.) pimientos, cream, skim milk cheese, condensed rhey, certified cheese color, etc., the particular ingredients depending pon the type of process cheese being manufactured.

Swiss (Emmenthaler) Cheese.—The general procedures applied in he making of cheddar cheese are used in the production of all cheeses, ut variations in the procedure of manufacture and aging are responsible or the different texture and flavor characteristics of particular types of heese.

In the manufacture of Swiss cheese, either raw or pasteurized cow's lilk may be used. After the milk is warmed to 88°–94°F (31.1°–34.4°C), he starter cultures are added. Starter cultures may include small mounts of *Streptococcus lactis* and/or *Streptococcus cremoris* (when asteurized milk is used), but larger amounts of cultures containing

Streptococcus thermophilus, Lactobacillus bulgaricus, and especiall: *Propionibacterium shermani* are also added. After about 30 min o holding at the temperature indicated, the cultured milk is coagulated with the enzyme, rennet, and the curd is cut into small kernels, stirred for 20 to 30 min, then heated. During heating, the temperature is raised to 126°–130°F (52.2°–54.4°C), and the mixture is stirred. Heating i: continued until the curd is firm. The cooked curd is then drained and taken up in a cloth, then pressed in hoops or forms of a large enough siz: to produce wheels of cheese weighing 125–210 lb (56.8–95.3 kg) o: blocks weighing 24–90 lb (11.4–40.9 kg). Pressing lasts for a period of 2: hr. The wheels or blocks of cheese are then immersed and held in a sal solution for a period of 1–4 days, then removed and held at 50°–60°F (10°–15.6°C) for 5 to 10 days. Swiss cheese is aged at 55°–70°F (12.8°– 21.1°C) for a period of 3 to 6 months.

Swiss cheese contains not more than 41% moisture and the solid contain not less than 43% butterfat. During aging, the lactobacilli and/o streptococci convert milk sugar (lactose) to lactic acid and the *Pro pionibacterium* group converts the lactic acid to propionic acid, aceti acid, carbon dioxide and water. The propionic acid provides the typica flavor and the carbon dioxide (a gas) forms the holes or "eyes" of th cheese. The "eyes" are usually 0.3–1.0 in. (0.8–2.5 cm) in diameter an: about 1–3 in. (2.5–7.6 cm) apart.

Some defects may develop in Swiss cheese. If the growth of th propionibacteria is not at optimum, when unsuitable conditions exist e.g., incorrect aging temperature, presence of antibiotic in the milk, etc. the bacteria may produce too little or too much carbon dioxide, resultin: in "eyes" which are too small or too large, and improperly spaced. Th cheese may also be crumbly or not sufficiently elastic if improperl: cultured or aged. Bitterness rarely occurs in Swiss cheese, but when thi happens, it is believed to be caused by the growth of a penicillin resistan *Streptococcus* that survives in milk containing a residual of the anti biotic, or by a starter culture which does not produce a sufficient numbe of peptidases (enzymes) to decompose bitter peptides.

Roquefort, Gorgonzola and Blue Cheese.—The main differenc among these cheeses is that Roquefort is made from sheep's milk whil Gorgonzola and Blue are made from cow's milk. Also, to be so labeled Roquefort cheese must be made in France.

Blue cheese may be made from raw, heated or pasterized whole milk o from skim milk and cream mixtures, but the butterfat content should b about 3.5%. Raw milk or milk which has been heated at temperature lower than those used for pasteurization is preferred, since lipase actio:

s required for the ripening of this type of cheese (lipase is an enzyme which splits fats into glycerin and fatty acids) and heating at pasteurization temperatures inactivates lipase. If skim milk and cream are used as the main ingredients, and the cream is too yellow in color, it may be bleached by treating with benzoyl peroxide. If whole milk is used, the temperature is adjusted to 85°F (29.4°C) and the milk is homogenized. If skim milk and cream are used, the cream is homogenized. After homogenization, the temperature of the product is raised to 90°F (32.2°C), a lactic acid starter culture is added, and the product is held at 90°F (32.2°C) for a period of about 1 hr. The enzyme rennet is then added to coagulate the mixture, which is allowed to stand for another 45 to 60 min. The curd is then cut into 1/2 in. (1.3 cm) cubes, after which it is stirred for 15 min while being held at the incubation temperature. The whey is then drained, and the curd is mixed with about 1% salt, then placed in racks lined with cheesecloth and allowed to drain. After draining, the curd is placed in sterilized hoops, and, as the hoops are filled, the curd is mixed with bread crumbs on which a culture of the mold *Penicillium roqueforti* has been inoculated and allowed to grow. The hoops containing the curd are held at 65°–68°F (18.3°–20°C) for part of the day, after which the product is placed in a room at 50°–55°F (10°–12.8°C) where salt is applied to the surface of the cheeses daily until the salt content reaches 4 to 4.5%. The cheeses are then removed to a ripening room where they are held for 2–3 months at 50°–55°F (10°–12.8°C) and a relative humidity of 95%. During ripening, the cheeses are mechanically pierced through the flat surface side with wire needles. The latter procedure permits air to enter the product so that the mold, which requires oxygen, will grow. After curing, the surfaces of the cheeses are scraped, and the cheeses are then cut into small wedges and wrapped in plastic or aluminum foil. Some blue cheese is packaged in plastic cups.

The flavor of blue-type cheeses is due to a blend of fatty acids (butyric, caproic, caprylic and capric) produced by the action of lipase on butterfat, and methyl ketones, formed from fatty acids, such as caprylic, by oxidative enzymes produced by the mold, *Penicillium roqueforti.*

Blue-type cheese should contain 46% moisture, or less, and the milk solids should consist of not less than 50% butterfat.

Camembert Cheese.—Camembert cheese is made from pasteurized cow's milk having a butterfat content of 3.5 to 3.7%. After pasteurization, the milk is cooled to 85°–86°F (29.4°–30°C) and inoculated with a starter culture of lactic-acid-producing streptococci. The milk mixture is then incubated for 1 hr at the stated temperature after which the enzyme rennet is added, and the material is allowed to stand for another hour while the incubation temperature is maintained. The coagulated material is then cut into 1/4 in. (0.6 cm) cubes, and the product is stirred until the curd is firm, after which the whey is drained. The drained curd is then filled into small, perforated forms 4.5-5 in. (11.4-12.7 cm) in diameter and 1.0-1.5 in. (2.5-3.8 cm) deep. After draining for 18 to 24 h (without pressure), the small cheeses are removed from the forms and salted daily until the salt content is about 2.5%. After salting, the surfaces of the cheeses are inoculated with spores of the mold *Penicillium camemberti*. The mold inoculum is prepared from cultures grown on crackers, and inoculation of the cheeses is carried out by spraying with, or dipping in, a water suspension of the mold culture. The cheeses are ripened or cured over a period of several months in a room held at a temperature of 55°-58°F (12.8°-14.4°C) and a relative humidity of 85 to 90%.

During ripening, mold grows on the surfaces of Camembert cheese and lactic acid-producing streptococci continue to increase in the product. The enzymes produced by these microorganisms convert much of the milk proteins to water soluble polypeptides and amino acids. Ammonia may be formed from amino acids during long periods of ripening. The moisture content of Camembert cheese is about 50% and at least 50% of the solids present must be butterfat.

Limburger Cheese.—Limburger cheese is made from raw or pasteurized whole cows' milk. The milk is brought to a temperature of 85°–90°F (29.4°–32.2°C), inoculated with a starter culture of lactic acid-producing streptococci, and held at the stated temperature for a period of 1 hr. The enzyme rennet is then added, and the temperature is maintained while holding the product for an additional hour. The curd is then cut, stirred, and the whey drained off. It is then placed in forms to make individual cheeses, but the hooped cheese is not pressed. The cheeses are salted by rubbing the surfaces with dry salt or by dipping them in a solution of salt, then aged for several months at a temperature of 60°-61°F (15.6°-16.1°C). During aging, a bacterium, *Bacterium linens,* grows in the cheese and produces the particular flavor components of the product. Limburger cheese is highly flavored and is enjoyed by some people but disliked by many. The moisture content of the cured product is about 42%.

Parmesan Cheese.—Parmesan cheese is made from cow's milk which has had part of the butterfat removed. The milk, pasteurized or not, is brought to a temperature of 90° to 100°F (32.2° to 37.8°C) and in-oculated with a starter culture of lactic acid-producing streptococci which grow at comparatively high temperatures. The inoculated product

held at the stated temperature for 1 hr. The enzyme rennet is then
added, the temperature being maintained until the mixture has set. The
curd is then cut and stirred, after which the whey is drained off. The curd
is salted and saffron may be added as a colorant. After salting, it is
placed in forms and allowed to drain, without pressing, over a period of
at least one week. The drained cheeses are aged in a room held at about
50°F (10°C) for a period of 1 year or more. During aging, the cheeses are
periodically cleaned and rubbed with oil.

Parmesan cheese is of the hard variety and has a moisture content of
about 34%. This type of cheese is ordinarily grated and sprinkled on
various types of prepared food.

Other Cheeses.—There are several hundred types of cheeses. No
attempt will be made to discuss them all. Certain cheeses not already
discussed will be briefly described.

Brie Cheese.—This is a high moisture content, highly salted, creamed
cottage cheese.

Edam Cheese.—This cheese is of the hard variety and is made from
whole cow's milk. It is shaped into a round form, and coated with
paraffin colored with a red dye. The moisture content of this cheese is
about 33%.

Munster Cheese.—Munster cheese is manufactured in much the
same way as Limburger cheese. However, it is aged at low temperatures
and hence has a mild flavor. The moisture content of Munster cheese is
about 52%.

Neufchatel Cheese.—This is a type of creamed cottage cheese which
is formed into small molds. The curd is not heated or cooked after
cutting. It has a moisture content of about 57%.

Stilton Cheese.—Stilton cheese is made from whole cow's milk, with
or without the addition of cream. The curd is not pressed but is well
drained. During aging, a blue-green mold grows in the product, and the
cheese takes on a marbled appearance. It has a final moisture content of
about 33%.

General.—The main factors which affect the unique flavor and/or
texture of various types of cheese are listed below:

(1) The type of milk used. This may be cow's milk or sheep's milk. The
    butterfat content of the milk may be high, normal or low (as in
    skim milk or partially skimmed milk).

(2) The degree to which the milk is heated prior to or during processing. If the milk is pasteurized, the enzyme lipase will have been inactivated, hence free fatty acids probably will not be formed in the cheese during aging. The application of comparatively high temperatures during processing will destroy some types of bacteria used as starter cultures but will allow the survival of others. Therefore, bacterial end-products produced in cheese during aging are affected by the temperatures used during processing; an example is Swiss cheese.

(3) The species of bacteria used to culture the milk. This will determine the types of bacteria which survive during processing and so, at least to some extent, the types of bacteria which grow and modify texture and flavor during aging.

(4) Whether or not the curds or the cheeses are inoculated with particular species of molds. This affects both texture and flavor changes which take place in cheeses during aging.

(5) The degree to which the curd is salted before or after draining or pressing. This, too, affects the particular types of bacteria which grow in the cheese during aging.

(6) The degree to which the curd is drained and/or pressed during processing. Both bacteria and molds grow best at comparatively high moisture content, and some types grow faster at a particular moisture content than others. The particular compounds produced in cheese during aging are, therefore, affected by the moisture content of the cheese.

(7) The temperature and relative humidity at which the cheese is aged. Some bacteria which are present in cheese do not grow or grow slowly at temperatures below 50°F (10°C); hence, cheese aged at temperatures above 50°F (10°C), especially those aged at 60°F (15.6°C) and above, are usually more highly flavored. Relative humidity affects the moisture content of cheese due to moisture loss during aging, and thus has a selective effect on the species of bacteria which grow during curing. Microbial growth, on the surface, especially of mold, is affected by the relative humidity at which the cheese is aged.

(8) The length of time during which the cheese is aged. Whereas the bacteria in cheese only grow slowly during aging at low temperatures, if curing or aging is carried out over a long period of time, significant amounts of bacterial end-products may be produced. Also, if milk enzymes are present they may, during long

time low-temperature aging, form significant amounts of reaction products in cheese. Parmesan cheese is an example of this effect.

BUTTER

Butter is made from cream having a butterfat content of 25 to 40%. The cream used is ordinarily pasteurized and cooled and may be sweet or unsoured or may be cultured with *Streptococcus diacetyllactis* (a strain of *Streptococcus lactis*) and allowed to ripen at 70°F (21.1°C). Sweet cream, to which a culture of *Streptococcus diacetyllactis* has been added, then held at temperatures below that at which the organism will grow, is sometimes used to produce butter, and a culture of the organism may be added directly to butter, made from uncultured cream. The purpose of adding bacterial cultures to the cream from which butter is made is to produce diacetyl, the main flavor component of butter. Prior to pasteurization, the cream is neutralized with alkali, if necessary, to regulate the acidity to about 0.4%, since excess acidity may hinder the churning process and may also cause the development of off-flavors in the butter during storage. Pasteurization is carried out by heating the cream to 160°–170°F (71.1°–76.7°C) and holding at this temperature for 30 min or by quickly heating the cream to 185°F (85°C) and holding at this temperature for 1 min. After pasteurization, the cream is cooled to 40°–50°F (4.4°–10°C) and held at this temperature for several hours prior to churning.

By the conventional method of churning, the temperature of the cream is raised to 50°–58°F (10°–14.4°C), the lower temperature being used when ambient temperatures are high. The cream is added to half-fill the churn—a metal cylinder which can be rotated and which has paddles that can be rotated. After the first few rotations, the churn is stopped briefly, allowing gas to escape from the cream. It is then rotated for a period of 30 to 45 min, during which time the fat globules adhere to each other, and form lumps about the size of a corn kernel. The small lumps eventually adhere to each other. After churning, the buttermilk is drained off, water at 55°–60°F (12.8°–15.6°C) is added and the churn rotated several times to wash the butter. The water is then drained off. Washing may be repeated. Salt may then be sprinkled over the butter in quantities of 1 to 2.5%, after which the churn is again rotated to work the butter. Working distributes the moisture, and salt, if added, evenly throughout the butter. The fat content of butter must be at least 80% and the moisture content not more than 16%. After working, the butter may be extruded into 1/4 lb (113.5 g) sticks or 1 lb (454 g) blocks, wrapped

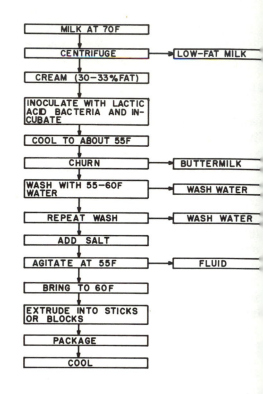

FIG. 16.6. MANUFACTURE OF
BUTTER
See metric conversion tables in
Appendix.

with parchment, then packaged, cased and stored, or it may be added to
parchment-lined boxes holding 50 lb (22.7 kg) of the product and stored
at 35°–40°F (1.7°–4.4°C) to be formed and packaged at some future date.
Butter which is to be held for significant periods of time is usually
formed, packaged, placed in shipping cases and stored at 0° to − 10°F
(− 17.8° to − 23.3°C). At this temperature, it has an almost indefinite
storage life.

Today, butter is produced by a continuous, rather than by the
conventional, churning method. In continuous methods, the cream is
first concentrated to 80% butterfat by centrifugation, after which it may
be further concentrated to more than 90% butterfat. This butter oil is
then pasteurized and partially cooled, after which specific amounts of
non-fat milk solids and water are added and mixed. The mixture is then
solidified under controlled cooling temperatures and extruded into the
required shape or form.

Since several species of bacteria may grow in the water present in
butter, the product must be handled as perishable. The water in butter

present as very small droplets; therefore this product may be considered a water-in-oil emulsion. While yeasty, tallowy or otherwise off-
flavors are rarely present in butter, the most common type of deterioration is the development of a strong odor and flavor which is wrongly
called "rancidity." This type of off-flavor is caused by the growth of
certain bacteria in the water droplets present in the product. These
bacteria produce the enzyme, lipase, which splits off fatty acids from the
butterfat. Butterfat contains short chain fatty acids, especially butyric
acid $(CH_3—CH_2—CH_2—COOH)$, which are very potent in odor and
flavor and which cause the strong off-flavor which may develop.

Salt is added in concentrations of 1–2.5% by weight to much of the
butter produced. Since the moisture content of butter is only about 16%,
the water phase of the salted product contains about 6 to 15% of salt. Salt
adds flavor to butter, but it is added primarily to inhibit the growth of
bacteria which might produce lipase, thus causing off-flavors. Although
the addition of salt will not indefinitely prolong the storage life of butter
at temperatures above freezing, it will extend it to a significant degree.
For this reason, sweet butter, which contains no salt, is more perishable
than the salted product when held at 32°F (0°C) and above.

DAIRY-PRODUCT SUBSTITUTES

While it is not the purpose of this book to cover dairy-product
imitations and variations, they appear to be growing in importance and
deserve some mention. Factors which have favored the growth of dairy
substitutes include their relatively low cost, adverse publicity associated
with the saturated fat content of butterfat, low caloric value, and the
freedom of substitutes from the strict standards that apply to dairy
products. These factors have permitted the formation of a broad spectrum of imitation products, each having a unique identification.

Filled Milks

Filled milks are products which resemble whole milk but which
contain no fat that is derived from milk. They contain vegetable fats
which are combined with nonfat milk solids or with a mixture of nonfat
milk solids and nonfat solids, derived from sources other than milk.

Imitation Milks

These products resemble whole milk but contain no actual milk
ingredients. They generally contain water, vegetable fats, corn sugar,

starch, vegetable protein, sodium caseinate, vitamins, minerals, an
stabilizers, such as gums or alginates. Imitation milks usually do n
taste like whole milk and often require the addition of flavorings.

Both imitation and filled milks are used as a basis for the formulatio
of synthetic and semi-synthetic flavored milk drinks, ice cream and oth
frozen desserts, butter, cream cheese, coffee cream, whipped cream an
other imitation dairy products. All these products are processed, store
and distributed in a manner similar to that of the dairy products th
they resemble.

WHEY

Whey is the fluid by-product of cheese manufacture. It is produced
far greater volume than cheese, the ratio of whey to cheese being abo
10:1. For numerous reasons, whey is underutilized, and not more th
half of the over 35 billion lb (15.75 billion kg) produced annually in t
United States is used. The other half represents a waste dispos
problem. Considering the growing rate of cheese production and the ev
tightening constraints on the disposal of processing plant effluents, t
problem of what to do with whey is one of major proportions. Wh
comprises about 5% lactose, 2% other milk components, and 93% wat

The utilization of whey is impeded mainly by the fact that its ma
solids component, lactose, is not easily digested by a large part of t
world's population, is not fermented by many microorganisms, and
only about $\frac{1}{3}$ as sweet as sucrose. Therefore, to obtain a particul
sweetness, it is required in larger amounts than other sugars. Whey c
be made sweeter by hydrolyzing the lactose with lactase produci
glucose and galactose.

$$C_{12}H_{22}O_{11} \xrightarrow[+H_2O]{\text{Lactase}} C_6H_{12}O_6 + C_6H_{12}O_6$$

Lactose   Glucose   Galactose

Glucose and galactose are sweeter than lactose, therefore the result
whey is a more effective sweetener. With hydrolysis of the lactose, t
resulting sugars are metabolizable by that segment of the populat
that cannot tolerate lactose in its diet. Of the few microorganisms th
can ferment lactose, *Kluyveromyces fragilis* has been reported to be t
most efficient. Many typical fermentation organisms are unable
ferment lactose.

Glucose is readily utilized by fermenting organisms (such as *Saccharomyces cerevisiae*). Since only the glucose is readily fermentable, in the production of alcoholic beverages from whey, it is necessary to add sugar even when the lactose has been hydrolyzed (galactose is not readily fermented). If the lactose has not been hydrolyzed, the sugar requirement is even higher. A factor hindering the use of whey for the production of certain beverages is its protein, which tends to produce an undesirable cloud.

Attempts to find uses for whey have produced numerous practical applications. It has been used in the manufacture of liquid breakfasts, snack drinks, alcoholic drinks, imitation milks, soft drinks, baked goods, lactic acid, vinegar, ice cream, sherbet, ice pops, fudge, candy caramel and other confections. One of its functions in many applications is as a substitute for nonfat dry milk. Whey can be used to produce a sweet syrup that, while not economically competitive with corn syrup, has potential value because of its properties as a humectant and a texture enhancer. Current investigations indicate that whey may be used in the production of wine.

The use of whey is noticeably associated with relatively high production costs and, as yet, unsolved technological problems. However, it is possible that the economics of its utilization may improve, if the costs of its disposal undergo prohibitive increases and, especially, when the regulation to ban the discharge of pollutants into navigable waterways goes into effect in 1985, according to the present position of the Environmental Protection Agency.

17

Poultry and Eggs

In the United States, most poultry used for food are chicken and turkey. Some ducks and geese are consumed, but they are relatively insignificant as food sources. The per capita consumption of poultry in the United States has been increasing at the rate of about 2 lb (0.91 kg) per year and is now at more than 40 lb (18.2 kg) per year. On the other hand, egg consumption has started to decrease from a high of about 400 eggs per capita per year in the 1950s to about 300 eggs per capita per year in the 1970s. Projections in food consumption trends indicate that poultry consumption will continue to increase while egg consumption will undergo additional slight decreases.

The only explanation to account for the decrease in egg consumption is the fact that eggs contain significant amounts of cholesterol which is now either avoided or held to a limit in the diets of a portion of the population.

In addition to its palatability, poultry costs significantly less than red meats, which accounts in part for the dramatic increase in its consumption in the United States.

Poultry is more efficient to raise than cattle and hogs in terms of the amount of meat produced per unit of feed consumed. This is why poultry is cheaper than most other meats. In order to be grown to market size, about 2.4 lb (1.1 kg) of feed are required per pound (0.45 kg) weight of bird over a 3 month period.

Today's modern technology, by manipulating genetics, allows poultry grown for meat purposes to grow rapidly, to be disease-resistant, and have good meat qualities including a tender texture, good flavor, and a light color. Chickens having white feathers are preferred over other types because there are no dark pin feathers which, if not removed, detract from the appearance. Also, the skin of white-feathered birds is much lighter and more desirable.

POULTRY

Present poultry breeds have been developed from wild birds, jungle fowl of southeast Asia and wild turkey of North America. Capons, roasters and broilers are mostly mixed breeds or hybrids.

Chicken

Male parents are usually selected from silver or dominant white Cornish varieties, because they develop meaty breasts and legs. Female parents are usually selected from Cornish strains with white plumage.

For meat production, almost all flocks are started from 1-day-old chicks, and, ordinarily, "straight run chicks" (about half male, half female) are used. Broiler flocks of this type are usually raised in housing system which provides 0.5 ft^2 (464.5 cm^2) per bird until they are 2 weeks old and 1.0 ft^2 (929 cm^2) per bird between 2 and 10 weeks old. At the end of 10 weeks, they will be removed for slaughter. For capons and roasters that are 10–20 weeks old, 2–3 ft^2 (1858–2787 cm^2) of floor space per bird is used. Older birds (fowl) require 4–5 ft^2 (0.36–0.46 m^2) per bird.

Commercial growers commonly raise at least 4 flocks of broilers per year.

In raising birds for meat, the floor of the chicken house, which may be soil, is first cleaned and scraped out with a bulldozer, and the floor and walls are cleaned and disinfected. Fresh litter (shavings) is then put down on the floor. The day-old chicks, which may have been debeaked to prevent cannibalism, are introduced. Lights are kept on continually, and the temperature of the brooders is maintained at 95°F (35°C) in cold weather and 90°F (32.2°C) in hot weather. The temperature of the brooders is lowered 5°F (2.8°C) weekly until 75°F (23.9°C) is reached. Heating is continued until the birds are well feathered, in winter for weeks and in late spring and summer for 4–6 weeks.

Troughs or hanging feeders may be used to hold food for meat-type birds, and hoppers for grit and calcium supplement may be used. Feeders, which must be adequate in size and number, may be filled automatically. Water troughs or hanging waterers may be used (16 ft [4.88 m] of watering space per 200 birds at 75°F [23.9°C], 20 ft [6.1 m] per 200 birds when the temperature goes above 80°F [26.7°C]). Feeder

ıd waterers should be kept clean, and waterers should refill auto
ıatically.

Food for meat-type birds, a complex mixture, is obtained from feed
ımpanies that are expert in formulating such rations. Recommended
arter feeds contain corn, fish meal, poultry by-product meal, corn
uten meal, soybean meal, alfalfa meal, dried distiller's solubles, small
nounts of calcium and phosphorus salts, iodized salt, and A, D, E, K,
ıd several B vitamins. Trace amounts of antibiotics are usually added
 feeds to prevent diseases. Among the above ingredients, corn and
·ybean meals constitute the major portion of the starter feeds. Finishing
eds, which may also include dried whey and steamed bone meal, are
ıed for birds from 6 weeks old to marketing sizes. Capons and roasters
·e fed increasing amounts of corn after 12–15 weeks of age.

Predators, such as rats and mice, must be kept out of poultry houses.
·oors, litter, walls, roosts, and even the birds themselves may have to be
·eated with malathion insecticide to get rid of mites, lice or ticks.
Birds grown for meat are the same age, and once marketing size is
·ached, they are placed in cages and removed from the growing house.
ıey are slaughtered and processed elsewhere. Prior to installing new
·cks, the house is cleaned and disinfected. In order for a growing
·eration to be economical, the number of birds in roaster and capon
·cks must be no less than 2000, and in broiler flocks, no less than 6000.
Some poultry processing plants are small, but the trend is to process
·ultry in plants which are capable of handling at least 10,000 birds per
·ur. These plants are usually divided into at least two rooms which
·parate bleeding, scalding and defeathering from the eviscerating and
·illing operations (see Fig. 17.1).

The birds are not fed for about 12 hr before they are to be slaughtered
 order that their crops will be empty. This is important because it
ıkes the operation much cleaner. The birds, shackled by their feet, are
·rried in the upside down position by conveyors from one operation to
·other. After shackling, they are slaughtered by slitting one or both of
·e jugular veins in the neck. An electrified knife or a stationary electric
ınner may be used to render them unconscious prior to bleeding.
·ndering birds unconscious prevents broken wings and bruising due to
·e flopping around during bleeding. After their jugular veins are
·vered, they are allowed to bleed from one to several minutes. Still
·ached to the conveyor, they next pass through the scalding tank,
·ntaining water at 135°–140°F (57.2°–60°C) for larger birds or
·2°–128°F (50°–53.3°C) for broilers. Immersion times vary with the size
 birds, but several minutes in the scalding tank are required even for
·oilers. If scalding time or temperature is too high, the skin may be
·maged.

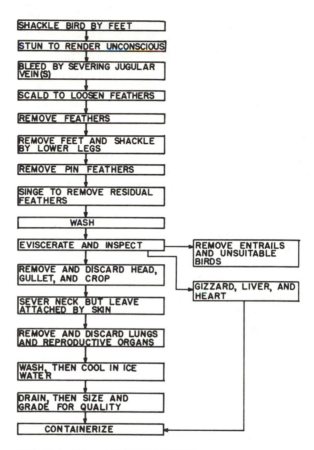

FIG. 17.1. CHICKEN PROCESSING

Automatic picking machines are used to remove the feathers fro[m]
poultry. In some systems, the birds are beaten by flexible rubber finge[rs]
as they pass through the machine. In other systems, they are dropp[ed]
into baskets where feathers are removed by flexible rubber finge[rs]
rotating on a central shaft.

After the birds are defeathered, their feet are cut off, and they a[re]
rehung on the moving shackles by the lower legs. They then pass along [a]
line where workers remove pin feathers by hand with the aid of a kni[fe.]
Then they pass through a gas flame for singeing residual pin feather[s.]
They are washed externally by water sprays as they pass along t[he]
conveyor.

Evisceration is usually carried out as the birds pass along the conveyo[r.]
The oil gland[1] may be cut out before or after evisceration. A circular c[ut]

---

[1]Oil gland: A gland that secretes oil, especially the one at the rump of a bird from whic[h]
it takes oil for preening its feathers.

made around the vent and the intestine is then pulled out a few inches. nother cut is then made through the abdominal wall from the vent ward the breastbone for broilers. For larger specimens a horizontal cut made. The gizzard, liver, heart and intestines are pulled out and llowed to hang so that they may be examined by a government inspector a veterinarian) for signs of disease. Diseased birds are removed and estroyed. Gizzards that pass inspection are opened, emptied, peeled nd washed, and then packaged together with the livers and hearts, ther for insertion within the bird or for holding separately. Intestines re not used.

The heads, gullets and crops are removed and discarded, and the neck ay be cut off and allowed to hang by the skin. Suction tubes are used remove lungs and traces of reproductive organs or these are scraped ut by hand. Both the interior and exterior of the bird are thoroughly ashed. The neck and neck skin are placed in the body cavity.

The birds are next cooled, either in an air-agitated ice-water slush in nks, in continuous ice-water chillers or in moving refrigerated air. uring chilling, the temperature of the birds, which may be 80°–95°F 26.7°–35°C), is lowered to 35°F (1.7°C), and they lose some residual lood and pick up a few percent of moisture from the chilled water. After hilling, they are drained, sized according to weight and graded for uality. Grading is based on conformation, fleshing, covering with fat, ie presence of pin feathers, torn skin, bruises, etc.

Graded birds are packed in wooden or waxed-fiber-board boxes. They hould be surrounded with crushed ice and held at temperatures below)°F (4.4°C). When the product is held at 28°F (−2.2°C), a significant xtension in the shelf-life may be realized.

Some poultry is bagged or wrapped with plastic and frozen, either by old air blast, by contact with refrigerated plates, by immersion in liquid itrogen at −320°F (−195.6°C) or by spraying with liquid freon-12 -21.7°F [−29.7°C]).

Chickens are classified, according to age and condition, as broilers or yers, roasters, capons (males castrated prior to maturity), stags (young ncastrated males), hens, or stewing chickens or fowl (hens older than 20 eeks) and cocks or old roosters. Poultry may also be graded for quality s A, B or C.

Some chicken products are battered and breaded, deep-fat fried and ozen. The high-quality storage life of uncooked chicken at 0°F -17.8°C) is more than two years if adequately packaged, while the high- uality storage life of fried chicken is less than 3 months at this mperature.

Some chicken is also precooked, deboned, and heat processed in glass rs or cans. Some deboned poultry is used as an ingredient of canned oups.

Chicken is also precooked, deboned, cut into cubes or small portions frozen, and then freeze dried. This product is used as an ingredient o dried soups.

The edible portion of chicken is about 54%. This includes about 39% meat and about 15% skin and parts of the viscera which are also edible The protein content of the chicken is about 20%, the fat content is abou 14%, the ash is about 1% and the remaining 65% is water.

Turkey

Turkey is the second most important poultry in the United States. I recent years, there has been a movement to develop strains whic produce more meat, especially breast meat. This development has mostl been limited to white and bronze feathered birds.

Turkey raising requires the same basic conditions as does chicke: raising. For birds that are 1–3 weeks old, 1 ft^2 (0.09 m^2) of floor space i required. Birds that are 4–8 weeks of age require 1.5 ft^2 (0.14 m^2) per bird and birds that are 8–15 weeks old require 2 ft^2 (0.18 m^2) per bird. Turkey are vaccinated to prevent erysipelas, salmonellosis and other diseases.

In growing turkeys, a serious disease known as "blackhead" has bee: known to break out. It was eventually learned that the disease wa caused by a bacterial infection which was carried by chickens, and it wa soon realized that if turkey flocks were going to be kept healthy, that the would have to be kept away from chickens or from places where chicken: had been, since the bacteria were found in heavy concentrations i: chicken droppings and in areas that had been contaminated by chicke: droppings.

Turkeys require a higher protein and vitamin content in their foo: than chickens, although, as with chickens, the protein content of the foo is gradually reduced as the birds mature.

Turkeys may be marketed as broilers which are 12–15 weeks old or a mature roasting birds which are 20–26 weeks of age.

Turkeys are slaughtered, defeathered, eviscerated and further pro cessed in much the same manner as chickens, except for some dif ferences. Longer bleeding times are required for turkeys. They must als be scalded for longer times and at higher temperatures to facilitate th removal of feathers. With turkeys, the tendons in the legs are pulled ou after defeathering and removing the feet. Further processing is essen tially the same as for chickens. During cooling, turkeys pick up 4.5–8% water, the smaller birds absorbing the highest proportional amounts Grading of turkeys is based on the same characteristics as those fo chickens, and they may be graded for quality as A, B or C.

Larger percentages of turkeys are frozen and sold to the consumer in the frozen state than is the case with chickens. However, a less proportional amount of turkey meat than chicken meat is canned or dried. Some turkey is frozen, to be used later as an ingredient of pies or dinners. While the meat appears to be quite stable at 0°F (−17.8°C) as an ingredient of pies, the meat alone is somewhat less stable than chicken meat in frozen storage.

Ducks

Ducks used for human consumption have been developed from wild types. The White Pekin, developed in China, the Aylesbury, developed in England, and the Muscovy, which originated in South America, are mostly used in the United States as meat-type birds.

Ducks may be housed in buildings, as are chickens. Cement floors covered with litter or wire floors [welded ¾ in. (1.9 cm) mesh] about 4 in. (10.2 cm) above the cement may be used. Floor space requirements are 0.5 ft² (0.045 m²) per bird on wire and 1 ft² (0.09 m²) per bird on litter. This should be increased to 2.5 ft² (0.23 m²) per bird at 7 weeks of age.

The brooder temperature for ducks is kept at 85°–90°F (29.4°–32.2°C) for the first week and then reduced 5°F (2.8°C) per week for a period of 4 weeks. As with chickens, ducklings should be supplied with clean water and adequate food. Food requirements are similar to those of chickens. After 4 weeks of age, ducklings are sufficiently feathered to be allowed to range outdoors in all but extremely cold weather. Swimming water for ducks is often provided on outside range areas.

Ducks, like chickens, are subject to a number of diseases, some of which may be prevented by the inclusion of antibiotics in the feed. Other diseases can be prevented by vaccination.

Pekin type ducks are ready for marketing at the age of 7−8 weeks and Muscovy types at 10−17 weeks of age.

Slaughtering, defeathering, eviscerating and cooling of ducks are carried out in much the same manner as for chickens. Ducks may be dry-picked or defeathered, scalded and machine-picked, and, in some cases, after picking, dipped in molten wax. The wax is then cooled and stripped off to remove residual feathers. After evisceration, most of the ducklings are packaged in plastic pouches and frozen, and sold in this condition. Uncooked ducks have a storage life of well over 1 year at 0°F (−17.8°C). Classes of ducks are: broiler or fryer ducklings, roaster ducklings, and mature or old ducks. As in the case of chickens, ducklings may be graded for quality as A, B or C.

Geese

Some geese are raised for human consumption, but the volume is comparatively small and relatively unimportant.

EGGS

Except for some variations due to breeds, and sometimes even individual hens, the chemical composition of eggs is fairly constant. Of the total weight of the egg, the shell is about 10.25%, the white is about 59.50% and the yolk is about 30.25%. The shell, largely calcium carbonate, has an outer coating (the cuticle) that protects the pores of the main part of the shell, as long as it remains intact. Inside the shell there are two membranes, the one next to the shell being thicker and tougher than the one covering the contents of the egg.

The white of the egg is about 10.5% protein, about 88% water, less than 1% ash. Small amounts of fat, sugar, carbon dioxide and other constituents are present. The white is in two distinct parts: the thick, jelly-like part that surrounds the yolk, and a less viscous "thin white" which spreads out when the egg is broken out of its shell. However, the thick white will also spread if it is cut, since its inner part is of thin consistency.

The yolk is about 15.5% protein, about 49.5% water, about 33.5% fat, about 1% ash. Small quantities of numerous other constituents, including vitamins, are also present. It is surrounded by the vitellin membrane which when ruptured or cut causes the yolk to spread when the egg is broken out of the shell. The yolk is much more complex chemically than the white, and accounts for the major nutritional composition of the egg, one of the most complete foods available to man.

The egg contains a small air pocket that develops after the egg is laid when it starts to cool from the body temperature of the hen, contracting the contents of the egg and pulling the inner membrane in with the egg contents. Since the eggshell is actually porous, the space formed by the contraction of the contents is soon filled by air that can be taken in through the shell.

The raising of chickens for the production of eggs has similar requirements to those of raising chickens for meat, but there are some differences. The chief breeds used for egg production are white leghorns which lay white-shelled eggs and New Hampshires, Plymouth Rocks, and Rhode Island Reds, which lay brown eggs. Some cross breeds are used.

In egg-laying flocks, only one age of bird is used, ordinarily. Chicks are usually received at an age of one day, although they may be obtained as starter birds at an age of 6–8 weeks. For an egg-producing operation to be economically feasible, the minimum number of egg-laying birds required is at least 2500.

Space requirements for egg-laying birds are similar to those for meat-type birds. An interior temperature in the range 45°–80°F (7.2°–26.7°C) is considered satisfactory for egg-laying flocks. Floors and walls should be easy to clean. Feeders and waterers should be provided, as in the case of meat-type birds.

Laying nests may be constructed of metal or wood. Roll-away floors with egg trays are desirable, since this type of nest minimizes the number of dirty eggs. With this arrangement, once the egg has been laid and the hen leaves the nest, the egg rolls away from the area where it is laid to a collecting area, where it will not be dirtied by droppings. One nest for each four birds is considered adequate. An individual nest should be about 10–12 in. (25.4–30.5 cm) wide, 12–14 in. (30.5–35.6 cm) high and 12 in. (30.5 cm) deep. To keep the nest clean, a perch is provided below the entrance of the nest.

Roosts [8−10 in. (20.3−25.4 cm) of space per bird] are 13−15 in. 33−38.1 cm) apart above dropping pits. Such pits should be cleaned periodically and should be constructed to facilitate cleaning.

The recommended food requirements for laying birds seem to be somewhat less complex than for meat-type birds, although the vitamin requirements are somewhat more complex. Antibiotics are included, and ground limestones or ground oyster shells are added as a source of necessary egg-shell ingredients. Feed requirements amount to 85–115 lb 38.6–52.2 kg) per bird per year depending upon bird size.

Poultrymen near cities may sell eggs directly to consumers via home deliveries, or they may sell to produce dealers, cooperatives, shippers or hucksters.

Eggs should be collected from laying houses at least three times daily in cool weather and 4–5 times daily in hot weather. They should be gathered in plastic or rubber-coated wire mesh baskets. Nests and baskets should be kept clean, since the interior of dirty eggs soon becomes contaminated and subject to spoilage. Once eggs have been gathered, the baskets should be placed in storage at 40°–45°F (4.4°–7.2°C) (relative humidity 70%). No other materials, including foods, should be stored in egg storage rooms, since pungent odors are readily absorbed by eggs.

Dirty eggs can be cleaned by buffing or washing, but unless done under rigidly controlled conditions, the interior may become contaminated with bacteria during cleaning. It is desirable, therefore, to emphasize production of clean eggs.

The egg shell presents a barrier to the entrance of microorganisms, but there are pores in the shell large enough to allow the entrance of bacteria and even molds. The number of pores that are found in the shell vary in the range 100–200 per cm². When just laid, the pores of the shell are sealed by a thin layer of protein, the cuticle. If buffed or washed, this protein coat is removed; during washing under improperly controlled conditions, contaminated water may enter the egg.

The logical reason for the porosity of the shell is to allow for the flow of gases in and out to the developing embryo in case the egg has been fertilized. The inner membranes also tend to prevent the entrance of microorganisms. In egg white, there is an enzyme (lysozyme), which tends to lyse or disintegrate some bacteria. There is also a substance in raw eggs, avidin, which ties up biotin, a required factor for the growth of some microorganisms. Finally, there is a material in fresh white which binds with iron, making it unavailable to several species of *Pseudomonas* bacteria, which are responsible for more than 80% of the egg spoilage.

The cuticle may be lost not only by washing; it can also be dissolved by droppings. In any case, about 3 weeks after the egg is laid, the cuticle becomes brittle and particles chip off. Some bacteria are not affected by lysozyme and require little or no biotin. As they are held in storage enzymes within the eggs cause chemical changes which deteriorate the iron-binding properties of the egg white. The defense mechanism against spoilage is, therefore, eventually lost, so that if microorganisms penetrate the shell, spoilage will occur. One of the most satisfactory treatments has been the application of mineral oil which has increased the shelf-life of the egg by reducing the contamination by bacteria and molds through the shell pores.

After the eggs are cooled, they should be packed in clean, odorless containers (usually fiber or wooden cases holding 30 dozen), which are held at 45°–50°F (7.2°–10°C) prior to shipment. They should be packed with their large ends up. All eggs should be candled (examined while in the shell under proper lighting) before selling to consumers. This should be done on the farm if they are sold directly to the consumer or by the distributor or retailer when marketed through retail channels. Candling is done to cull out specimens with such defects as blood spots, blood rings, meat spots, and germ spots (in fertile eggs).

Eggs are classified according to size as jumbo [30 oz (851 g) per dozen] extra large [27 oz (766 g) per dozen], large [24 oz (680 g) per dozen] medium [21 oz (595 g) per dozen], small [18 oz (510 g) per dozen] and peewee [16 oz (425 g) per dozen].

Eggs may be graded according to the interior quality and the condition and appearance of the shell (see Fig. 17.2). For grade AA, the shell must

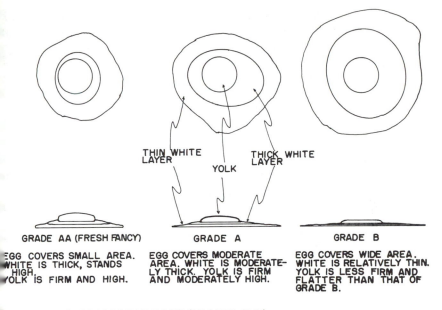

GRADE AA (FRESH FANCY)
EGG COVERS SMALL AREA. WHITE IS THICK, STANDS HIGH. YOLK IS FIRM AND HIGH.

GRADE A
EGG COVERS MODERATE AREA. WHITE IS MODERATELY THICK. YOLK IS FIRM AND MODERATELY HIGH.

GRADE B
EGG COVERS WIDE AREA. WHITE IS RELATIVELY THIN. YOLK IS LESS FIRM AND FLATTER THAN THAT OF GRADE B.

FIG. 17.2. U.S. GRADES FOR EGGS (BROKEN OUT)

be clean and unbroken, and when broken out of the shell, the egg must cover a comparatively small area, the white must be thick and stand high and the yolk must be firm and stand high. For grade A, the shell must be clean and unbroken, the broken out egg must cover a comparatively moderate area, the white must be reasonably thick and stand fairly high and the yolk must be firm and high. For grade B, the shell must be clean and unbroken, the broken out egg covering a wide area with only a small amount of white that can be considered thick, and the yolk somewhat flattened, covering a comparatively large area. Dirty or broken eggs may not be graded.

Processing of Eggs

Almost all eggs used in bakeries have been preserved either by freezing or by drying. The eggs should first be candled to eliminate rots, blood rings, etc., and then washed before they are broken out. At one time, eggs were broken out by hand. Today, this is done almost entirely by machine. Machines break the egg out into cups, which may or may not also separate whites from yolks, depending upon whether the process is to produce whites, yolks, or whole egg magma (mixed white and yolk). While the separation of the egg is such that the white can be separated so that it is free of yolk, the yolk cannot be separated so that it is entirely free of the white. In fact, standards or definitions concerning yolk allow as much as 20% white which is impossible to remove in the separation. As the eggs in cups are carried along by an endless conveyor, they are examined by inspectors. If a particular cup contains a bad specimen, the cup with the egg is removed and replaced with a clean, chemically sanitized cup. The egg from the removed cup is then discarded and the cup washed and sanitized by rinsing in a chemical solution (usually 50 ppm or more of chlorine). As the conveyor moves along, the contents of each cup are emptied into one container when eggs are to be kept whole or the whites and yolks are emptied into separate containers when separation is desired.

Egg products produced in the United States must now be pasteurized prior to freezing or drying, in order to destroy *Salmonella* bacteria, since in the past the disease salmonellosis has been traced to contaminated eggs. After screening to remove chalaza, shell fragments, etc.—eggs are pasteurized by passing them through a plate-type heat exchanger. The product is heated to 140°–145°F (60°–62.8°C) and held at this temperature for 1–4 min prior to cooling. Cooling may be carried out in tanks provided with cooling coils and paddles which agitate the product to facilitate cooling, or in thin film heat-exchangers.

After cooling to 40°F (4.4°C), the pasteurized eggs may be placed in metal cans holding about 30 lb (13.6 kg) of product. The filled cans are then placed in a cold room at 0° to −20°F (−17.8° to −28.9° C) until the product is frozen, after which it will be held at 0°F (−17.8°C) or lower until shipped out to the distributor or to the point of utilization.

Frozen whole egg magma and frozen yolks are subject to deterioration during frozen storage. Ingredients in the yolk tend to form a gummy mass during frozen storage. In order to prevent this, 5–7.5% salt or glycerin or 5–10% sugar may be added and mixed with the product.

Whites, yolks or whole egg magma may be spray-dried by being forced through a nozzle (to form droplets) into a chamber of heated air where most of the moisture is removed from the droplets to the heated air which is vented to the outside. The dried product falls to the bottom of the drier and is collected. Spray-dried eggs have a moisture content of about 5%. This moisture content is not sufficiently low to prevent nonenzymatic browning during storage. Nonenzymatic browning involves sugars; thus it can be prevented by removing sugars from eggs when they are allowed to undergo a natural fermentation at 70°F (21.1°C) over a period of several days. This is done when bacteria are allowed to grow and ferment out the sugar. While this method is effective in removing sugars, it is considered to be unsanitary, since disease bacteria may also grow during fermentation.

Sugars may be removed from egg products by adding yeasts, which utilize sugars, and holding the product at temperatures and for periods which will allow for an adequate growth of these microorganisms. The use of yeasts to ferment sugars, however, may produce undesirable flavors in egg products.

A mixture of two enzymes, glucose oxidase and catalase, may be used to remove sugars from eggs. By this method, the sugar (glucose) is oxidized to gluconic acid and hydrogen peroxide by glucose oxidase, and the hydrogen peroxide, which is undesirable, is decomposed to water and oxygen by catalase (see Chapter 9). The enzyme method is probably the most satisfactory means of removing sugar from egg products. Treatment of these products to remove sugar must be carried out prior to drying.

The heating encountered during the pasteurization treatment and the physical forces encountered during spray drying have some effect on the functional characteristics (whipping quality, etc.) of egg products, especially those made from whites. Therefore, in some countries, egg white is allowed to undergo a natural fermentation and is then dried in cabinets on trays. In such instances, to eliminate disease-causing bacteria which may be present (since the product was not pasteurized prior to drying), the dried product is held at 130°F (54.4°C) or at higher temperatures for several days. This heat treatment after drying is said to destroy disease-causing bacteria.

18

Fish and Shellfish

Although the word "fish" is used to classify one type of food much the same as meat, poultry and cheese, varieties of fish are much greater in number than those of other foods. In the United States alone, at least 50 species of fish and shellfish are used as food for humans. Considering that the variations among aquatic species are relatively greater than those among species of meat animals, we can appreciate the magnitude of the time, space and effort required to give even minimum coverage to fish as food.

The commercial fishermen of the United States land about 2.5 million U.S. tons (2.3 MMT) of fish and shellfish per year. This catch has a total value of about $1.4 billion. The state that lands the most fish and shellfish is Louisiana, harvesting over 613,000 U.S. tons (556,000 MT) annually. Alaska leads in the value of fish and shellfish landed—about $225 million annually. The principal United States fishing port is San Pedro, California. It accounts for nearly 325,000 U.S. tons (295,000 MT) per year. Menhaden is the leading species in amount harvested—about 1 million U.S. tons (907,000 MT) per year.

United States fishermen supply only about 37% of this country's requirements for edible fish and shellfish. Therefore, about 63% of the edible fishery products used must be imported.

In some countries, fish and shellfish do not supply much animal protein; for example, in the United States, the per capita consumption (consumption per person per year) of fish and shellfish is only about 12.5 lb (5.7 kg). This accounts for less than 5% of the animal proteins (beef, pork, chicken, dairy products, eggs and lamb) which are utilized. In Scandinavian countries, the per capita consumption of fish and shellfish is more than three times greater than in the United States, and in Japan, per capita consumption of seafoods is even larger. In some parts of Southeast Asia, fish and shellfish make up a much greater part of the animal protein supplies, in some areas accounting for essentially all of the complete protein type food.

One of the chief reasons that fish and shellfish are of lesser importance in the food supplies of many countries than are other animal proteins is that, by and large, we are still hunting them as wild animals, even though we depend no longer on wild deer, buffalo and other warm-blooded animals as a source of meat. Little attempt has been made to cultivate fish and shellfish. If some of the time, effort and money expended on the raising of land animals and agricultural products had been applied to the raising of fish and shellfish, these products might now play a more important part in our diet. The advantages of fish farming over conventional fishing are many:

(1) Harvest is proportional to effort, simpler and safer;

(2) Conditions can be largely controlled (contaminants, disease temperature, salinity, etc.);

(3) Size of crop can be predicted and stocks easily and reliably assessed;

(4) Genetics can be manipulated to improve yield, improve resistance to disease, shorten generation times, etc.;

(5) Habits and life processes can be studied;

(6) Feeding of fish can be controlled;

(7) Operation not vulnerable to overfishing;

(8) Requires no expensive fishing gear, ships, ship maintenance or ship insurance;

(9) Requires no sailing energy and time;

(10) Is not dependent on weather;

(11) Time between slaughter and process is very short, ensuring top quality;

(12) No need for international agreement.

Fish are the most efficient converters of food, having a conversion ratio of about 1.5 to 1. Fish also require much less space than other animals (e.g., catfish space requirements are about 2500 lb per acre (2750 kg per ha); silo systems can reportedly produce about 1 million lb of fish per acre [1.1 million kg per ha]). These facts suggest that fish for human use will eventually be produced largely by fish farming. It may be that the situation is beginning to change. Methods of culturing oysters, clams, mussels, abalone, shrimp, crawfish, crabs, northern lobsters, salmon, catfish, carp, buffalofish, milkfish, tilapia, shad, striped bass, trout, mullet and plaice are being investigated in various countries. In some countries, several freshwater species (catfish, carp, trout and tilapia) have been raised as a commercial enterprise for some years and milkfish have been raised (from the captured young fish) for many years in the Philippines. Oysters are now grown commercially in some areas, and the raising of shrimp in Japan is already commercialized and may eventually become a thriving industry.

As the availability of other animal protein decreases, a situation now existing even in affluent countries, it may be that greater efforts will be applied to the culturing of marine and freshwater species of fish and shellfish, and eventually these species may play a much more important part as a worldwide supply of animal protein.

Fish are the most efficient converters of feed when compared to other animals that comprise man's complete protein sources. It takes about 1 1/2 lb (680 g) of feed to produce 1 lb (454 g) of fish, whereas it takes about 2 1/2 lb (1.14 kg) of feed to produce 1 lb (454 g) of poultry, 4 lb (1.82 kg) of feed to produce 1 lb (454 g) of hog and about 10 lb (4.54 kg) of feed to produce 1 lb (454 g) of cattle.

Fish require relatively less space than other high-protein sources. Catfish can grow with a space need of about 2500 lb per acre (2750 kg per ha). In more efficient culturing methods, trout have been produced with a space need of about 100,000 lb per acre (113.75 MT per ha). Using silo-type systems, recent production figures indicate that 1 acre (0.4 ha) provides enough space to produce about 1,000,000 lb (454 MT) of fish.

Fish flesh is readily digested, and it is subjected to highly active bacterial enzymes. Therefore, fish tends to deteriorate rapidly and cannot be held at temperatures above freezing for long periods. A simple principle that applies to all fresh food, and especially to fish, is the 3/H rule: *Handle the product under strict sanitary conditions* (to keep the microbial contamination at a minimum). *Handle the product at a cool temperature* (microbes multiply rapidly and spoilage reactions proceed rapidly at warm temperatures but both proceed slowly at cool temperatures). *Handle the product quickly* (fish deteriorate as a function of time as well as temperature). To give some idea of the importance of temperature, fresh caught fish will generally last about 12 days if held in ice (temperature at about 32°F or 0°C) whereas they will last only about 4 days at 46°F (7.8°C), a temperature sometimes found in domestic refrigerators. There are at least three reasons why fish spoil so rapidly at refrigerator temperatures. Primarily, because they are readily digestible;

secondly, because the muscle glycogen is nearly depleted during harvesting, leaving little to be converted to acid which would act as a preservative; finally, because the bacteria which are found on fish are psychrophiles—that is, they can grow well at low temperatures, and their enzymes are functional at low temperatures. Even among psychrophiles there is a range of optimum growth temperatures for different species, and it is known that some of the psychrophilic bacteria found naturally on fish grow at such low temperatures that they are not reliably detected by standard bacteriological plating techniques.

FISHING METHODS AND EQUIPMENT

Hook and Line Gear

Hand Lines.—The hand line (see Fig. 18.1) is one of the simpler types of equipment used to catch fish. It has a baited hook attached to the end of a line and a weight or "sinker" fixed to the line in a position above or below the hook. The hook is barbed so that, once caught, the fish cannot escape. The "sinker" must be heavy enough to keep the line more or less vertical in the water. The line is let down to or near the bottom, and when the fish bites at the bait and becomes hooked it is pulled into the boat and removed from the hook. Some ground fish are caught in limited quantities with hand lines.

Pole Lines.—At one time, pole lines were used extensively to catch tuna and may still be used to a limited extent for this purpose. Pole lines

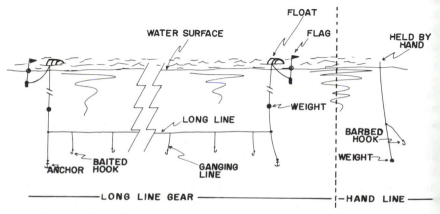

FIG. 18.1. LINE FISHING

have a nylon hoop attached to a short bamboo pole. A heavy line is fastened to the hoop. The line is short and is fixed to a wire leader at the end of which a barbless hook and feathered "jig" is secured. The hook is thrown into the water; the "jig" attracts the fish which tries to take it, becomes hooked in the process, and is then swung out of the water and over the deck. Since the hook is barbless, the fish falls off the hook and into the deck and the hook and "jig" are returned to the water. For this type of operation, fishermen stand on a platform over the side of the boat in a position just above the water.

Long Lines.—Long lines (see Fig. 18.1) are used to catch halibut and, in some instances, cod and haddock. The long line has a comparatively heavy central line to which short lines or "gangings" are attached at right angles to the main line and at distances of every few feet (1 ft is about 0.5 cm). A barbed hook, which is baited, is fastened to the end of each ganging." An anchored line carrying a float and flag is fixed to each end of the central line, the anchor holding the gear on the bottom and the flag, which rests above the water surface, serving to indicate the position of the gear. In setting the long line, the float and anchor at one end are thrown overboard, and the central line, with "gangings" and baited hooks, is run out and allowed to sink to the bottom. The anchor and float at the other end of the line are then thrown overboard. After the set has been allowed to lie for a period of several hours, the float and anchor at one end are taken in, and the line is pulled into the boat with (for halibut) or without the aid of a rotating block or cylinder. As the fish reach the boat they are flung onto the deck, or removed from the hook by hand with or without the aid of a "gaff" (large unbarbed hook with a short handle). As the line, with fish removed, is brought into the boat, it is coiled into tubs so that the hooks can be baited without difficulty. Long lines may be set over a distance of one mile (1 mi is about 1.6 km) or more.

Trot Lines.—Trot lines are sometimes used to catch the blue crab. They are similar to long lines, consisting of a central line, with "gangings," attached at both ends to an anchor and a float. They differ from long lines in that the ends of the gangings are baited but contain no hook. As the line lies on the bottom, the crab grabs the "ganging" with its biting claw to feed on the bait and hangs on even if pulled out of the water. The trot line is allowed to lie on the bottom for a period of time, then pulled into the boat through a metal ring beneath which a small net is positioned. As the crab hits the ring, it releases its hold on the "ganging" and falls into the net from which it is transferred to the boat.

Troll Lines.—Troll lines (see Fig. 18.2) are used to catch certain species of salmon and may be used occasionally to catch other fish which are found near the ocean surface. The lines are strung from poles or masts on auxiliary powered boats. The terminal end of each line contains a hook which may be baited but usually has a metal "spoon" attached to the line and to the hook by swivels. As the boat moves along, the spoon rotates and flashes in the water, attracting the fish, which bite at it and become hooked. A mechanical attachment to the line may be used to pull the hooked fish automatically to the side of the boat.

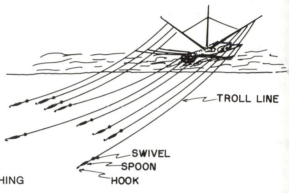

FIG. 18.2.   TROLL LINE FISHING

Nets

Gill Nets.—Gill nets (see Fig. 18.3) are used to catch salmon and shad and sometimes to catch herring, mackerel, cod and haddock. Gill nets

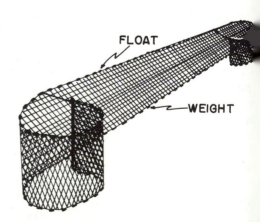

FIG. 18.3.   DRIFT GILL NET

are constructed of twine, the mesh size of which is large enough to allow the fish (of a particular species) to swim through until the thickest part of its body stops it from swimming further. When the fish tries to back out of the mesh, it is prevented from doing so by its gill covers which must be opened in order for the fish to pump water through the gills for the purpose of obtaining oxygen. Drift nets have floats which keep the top of the net at the surface of the water and weights or "sinkers" attached to the bottom which keep the net extended vertically in the water. In fishing, the net is let out to extend at right angles from the boat, which is allowed to drift. Eventually the net is removed from the water, and the fish are taken from the net. Anchored gill nets may be used to catch some types of fish.

Otter Trawls.—Otter trawls (see Fig. 18.4) are used to catch cod, haddock, flounder and other bottom fish. The otter trawl is a large, cone-

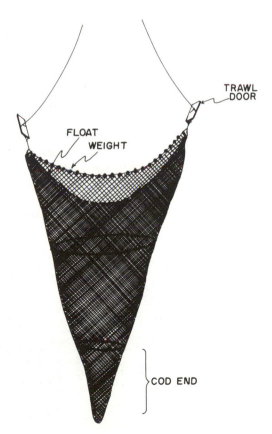

FIG. 18.4. OTTER TRAWL

shaped net which is towed behind the fishing boat along or just over the bottom. The mouth of the net is fitted with floats at the top and weights at the bottom which serve to keep it open vertically. Attached to the towing lines, near each side of the mouth of the net, are "doors" or large rectangular wooden frames which keep the mouth of the net open in the horizontal direction. The far end (cod end) of the otter trawl has a mesh small enough to retain fish of edible size but large enough to allow very small fish to escape. After the net has been towed for some time, it is pulled up to the boat and the "cod end" is tied off with a strap. This portion of the net is then hoisted out of the water to a position over the deck, and the bottom end is opened by a line attached to a special closing mechanism, which allows the fish to fall onto the deck of the boat.

Purse Seines.—More fish are caught with purse seines (see Fig. 18.5) than by any other method. The purse seine is a long, deep, fine-meshed net which has floats at the top and weights at the bottom which keep the net, when let out, in a vertical position in the water. Along the bottom and ends (vertically), rings are attached through which lines are run to allow for closing the bottom of the net when a school of fish has been surrounded. Purse seines are used to catch fish which swim together in large groups (schools) near the surface of the water. Menhaden, tuna, salmon, herring, mackerel and other fish are caught with purse seines. When a "school" of fish is sighted, one end of the net is attached to a small power boat which encircles the fish, paying out the net as it goes, after which the purse lines are drawn, closing the bottom of the net. Portions of the net are then brought aboard the fishing vessel with a power block, concentrating the fish in the "bunt," a portion of the net

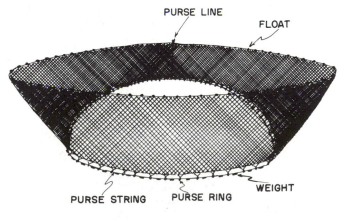

FIG. 18.5. PURSE SEINE

which is constructed of stronger twine. Once in the "bunt," the fish are pumped into the vessel or a carrier boat with a large suction hose attached to a centrifugal pump. The fish may also be removed manually with a "brail," a large dip net attached to a boom which can be guided by a long handle. A release mechanism attached to a line allows the opening of the bottom of the "brail" to discharge the fish and provides for closing it once the fish have been emptied from the net.

Traps.—Pound traps (see Fig. 18.6) are used less frequently than formerly to catch fish but are still sometimes used to catch salmon, herring, mackerel and other species. Pound traps are constructed of twine and have a small mesh size. They differ somewhat in construction, but a typical configuration is a leader, running outward from shore to a V-shaped section called the heart (there may be another V-shaped section farther from shore called the inner heart), the outer heart (or inner heart when used) is connected to a rectangular section, called the pot, by a narrow, funnel-like entrance. This may have a rectangular section (the spiller) connected to it by a funnel-like entrance. The pot, or spiller when used, is located farthest from shore. The netting of the leader, outer heart and inner heart, when used, extends from the water surface to the bottom. The netting of the pot and spiller, if used, extends from the surface to the bottom, and the bottom is enclosed with netting. The various parts of the pound trap are attached by ropes to poles anchored to the bottom or driven into the bottom mud.

In operation of the pound trap, fish swimming near the surface contact the leader and follow it into the heart section and thence into the pot and into the spiller, when used. The fish are removed from the pot or spiller

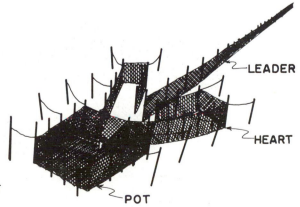

FIG. 18.6. SINGLE-HEART POUND NET

with a powered brail with or without concentrating the fish in a section of the netting by loosing the rope fastenings and pulling part of the netting into or under the boat.

Pots

Pots (see Fig. 18.7) are used mainly for catching crabs, lobsters, and, in some instances, certain freshwater fish.

Lobsters are caught with pots that are constructed from wooden laths spaced about 1 in. (2.54 cm) apart or from wire mesh covered with plastic. The pot usually has several chambers and is weighted at the bottom to keep it upright on the bottom. The bottom of the pot is flat while the body is semicircular in shape. Entrances to the pot are provided by circular openings attached to the frame with netting. In fishing, the pot is baited with fish portions in a mesh container fixed to a spindle. The pot is lowered to the bottom, attached to a rope which has a buoy or float at the upper end which rests on the water surface. After a period of 24 hr or more, the pot is raised from the bottom (usually with the aid of a rotating power block), the lobsters are removed and the pot is rebaited and again lowered to the bottom. Pots may be operated singly or strung out in succession, one attached to another with rope.

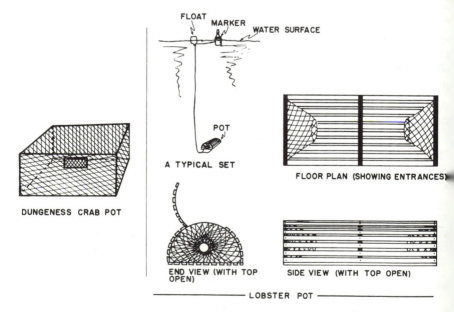

FIG. 18.7. ENTRAPMENT DEVICES

Crabs are caught with circular or rectangular shaped pots constructed from wire mesh attached to a metal frame. The wire may or may not be covered with plastic. Circular pots used to catch blue crabs are about 42 in. (107 cm) in diameter and 14 in. (35.6 cm) deep. Rectangular pots used for this purpose are about 30 in. (76 cm) square and 14 in. (35.6 cm) in depth. The circular pot has two entrances, and the rectangular type has four entrances. This type of gear is weighted or anchored to the bottom to keep it upright, and a line is connected to the pot and to a buoy or float which lies on the water surface. Crab pots are baited with shucked clams or dead fish and are pulled up to remove crabs 8 to 24 hr after setting. The pot is then rebaited and again lowered to the bottom.

Pots are used on the West Coast of North America to catch the Dungeness, the snow and the King crabs. Except for its larger size, this gear is of similar construction to that used to catch the blue crab, and it is used in a similar manner as that used for blue crabs. Circular pots are used for deep-water fishing and rectangular pots for shallow-water fishing for Dungeness crabs. King and snow crabs are caught with rectangular pots which are about 7 ft (about 2 m) square and 2.5 ft (76.2 cm) deep.

Dredges

Dredges (see Fig. 18.8) are used to harvest scallops, surf clams, hard-shell clams, soft-shell clams and oysters. Scallops, oysters and hard-shell clams may be harvested with dredges constructed with a metal-mesh bag positioned behind a toothed metal bar. The dredge is dragged along the bottom, the bar penetrating the bottom mud far enough to remove hard-shell clams and other shellfish and depositing them in the metal-mesh bag. After dragging for a period of time, the dredge is brought to the deck of the boat and emptied onto a platform where the shellfish can be separated from the detritus.

Suction dredges, using a water-jet vacuum to remove shellfish and deposit them onto conveyors which carry them to the deck of the boat, may be used to harvest oysters. Water-jet dredges, with or without an escalator, may be used to harvest hard-shell clams. Hydraulic (water-jet) dredges, together with escalators, are used to harvest surf clams and to obtain soft-shell clams from the Chesapeake Bay area.

Tongs, Rakes and Forks

These implements (see Fig. 18.8) are used to harvest oysters, hard-shell clams and soft-shell clams. Tongs have a two-sided chamber

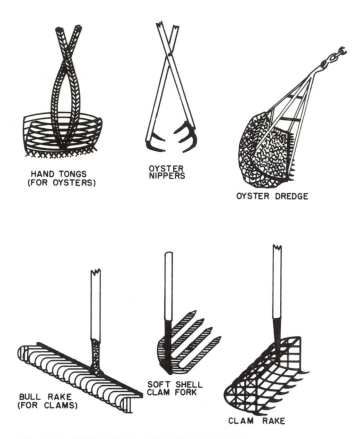

HAND TONGS
(FOR OYSTERS)

OYSTER
NIPPERS

OYSTER DREDGE

BULL RAKE
(FOR CLAMS)

SOFT SHELL
CLAM FORK

CLAM RAKE

FIG. 18.8. SHELLFISH HARVESTING DEVICES

constructed of metal strips approximately 1 in. (2.54 cm) apart. T
bottom part of each half-chamber is fronted by a toothed metal bar. T
half-chambers can be made to open or close like the blades of a scissor
means of two long wooden handles connected to a pivot. In fishing, t
tongs are lowered to the bottom in the open position, then close
scraping up whatever is in the path of the closing tongs. They are th
raised to the boat where the shellfish are removed and the detrit
discarded. Tongs are used in comparatively shallow water.

Rakes are sometimes used in shallow water to harvest oysters, har
shell clams or bay scallops. The rake has a basket-like chamber co
structed of metal strips or metal wire fronted by a toothed-metal ba
The basket is attached to a long, wooden handle. In fishing, the rake
lowered to and pulled over the bottom as the operator walks from one e
of a small boat to the other. In shallow water, the rake is simply reach

ut as far as possible and pulled toward the operator. After raising the ake to the boat, the shellfish are removed and the waste is discarded.

In some areas, soft-shell clams are harvested with the clam hoe or fork. 'his is a short-handled, four-tined fork. The tines are flat and fastened o the handle at an angle of about 60 degrees. An area where clam holes re present is selected and the fork is inserted into the mud and pulled up o remove the clam with the mud while preventing breakage of the clam r clams. The clams are removed from the mud by hand and placed in a ail or other similar small tote-type container.

MPORTANT FAMILIES OF FISH AND SHELLFISH

'he Herring Family (Clupeidae)

Among the various categories of fish which are used by man for his)od, one of the most important is the family Clupeidae. This is also the 1ost prolific group of fish to be found in the ocean. The various species ave rounded bellies with or without sharp edges. In color they are dark reenish or brownish blue above and silvery on the sides and belly. Some ave dark spots on the sides, just behind the head. The tail is deeply)rked. The dorsal (back) fin is located about midway of the body just bove the small ventral (belly) fin. There are small pectoral fins (on the elly side just behind the head), and an anal fin is located ventrally near 1e tail. The scales are wide and loosely attached. The sea herring is up) 17 in. (43 cm) in length and over 1 lb (454 g) in weight. Menhaden ach 15 in. (38 cm) in length and 12 oz (340 g) in weight. Alewives reach ft (30.5 cm) in length and more than 8 oz (227 g) in weight, and shad 1ay be 2 1/2 ft (76 cm) in length and 8 lb (3.6 kg) in weight. The lupeidae are pelagic, traveling in groups or schools near the ocean 1rface.

Sea Herring.—In the United States sea herring (see Fig. 18.9) are)und in ocean waters from Alaska to the state of Washington on the

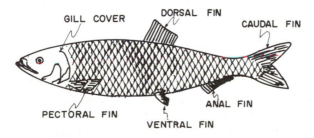

FIG. 18.9. HERRING *(CLUPEA HARENGUS)*

West Coast, and from Labrador to Cape Hatteras on the East Coast. Sea herring are plankton feeders, eating the various microscopic plants and animals (diatoms, larvae of various shellfish, etc.) when very young and as adults, eating small shrimp, small fish, etc. When unmolested, the sea herring may live to an age of 20 years or more.

Most sea herring are caught with purse seines, but some are caught with pound traps or weirs (similar to pound traps but constructed with poles and brush). Gill nets are sometimes used to catch these fish.

The larger fish may be used for export, as there is a foreign market for this species where it is used for human food. The smaller herring, which are canned as sardines, must be held in the purse seine (sometimes for longer than 24 hr) until the stomach is free from feed. This is done to prevent excessive enzyme action, which, after removal of the fish from the water and prior to processing it, might digest parts of the skin and flesh and cause the canned product to be inferior in quality. During the removal of herring from seines and during removal from the boat (a hose and centrifugal pump are ordinarily used for this purpose), most of the scales are removed from the fish. These scales are not thrown away but are separated and sold, since they are a source of pearl essence.

Some salt is added to herring aboard the boat if they are to be processed as sardines. At the sardine plant, they are unloaded into tanks, some salt again being added, where they are held until processed. In processing, they are first headed and gutted by hand with the aid of a knife, without splitting the belly. Some processing methods require the precooking of the fish using steam or hot oil before packing into cans. The usual procedure is to pack the raw fish into small rectangular cans by hand, place the open cans on racks, heat the product in free flowing steam for 18 to 20 min, then invert the racks to drain off the liquid formed during heating. Just before sealing the cans, vegetable oil (hot or cold), a tomato sauce mixture or a mustard sauce mixture (about 2 oz [56.7 g] per 4 oz [113.4 g] of fish is added). The cans are then sealed and heat processed at 240°F (115.6°C) for 20 min, then cooled in the retort. Some of the larger herring are canned (after heading and gutting) in larger oval cans with the addition of oil or sauce, and some are steaked (cut across the body) into pieces about 1 in. (2.54 cm) thick and packed in sauce in the 4 oz (113.5 g) cans.

Some of the larger herring taken on the east coast are filleted or gutted and split, packed into cartons holding 10 kg (22 lb), frozen between refrigerated plates and sold to the European trade. Mature herring are also prepared as the pickled product. The herring are headed and gutted and the bellies are split. The fish may then be packed in barrels in lternating layers of salt and fish, and the barrels capped. After about 3 eeks, part of the brine is removed through the bung and the fish from ne of the barrels are distributed among other barrels to compensate for hrinkage. The old brine may be discarded and new brine (saturated salt olution) added or old (used) brine may be added to fill the barrels. Fish repared in this manner may be shipped directly to the pickling plant. nother method of preparing herring for the pickling plant is to cure the eaded and gutted fish for 3 to 7 days in 80 to 90% saturated salt solution ontaining 2 1/2% of vinegar (this vinegar has 2% of acetic acid). At the ickling plant, the fish may be repacked in 35% salt solution containing me vinegar, then held at 34°F (1.1°C) until used for further processing. or the final processing, the fish are soaked in running water overnight or r sufficient time to remove almost all the salt, then held in a 6% salt d 3% vinegar (this vinegar contains 5% of acetic acid) solution until oss cut and packed in glass jars. When packed in jars, the cut herring ay be packed in the 6% salt, 3% vinegar solution with spices or in a lution containing 2 1/2% acetic acid (as vinegar), about 1% sugar and 5% salt. Soured cream, onions or other flavoring material may be used ingredients in the glass-packed pickled fish. Also, rollmops may be epared as pickled herring fillets wrapped around a piece of pickle or ion and packed in one of the pickling solutions.

During salting and subsequent storage, herring should be held at 60°F 5.6°C) or below, otherwise spoilage (putrefaction) may occur, bacterial owth taking place before the salt has thoroughly penetrated the flesh. lso, the pickled product is not stable at room temperature and must be ld under refrigeration, preferably at 40°F (4.4°C) or below.

In Scandinavian countries, freshly caught herring are placed in barrels ithout gutting, with the addition of about 10% salt and 10% sugar (by eight). They are held in this manner at 50°F (10°C) below until filtered gutted and crosscut and packed into tins with the addition of salt lutions containing any of a number of different sauce mixtures. It is nsidered that proteolytic enzymes present in the fish are responsible r the development of desirable flavors in these products.

In some areas, mature herring are split, gutted, washed, soaked in salt lution (90% saturated) for 1–2 hr, then lightly smoked without heating. is product may be held under refrigeration and sold as kippered rring, or it may be filleted, packed in aluminum cans, usually holding 5 oz (99.2 g), and heat-processed to provide a commercially sterile oduct. Small amounts of herring are highly salted and highly smoked d sold in this form, in which condition they are stable at room mperature.

In Alaska, and to some extent in eastern United States, the roe (eggs) is taken from herring approaching the spawning stage, salted in containers and sold at high prices to certain Asian countries.

Large quantities of sea herring harvested in some countries are converted into fish meal, which is used as a protein supplement for cattle and poultry as a portion of the feed. When used for this purpose, the fish are sometimes brought to port in unrefrigerated vessels, although some vessels, especially those operating in southern waters, are refrigerated. At the processing plant, the fish is first cooked in live steam in a continuous cooker, then pressed in a screw-type continuous press. The press cake is dried with hot gases from oil-produced flames in a rotary drier, to a moisture content of 5 to 8%. The liquid from the pressed product is not discarded. It is first centrifuged to remove oil, which is collected and sold for industrial uses. The remaining liquid (stickwater), which contains proteins, peptides and amino acids, is then vacuum concentrated to a solids content of 50% and acidified to prevent spoilage. This product may be sold as a protein supplement or it may be added back to the presscake before the latter is dried.

Shad

Shad are anadromous fish (ascend rivers to spawn) that spend the greatest part of their lives in ocean waters as far as 50 mi (80.5 km) from shore. Shad are plankton eaters and are said not to eat fish. They range from the Gulf of St. Lawrence to Florida on the East Coast, but are caught in significant numbers only from New York southward. They were brought, some years ago, to the Pacific waters, and some are now caught in California. Shad are caught in rivers and their estuaries with drift gill nets. Aboard the boats, they are neither eviscerated nor iced since they are brought to shore shortly after removal from the water.

Shad are used almost entirely as the fresh product, with only small amounts being frozen. They contain many small bones, but can be filleted to eliminate most of them from the flesh. The roe (unfertilized eggs), which prior to spawning is held together by a thin membrane, is highly prized. It is sold fresh, or packaged in moisture vapor-proof material and frozen to be sold to the restaurant trade and, in this state, may be stored at 0°F (−17.8°C) for 6 to 8 months. Longer storage under these conditions usually results in a rancid product due to oxidation of the fats contained therein.

Menhaden

Four species of menhaden, sometimes called "pogy," "bunker" or "mossbunker" are found in the Western Atlantic. They range from North

Scotia to Brazil. Menhaden feed on microscopic plants and animals. They are caught with purse seines when schooling near the surface of the water and removed to the hold of the power boat or carrier vessel. Aboard the boat, they may be held without refrigeration, in which case they will be brought to port within a period of 24 hr after catching. Boats which keep these fish in refrigerated holds may remain at sea for several days prior to landing the catch.

Menhaden are not used for human food. They are processed to produce fish meal and oil in the manner described for herring. In the United States, larger quantities of menhaden are caught (several hundred thousand metric tons) than that of any other fish or shellfish.

The Anchovy

The anchovy (family Engraulidae) is a small herring-like fish found off the coasts of California and Mexico. It is caught in purse seines, and at one time was used as live bait for tuna fishing. It is now used mainly for the production of fish meal and oil.

The Anchoveta (Family Engraulidae)

This is a small herring-like fish found off the coast of Peru. It is usually caught in large quantities, as much as 9 billion lb (4,080,000 MT), by purse seines, and is used entirely for the manufacture of fish meal and oil.

Other Clupeidae

Pilchards are members of the herring family, and at one time were plentiful off the coast of California, where they were caught with purse seines. They were used for canning as sardines and also to produce fish meal and oil. Due to scarcity, fishing for this species has been discontinued in California, but pilchards are caught and canned by the South Africans.

THE COD FAMILY (GADIDAE)

The cod family (see Fig. 18.10) includes the cod, haddock, pollock, cusk and several species of hake. The members of this family vary greatly in size. The cod may reach a length of 6 ft (1.8 m) and a weight of 200 lb (91 kg) and averages 10–12 lb (4.5–5.5 kg). The haddock may reach 3 ft (1.5 cm) in length and 24 lb (10.9 kg) in weight, but the average is much

smaller than this. The pollock reaches a maximum of 3.5 ft (1.1 m) an
a weight of 25 lb (11.4 kg). The cusk and the hakes are smaller than thos
listed above.

In color, the members of the cod family vary greatly. The cod i
brownish, grayish or reddish above and whitish below, with the uppe
part being speckled with small rust colored spots; the haddock i
purplish-gray above, silvery gray below, with a dark shoulder patch
Pollock are olive or brownish-green above and silvery-green below.

Cod, pollock and haddock have three wide dorsal fins and two wid
anal fins (middle of the body to the tail). The pectoral fins (on each sid
just behind the head) are of moderate size, and the ventral fins are sma
and located on the underside just forward of the pectoral fins. The tail i
wide and not deeply forked. The cod has a small, fleshy barbel under th
front part of the lower jaw. The fin structure of the cusk and the hakes i
quite different from the other members of the group. They have on
(large) or two (one large, one small) dorsal fins and one large anal fin, an
the tail is rounded and not forked.

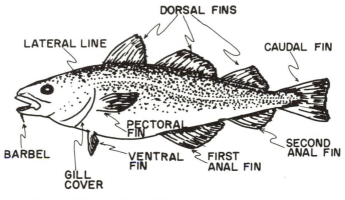

FIG. 18.10. COD *(GADUS MORHUA)*

The Cod

Cod (see Fig. 18.10) are found on both sides of the Atlantic and a
most plentiful around Norway, Iceland, Newfoundland, Nova Scotia ar
on Georges Bank off Cape Cod. In the Eastern Atlantic, they range fro
Greenland to North Carolina. Molluscs (clams, oysters, scallops, etc
are said to make up an important part of the diet of the cod, but cod al
eat small fish.

By far the largest quantities of cod are caught with otter trawls
waters ranging from 300 to 1500 ft (91 to 457 m) in depth. Sm

quantities of this species are caught with long lines, hand lines or gill nets. When caught with otter trawls, the fish are gutted and washed on the deck of the boat. During summer months, the gills must be removed, but the head is left intact. The fish are then stored in hold pens in the iced condition, layers of ice being alternated with layers of fish until the pen is filled. A shelf is inserted in the pen, once it has been filled about half-way, to prevent excess weight on the bottom fish. The proportion of fish to ice should be not more than 2:1. A more recent method of holding trawl-caught fish aboard boats is to store them iced in wooden, plastic or metal boxes. The boxes are stored in the hold in tiers, and the boxes are so constructed to allow the water from melting ice to drain away to the bilges (lower part of the boat) without running over fish stored in the lower boxes. The holding of fish in boxes facilitates both preservation and unloading of the product from the boat.

In port, for purposes of unloading, if boxes are not used, the fish are placed in canvas baskets holding about 100 lb (45 kg). The baskets are hoisted to the dock and dumped into a weighing box, then transferred to carts, barrels or boxes for transport to a nearby processing plant. If the fish are to be held for significant lengths of time prior to processing, they should be iced to keep their temperature as near 32°F (0°C) as possible, but under no circumstance above 40°F (4.4°C).

In processing cod, the fish are first washed under rotating brushes and heavy sprays of water. Fillets are then cut from both sides of the fish by hand with a knife or by machine. Fillets are usually skinned, and this may be done with a knife, by hand or by machine. If the fillets are to be sold with the skin, the fish must first be scaled by being passed under rotating serrated discs or stiff brushes. Cod fillets from fish taken from certain areas must be candled, since a parasite, a roundworm, may be present. In candling, the fillet is placed on a glass plate below which a strong light is positioned. Parasites show up as an opaque area, when present, and are cut out with a knife. The law requires the removal of roundworms from fish fillets. If the fillets are to be sold as fresh, they are packed in metal boxes holding 10, 20, or 30 lb (4.5, 9.1 or 13.6 kg) of product. Individual fillets are usually not wrapped with plastic. Some fillets, or portions, may be packaged on cardboard trays and over-wrapped with plastic for the retail trade. Products to be packaged in metal boxes are placed in wooden boxes or barrels and surrounded by ice, then shipped to market.

Fillets which are to be frozen for the retail trade are packed in 1 lb (454 g) waxed cartons with or without being first wrapped in moisture vapor-proof plastic. The fillets may have been passed through a weak brine (10 to 45% saturated salt solution) prior to packaging. For purposes of freezing, retail-size cartons of fillets may be placed on trays, the trays placed on racks, and the racks wheeled into a blast freezer where very cold air is blown over the product, or the cartons may be plate frozen by being placed on trays in contact with refrigerated plates at −28°F (−33.3°C). Some cod fillets are frozen in contact with refrigerated plates in waxed cartons holding 10 lb (4.5 kg) of product. The pressure of the plates causes the fillets to adhere to each other to form a block of frozen fish which can be sawed into 4 oz (113.4 g) portions or into the smaller sized fish sticks. The cut portions are covered with batter or with batter and breading, packaged, refrozen and stored at 0°F (−17.8°C) or below for the restaurant trade. Fish sticks are generally breaded and passed through hot oil at 375°F (190.6°C) for about 30 sec before packaging and refreezing. The batter used for fish portions and fish sticks varies with the manufacturer but generally contains some type of cereal flour, dried milk solids, egg solids, spices and flavoring.

Although very small amounts, if any, of dried salt cod are produced in the United States, some is consumed in this country, and is an important part of the diet of people living in certain tropical areas. In preparing dried salt cod, the fish are headed, then split open along the backbone to the tail. The triangular portion of the backbone is then removed with a knife. The fish are then washed, and the black lining covering a part of the body cavity is scrubbed off. The fish are then salted, either in large casks or in kench (piled together on a solid floor). In salting, the bottom layer of fish is placed on a bed of salt, skin side down. This is continued with first a layer of salt then a layer of fish until the top layer is reached and placed skin side up. Salting and holding should take place at ambient temperatures of about 60°F (15.6°C) or below to prevent bacterial spoilage. The salted fish may be held for 3 months or more prior to drying. Much of the salted cod is now dried in machine driers where the relative humidity is regulated to 70% or less, and the temperature is held at 75°F (23.9°C) or below. High drying temperatures may allow proteolytic enzymes to cause a softening of the product. Some salted cod is dried outdoors, exposed to the sun on slatted wooden frames, in areas where ambient temperatures and humidities are suitable. In such cases several days are required for drying, and the fish must be packed together in small piles and covered at night. Salted cod is dried to a moisture content of about 45% and is not entirely stable without refrigeration. Usually, after drying, the product is skinned and boned by hand, then packed in wooden boxes or plastic film in 1 lb (454 g) portions. Some salt cod is dried to a moisture content of about 30%, in which condition refrigeration is not required.

Fish cakes are prepared from salt cod by first cooking and freshening the fish to remove most of the salt, then mixing it with mashed cooked potatoes and small portions of oil, onions and pepper. A proportion of about 40% shredded cooked fish and 60% cooked potatoes is used. This product may be canned without forming or it may be formed into small cakes, deep-fat fried to brown the surface, and frozen, or sold in the refrigerated state. Some manufacturers of fish cakes are using fish sawdust, from fish stick processing, and broken fish sticks as the fish component in fish cakes.

Haddock

The haddock is the second most important member of the cod family. Haddock are found on both sides of the Atlantic from Norway to New Jersey but are most plentiful in waters off Nova Scotia and Cape Cod (Georges Bank). The mature fish feed on crustaceans (crabs, shrimp, etc.), molluscs (clams, etc.) and on small fish. In recent years, the stocks of haddock have been greatly depleted due to overfishing.

Haddock are usually caught in areas where the depth of water is 50–360 ft (45.7–109.7 m). They are caught, handled aboard the boat, and processed to produce fillets, fish blocks, fish sticks, etc., in the same manner as that described for cod. Haddock are not salted and dried. Some haddock are lightly salted and lightly smoked, without heat, to produce a product called "finnan haddie."

Pollock

Pollock are found on both sides of the Atlantic from Norway to the Chesapeake Bay but are most plentiful in waters off Nova Scotia, Cape Cod (Georges Bank) and in the Gulf of Maine. Pollock eat shrimp, crabs and other crustaceans, and small fish. They do not eat bivalve molluscs.

Pollock are caught in waters at levels between the surface and a depth of 450 ft (137 m). They are caught, handled aboard the boat, and processed in much the same manner as that described for cod. Small quantities of pollock are salted and dried.

Hakes

There are several species of hakes, the most important of which is the silver hake or whiting. The whiting is most abundant in waters off Nova Scotia. It is caught and handled in much the same manner as are cod. Some whiting are headed, gutted, washed and frozen in blocks without further cutting, for utilization as food.

With all members of the cod family, small fish, species not especially prized as food (red hake, etc.), fish frames (the portion remaining after the fillet has been cut away), etc., may be passed through mechanical meat/bone separators (machines which separate the flesh portion from bones and skin). This provides a significant yield of edible, ground fish flesh (resembling hamburger in texture) which may be used to produce frozen fish blocks to be further processed into fish portions, fish sticks, etc. Handled in conventional fashion, such products are not stable in frozen storage since the fat oxidizes and becomes rancid, and the tissues get tough at a faster rate than that of the corresponding fillets held under the same conditions. The faster oxidation of fats is probably due to a greater exposure to oxygen because of the great increase in surface area. The increased rate of toughening may be due to a wider distribution of the enzyme which decomposes trimethylamine oxide to form dimethyl amine and formaldehyde. The latter compound is known to denature proteins. The spoilage reactions may be slowed considerably by storing at lower temperatures, e.g., −20°F or −28.9°C. Rancidity can be prevented altogether by protecting the product with a wrapper of gas-impermeable plastic film, e.g., polyester, PVC, Nylon 11, aluminum laminate, etc.

THE MACKEREL FAMILY (SCOMBRIDAE)

The most important fisheries of the United States in terms of volume caught and value are those which process fish of the mackerel family. Included in this group are the various tuna, the Atlantic mackerel, the jack mackerel and the Spanish mackerel.

Tuna (see Fig. 18.11) are torpedo-shaped fish tapering to a pointed nose and a slender caudal peduncle (the portion near the tail). The first dorsal fin is high in front and low at the back end. The second dorsal fin

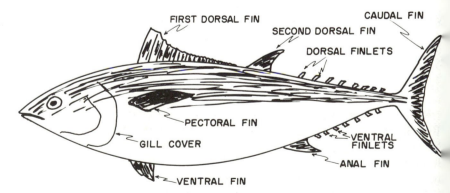

FIG. 18.11. BLUEFIN TUNA *(THUNNUS THYNNUS)*

situated quite close to the first, is lower in young and higher in old fish than the first. The anal fin originates under the rear of the second dorsal fin, and is similar to it in shape. There are a number of dorsal and ventral finlets extending from the second dorsal and anal fins to the tail. The tail is broad and crescent-shaped. Tuna vary in color from blue to black above and silvery below.

Among the important species of tuna are the yellow fin, which reaches 400 lb (181.6 kg) and averages 30 lb (13.6 kg) in weight; the bluefin, which attains a maximum weight of 1000 lb (454 kg) and averages 30 lb (13.6 kg); the skipjack, which varies in size 3–20 lb (1.4–9.1 kg); the albacore, which reaches a maximum of 80 lb (36.3 kg) and averages 25 lb (11.4 kg); and the yellowtail, which has an average weight of 20 lb (9.1 kg).

The Atlantic mackerel is shaped like the tuna and the fins are similar, but the tail is more deeply forked. In color, this fish is steely to greenish blue on top and silvery-white below. Mackerel are much smaller than tuna, the Atlantic mackerel reaching a length of 20 in. (50.8 cm) and a weight of about 3.5 lb (1.6 kg) but averaging only 13 in. (33 cm) in length and 12 oz (340 g) in weight.

The shape and fin structure of the Spanish mackerel is similar to that of the Atlantic mackerel. In color, this fish is blue-green along the back and silvery on the sides with orange or yellowish spots. The second dorsal and pectoral fins are yellowish. Spanish mackerel reach a maximum of 10 lb (4.5 kg) and average 2 lb (908 g) in weight.

The jack mackerel has a shape and fin structure like that of the Atlantic mackerel. The back of this fish is dark green in color, and the belly is silvery. Jack mackerel are somewhat smaller than the Atlantic mackerel in average size and belong to a different genus.

Tuna

The yellowfin tuna is found on the West Coast from Southern California to Southern Chile. Bluefin tuna range from Nova Scotia to Brazil on the East Coast and from Southern California to Northern Mexico on the West Coast. The skipjack is found in the Pacific Ocean from Southern California to Central and South America. The albacore ranges from Puget Sound (state of Washington) to lower California. The yellowtail is found in Pacific waters from Southern California to the coast of Mexico. Several other species of tuna are found elsewhere, especially in the Eastern Pacific Ocean.

At one time, fishing for tuna was carried out exclusively with pole lines. Today, tuna are mainly caught with purse seines which, due to the

size of the fish, are made with heavy twine. Some tuna may still be taken
with pole lines and some with troll lines.

Most tuna fishing boats make trips which last for several months, and
for this reason, the fish are frozen aboard the boat. In freezing, the
uneviscerated fish are placed in water-tight wells through which sea
water, refrigerated to about 28°F (−2.2°C), is circulated. When the fish
are cooled, this water is pumped overboard or to another well, and brine
at a temperature of 10°F (−12.2°C) or lower is circulated around the fish
in the well. After the fish are frozen (about 3 days), the brine is discarded
and the fish are held in the well in the frozen state by mechanical
refrigeration (cold air).

When the frozen fish are unloaded at the canning plant, they must be
thawed. This may be done by placing them in single layers in a room with
open air at ambient temperatures. The fish may be rinsed with water
from time to time. With this method, periods of 4–36 hr, depending upon
the size of the fish, are required to thaw the product. Tuna may also be
thawed in tanks of running water holding 1–5 tons (0.9–4.5 MT) of fish.
In this case, thawing times of 2 to 8 hr are required.

When thawed, the fish are dressed or butchered. The belly is cut open
with a knife, the entrails removed, and the fish and belly cavity washed
with water. With their heads intact, the fish are placed in wire baskets
which are wheeled into rectangular chambers where they are cooked in
steam at 216°–220°F (102.2°–104.4°C). At these temperatures, the
pressure is 1.2–2.5 lb/in.^{2} (84.5–176 g/cm^2). The length of time required
for cooking is 1.5–9 hr, depending upon the size of the fish. When cooked,
the fish are placed in a room at ambient temperature and allowed to cool
and drain. Cooling usually requires a period of 24–36 hr. Once cooled, the
fish are beheaded, the fins are removed and the skin is scraped away,
after which the flesh is separated into halves and the backbone and rib
bones are removed. The two halves of flesh are then split longitudinally,
and the dark meat (a V-shaped layer along the sides which is not
generally used as food for humans) is scraped away. The remaining white
or light meat (loins) is cleaned and placed on trays which are placed on
a belt that is flanked on either side by a vertical belt. The vertical belts
have links of semi-circular cross section facing inward and these gradually come together to form a hollow cylinder which squeezes the fish loin
to form a cylindrical shape as it moves along. When the loin reaches the
head of the shaper, the forward end is forced into a waiting can and a
circular knife cuts if off to fill the can. When packed in chunks, the loins
are cut into pieces and filled into cans from an adjustable filler.

Filled cans pass under automatic salters, then feeders, which automatically add oil, or broth containing hydrolyzed vegetable protein in
water. The covers may then be clinched on the cans which are then

heated in steam for about 3 min, then tightly sealed. The heating followed by cooling in the sealed can results in a partial vacuum. When jet can sealers are used, they remove sufficient air from the headspace in the can to provide the required vacuum. After sealing, the cans are washed, then heat processed. Temperatures ranging from 240°–250°F (115.6°–121.1°C) and times of 40 to 230 min are used in retorting, depending upon the size of the can used to hold the product. Cooling of the heat processed product is carried out in the retort and, if not lithographed, the cooled cans are labelled and stored in a warehouse.

Mackerel

Atlantic mackerel are found from the Gulf of St. Lawrence to Cape Hatteras in America, and from Norway to Spain in the Eastern Atlantic. Spanish mackerel range from Maine to Brazil in the Western Atlantic but are mostly caught in waters off the Carolinas and southward of these waters. The jack mackerel ranges from British Columbia to Mexico in the Pacific Ocean.

Mackerel may be taken in pound traps or with gill nets, but by far the greatest quantities are taken with purse seines. If the boat is to remain out of port after the fish are caught, they are held in ice in the round, uneviscerated state. Atlantic and Spanish mackerel are sold to retailers as the fresh product, either as fillets or as the round uncut fish. Some are frozen by placing the round fish in pans and holding at 0°F (−17.8°C), or below in rooms with or without circulating air. The fish, frozen as a block, are sprayed with water for purposes of glazing to prevent dehydration and held in the frozen state until defrosted for sale to restaurants or retail outlets.

Jack mackerel are canned in 1 lb (454 g) tall containers. The fish first pass on a conveyor belt under circular knives which cut off the heads and tails and also cut the fish to can-size lengths. The entrails are then removed, after which the fish are washed and flumed to a container feeding the packing table, where they are filled into cans by hand. The open cans are then heated in a steam box to raise the product temperature to 145°F (62.8°C), after which they are inverted to drain off liquid formed during heating. Oil, brine, tomato sauce or mustard sauce is then added to cover the fish, and the cans are then sealed and heat processed.

THE SALMON FAMILY (SALMONIDAE)

A number of commercially important species of the salmon family are found throughout the world. In the United States, the red, sockeye or blueback, the spring, king or chinook, the silversides or coho, the pink or humpback, the chum or dog salmon are of chief importance. The steelhead trout, which behaves like a salmon, is caught in some volume. All the above named fish are caught on the West Coast. Only small numbers of Atlantic salmon are caught on the American side of the Atlantic.

Salmon (see Fig. 18.12) have a deep body and a rounded belly. All have a wide, slightly lunate tail. The single dorsal fin (center of back) is fairly wide. The wide anal fin is located near the tail and the ventral fins are located just behind the midpoint of the body toward the tail. The moderately large pectoral fins are located just behind the head below the mid-section of the body. On the back, near the tail, there is a small fleshy (adipose) fin.

The various salmon differ in color from blue to green on the back. The sides and belly are silvery. The flesh of the sockeye salmon is deep red, that of the spring salmon is deep red to white. Coho salmon have a medium red flesh, pink salmon have a light red flesh, and the flesh of the chum salmon is pink to yellowish-white. The flesh of the steelhead trout is pale pink in color. The average weights of salmon are: sockeye 6.5 lb (3 kg); spring 20 lb (9.1 kg); coho 8 lb (3.6 kg); pink 4 lb (1.8 kg) and chum 8 lb (3.6 kg).

Pacific salmon range from waters off northeastern Asia and northern Alaska to California. In the United States, sockeye and pink salmon are mostly caught off the coast of Alaska.

Salmon eat small crustaceans and small fish when young and larger members of these groups as they become older. The Pacific salmon and steelhead trout are anadromous, ascending rivers sometimes more than 1000 mi (1609 km) to lay their eggs in the same rivers or streams in which they were hatched. These fish die soon after spawning.

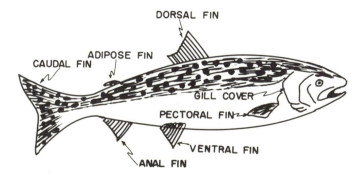

FIG. 18.12. PINK SALMON (*ONCORHYNCHUS GORBUSCHA*)

West Coast salmon are caught, in greatest quantities, with purse
eines in ocean waters near the coast. Purse seines cannot be operated in
ivers or their estuaries, but significant quantities of salmon are caught
n estuaries with drift nets. Salmon caught with purse seines or pound
raps are either transported to the cannery by boats or scows (in the
ineviscerated state) a few hours after harvesting or they may be held
board packer boats in refrigerated sea water in the uneviscerated state
or transportation to the cannery. Salmon caught in drift nets are picked
ip by packer boats on each tide and are neither eviscerated nor iced for
urposes of transportation.

Since, unlike other species of the group, coho and spring salmon are
eeding when they enter coastal waters, they are caught sometimes on
roll lines strung from small auxiliary powered boats. When caught in
his manner, they are eviscerated, washed and iced down in boxes aboard
he boat where they may be held for more than 24 hr before landing in
ort.

Some coho and spring salmon is sold at retail in the fresh state as
teaks. During the fishing season, these fish may be shipped to eastern
narkets packed in ice. These species of salmon are also frozen. In
reezing, the fish are beheaded and placed in racks in rooms held at 0°F
−17.8°C) or below until solid, then glazed by dipping in water. The fro-
en product is stored at 0°F (−17.8°C) until shipped out. In this form,
almon is not stable in the frozen state and the flesh becomes rancid in
. few months due to oxidation of the fat. To prevent rancidity, the fish
nust be protected from oxygen by being packed in gas-impermeable
ontainers.

By far the greatest quantities canned are West Coast salmon. The
ish are first conveyed to a machine, the "iron chink," which removes the
ead, tail, fins and viscera. They are next trimmed to remove extraneous
naterial left by the iron chink, after which they are washed. The fish
ext pass under rotary blades on a slotted conveyor where they are cut
ito can-sized lengths. The cut salmon pieces then pass on to a
olumetric filling machine where salt is added to the can (about 1.25% by
reight), and the cans are filled with fish. After filling, the covers are
linched on the cans which are then sealed under vacuum, or they may
e sealed without first clinching, a steam jet being used to remove air
rom the head space in the can. After sealing, the cans are washed, heat
rocessed to provide commercial sterility, then cooled in the retort. The
ans are then labeled, if not lithographed, packed in cases and stored in
warehouse until shipped out. Heat processing times and temperatures
re applied according to the weight of the product in the can. For
istance, at 240°F (115.6°C), cans holding 1 lb (454 g) of product should

be heated for 90 min, and those holding 1/2 lb (227 g) of product should be processed for 80 min.

Spring salmon are sometimes preserved by salting. The fish are split, trimmed, washed, covered with salt and packed in casks, after which the casks are filled with saturated salt solution and held at 35° to 40°F (1.7 to 4.4°C) for 30 days. This product is usually shipped to processors who smoke the fish. In smoking, the salted fish is first soaked in water to remove salt, then smoked at temperatures below 90°F (32.2°C) or hot smoked at a temperature of about 175°F (79.4°C).

THE FLATFISH FAMILY (PLEURONECTIDAE)

Many species of flatfish are utilized as food. On the East Coast of the United States, the halibut, the turbot, the sand dab, the fluke, the yellow tail flounder, the blackback flounder, the lemon sole, the plaice and other species are edible types. On the West Coast, the halibut, petrale sole, English sole, rex flounder, arrow tooth flounder or turbot, Dover sole, starry flounder, rock sole and other types are caught as edible fish. All the above are flounder; none are true sole.

In shape, flounder are flat, comparatively thin fish (see Fig. 18.13). The dorsal fin runs along the back from the head to the narrow part (caudal peduncle) near the tail. The anal fin runs from behind the head to the caudal peduncle. Both of these fins are somewhat wider at the midpoint. The ventral fins, located just below the head, are small, and the small pectoral fins are located just behind the head below the lateral line. The tail is either slightly lunate or moderately rounded, depending upon the species. Flounder are colored in various shades of black or brown on the upper side and various shades of white on the under side (the side that lies on the bottom).

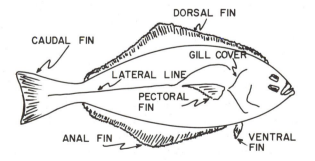

FIG. 18.13. CALIFORNIA HALIBUT *(PARALICHTHYS CALIFORNICUS)*

In the larval stage, the eyes of flounder are on either side of the head, as in other fish. Eventually, one eye migrates to the other side of the head, and the fish becomes reoriented so that the side having the eyes is uppermost. The fish swim or lie on the bottom with the eyed side uppermost. In some species, the left eye migrates to the right side, and in others, the right eye migrates to the left side.

Flatfish vary greatly in size. On the East Coast, halibut weighing up to 700 lb (318 kg) have been caught, but due to the depletion of this species, the average weight of those caught today is probably 30–50 lb (13.6–22.7 kg). Turbot and plaice average 5–10 and 10 lb (2.27–4.54 and 4.54 kg), respectively. Other flatfish are smaller, averaging 1/2 to 5 lb (0.23–2.3 kg). On the West Coast, halibut which weighed 470 lb (213 kg) have been caught, but the average is 20 to 60 lb (9 to 27 kg). Various other species may reach a maximum weight of 10–25 lb (4.5–11.3 kg), but average weights are much less.

On the East Coast, halibut are found from the Grand Banks (off Newfoundland) to the Gulf of St. Lawrence, and south to New York waters. The turbot is found on the Grand Banks and in Nova Scotian waters. Various other flounders are found from the Gulf of St. Lawrence to South Carolina. The yellowtail is taken only on Georges Bank. West Coast flounder are taken from waters which extend from California to North Alaska. The most important West Coast species, the halibut, is caught mostly in waters that extend from Northern British Columbia to Northwest Alaska.

Flounder are found in waters which vary in depth from less than 50 ft (15 m) to more than 1200 ft (366 m). Halibut, lemon sole and turbot are mostly found in deep water. The mature flounders feed on crabs, shrimp, worms, squid and other molluscs but halibut, turbot and dab are mostly fish eaters.

Small flounder are caught with otter trawls. Aboard fishing boats, they are held in pens or boxes in ice, as are cod, but these fish are not eviscerated prior to icing. On the West Coast, halibut are caught with long lines. The fish are eviscerated, the gills are removed, and they are placed in hold pens in ice much in the same manner as described for cod, but in this case, the "poke" (belly cavity) is also filled with ice.

Small flounders are usually sold as fresh or frozen fillets. The fish are filleted and skinned by hand. Fillets to be sold as fresh are packed in tins holding 10, 20 or 30 lb (4.5, 9.1 or 13.6 kg). The tins are surrounded with ice and shipped to retail outlets or to restaurants. When they are to be sold as frozen, the fillets are packed in retail-sized waxed paperboard cartons (with or without a plastic lining and waxed paper overwrap) and frozen between refrigerated plates. Such products should be held at 0°F (–17.8°C) or below at all times until sold for consumption.

Halibut are handled in the fresh or frozen state. As the fresh product the fish are beheaded, washed and packed in ice in wooden boxes. Then they are shipped from the West Coast to the Middle West or the East Coast in refrigerated trucks or railroad cars. If the fish are small, they may be sold by distributors to retailers as received. If the fish are large, they may be sold to retailers as portions. Frozen halibut are beheaded, washed and placed on racks in freezer rooms at 0°F (−17.8°C) or below. When frozen, the fish are glazed by dipping in water, and then stored at 0°F (−17.8°C) below until shipped to distributors in the frozen state. Small halibut, after freezing, may be sawed into steaks, trimmed and packaged in moisture/vapor-proof plastic film as 12, 14 or 16 oz (340, 39 or 454 g) portions.

OTHER FISH

Many species of fish have not been mentioned in this chapter. Some of these are: bluefish, butterfish, croaker, red and black drum, eels, groupers, mullet, ocean perch (fairly important fish of small size, caught with otter trawls, handled aboard boats in the iced uneviscerated state and processed as fresh or frozen fillets), pompano, rockfish, sablefish, sea trout, red and other snappers, spot, striped bass, swordfish and other marine species, as well as freshwater fish, such as buffalofish, carp, catfish, chubs, cisco, trout and whitefish.

Today, there is a considerable fish farming industry in the United States in which catfish and trout are grown in freshwater ponds. In many countries, carp and tilapia are grown in freshwater ponds and harvested as food for humans.

BIVALVE MOLLUSCS (CLASS PELYCOPODA)

There are other molluscs, besides bivalves, which are used in various countries as food for humans; squid is among them. In the United States bivalves, including clams, oysters (see Fig. 18.14) and scallops, are the chief types that are used for food. Bivalves have a calcareous outer shell which covers the living organism. The shell is lined on the inside with smooth enamel and varies in thickness and in roughness on the outside. The two halves of the shell are hinged and joined at a point on one side by a ligament which tends to force the shell open against the pull of the adductor muscle which is used to close it. The visceral mass (see Fig. 18.15), including the intestinal tract, liver or digestive gland, heart

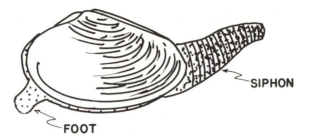

SIPHON

FOOT

SOFT SHELL CLAM *(Mya arenaria)*

AMERICAN OYSTER *(Crassostrea virginica)*

FIG. 18.14. TYPICAL BIVALVES (SHELLFISH)

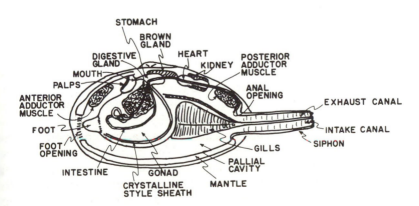

STOMACH

BROWN GLAND

DIGESTIVE GLAND

HEART

KIDNEY

POSTERIOR ADDUCTOR MUSCLE

MOUTH

PALPS

ANAL OPENING

ANTERIOR ADDUCTOR MUSCLE

EXHAUST CANAL

FOOT

INTAKE CANAL

FOOT OPENING

SIPHON

INTESTINE

GONAD

GILLS

PALLIAL CAVITY

CRYSTALLINE STYLE SHEATH

MANTLE

FIG. 18.15. INTERNAL STRUCTURE OF THE SOFT-SHELL CLAM *(MYA ARENARIA)*

reproductive and other organs, lies in an area from just beyond the center of the shell to near the hinge and is enclosed by the mantles, which are ribbon-like tissues extending along the margin of the shell from one end of the visceral mass to the other. The mantles which form the outer shell are lightly attached to the shell in areas away from the visceral mass. At the posterior end, the mantles are fused to form a projection called the neck which is small in some species, large in others. The neck has two tubes called syphons. Cilia (fine protoplasmic projections which move back and forth) lining these tubes cause water to be brought into the shell area through the inlet syphon, pumped over the visceral mass and forced to the outside through the outlet syphon. By this means, both food and oxygen are obtained for sustenance of life. The gill structure, which removes oxygen from the water to provide for respiration, is attached to the posterior end of the visceral mass and is composed of fine tubes lined with cilia. Bivalves cease to syphon water at a minimum temperature of about 40°F (4.4°C) and at a maximum temperature of 104°F (40°C). Besides their role as the breathing organ, the gills strain out microscopic plant life (mostly algae) which serve as food for these species. Another set of cilia moves the food material against the water currents within the shell and passes it along to flesh-like folds (the palps) which lie over the anterior visceral mass. Cilia in the palps move the food to the mouth parts. The palps also reject extraneous material, such as sand, allowing it to fall into the shell cavity. Feces from the intestine also fall into the shell cavity, and this and other extraneous materials are pumped out through the outlet syphon.

Some bivalves are either male or female throughout their life span, others change sex from male, in early stages, to female in later years. When spawning occurs, millions of eggs and sperm are shed into surrounding waters where fertilization takes place.

Oysters

There are five species of oysters in the United States, three on the East Coast, and two on the West Coast, one of which was introduced from Japan. About two weeks after the eggs hatch, the free swimming larvae attach and cement themselves to a hard surface (rock or shell) on the bottom. To provide for this attachment, oyster growers throw materials, such as the shells of quahogs (the clutch), into the water where spawning takes place. Some time after the set (attachment), the small bivalves may be removed to areas where tidal conditions provide a better supply of food. This also allows more room for growth.

Since oysters and other bivalves may be eaten raw or without sufficient ooking to destroy any disease-causing bacteria which might be present, nd since they are grown near the shore, often near highly populated reas, great care must be taken to make sure that bivalve growing areas re not polluted with even traces of human excrement. Control of bivalve arvesting areas is supervised by a division of the Food and Drug dministration but must be effected by state authorities. This control onsists of tests for disease-indicator bacteria on shellfish growing waters nd on shellfish meats, sanitary surveys to determine that traces of ewage are not reaching the growing areas, and the licensing of shellfish ealers who must record the areas from which the bivalves were taken, om whom they were purchased and to whom they were sold. Some ivalves may be taken from areas which do not meet the absolute pecifications for approved areas but which are not grossly polluted, rovided they are depurated in bacteriologically clean waters, either in he ocean or in tanks, under supervision by the state. The shellfish must e held in clean waters for periods long enough to allow them to purify hemselves or eliminate pathogenic bacteria by syphoning clean water.

Oysters are harvested with rakes, tongs, dredges, or with water-jet acuum dredges. Oysters and other bivalves, except scallops, are able to ve out of water at suitable temperatures for some time, since they can btain oxygen from that which is dissolved in the water retained within he shell in contact with the gills. Aboard boats, oysters and clams must e held under sanitary conditions away from the bilges. The boat used to arvest bivalves should be outfitted with a chemical toilet so that rowing areas will not be polluted with human discharges.

At the processing plant, oysters which are to be marketed in the shell re washed in sea water, which may be chlorinated, packed in sacks or arrels, cooled and shipped to restaurants. They should be held at mperatures between 32° and 40°F (0° and 4.4°C). Most oysters are nucked (meats removed from the shell) by hand with the aid of a knife. he meats are washed or agitated in fresh potable water by air blown into ne wash tank, graded for size and packed in glass or metal containers. he filled containers are cooled and shipped to market in crushed ice emperature of the product is about 33° to 34°F or 0.6° to 1.1°C).

Some shucked oysters in metal containers are frozen in moving air at 5°F (−20.6°C) and stored at 0°F (−17.8°C) or below until shipped to arket. Oysters may also be breaded, packed in waxed paperboard rtons holding 10 to 14 oz (284 to 397 g) of product, frozen between frigerated plates or in cold air, and stored at 0°F (−17.8°C) until nipped to market.

Oysters are eaten raw from the half shell, or in stews (lightly heated i milk with some butter) or breaded and deep-fat fried.

The Hard-shell Clam

The hard-shell clam is similar to the oyster in its internal structure The shell is rounded, symmetrical and relatively smooth on the outside coming to a gradual peak near the hinge. The shell is quite hard an thick. Once the larvae have developed into clams which are 1/8 to 1/4 ir (0.3 to 0.6 cm) in diameter, they burrow into the mud and remain jus below the surface of the ocean bottom. The hard-shell clam is found fro the Maine Coast to the Gulf of Mexico, but is most abundant off th Atlantic coast from Southern Massachusetts up to and including Vi ginia.

Hard-shell clams may be harvested by hand (feeling for them with th hands or feet and removing them by hand). They may also be remove from shallow water with clam rakes. The largest quantities of this clar are harvested with scratch rakes, with tongs similar to those used t remove oysters, or with dredges. Dredges are used in comparatively dee water and may be of the basket or water-jet type. Aboard boats, har shell clams should be handled in the manner described for oysters.

In preparing them for market, hard-shell clams are washed with se water, graded for size and cooled. Some clams may be taken from sem polluted waters, provided they are depurated (held in tanks of sea wate which has been chlorinated to destroy disease-causing bacteria, when th clams pass this water through their systems).

In the shell, hard-shell clams are marketed according to size. Th different sizes are: "chowders" (large size) which are used to prepar chowders, fritters or stuffed clams; "cherrystones" (medium size) whic are used for baking; and "littlenecks" (small size) which are used a steamed clams or for eating raw on the half shell. Hard-shell clams a neither canned nor frozen in significant quantities.

Hard-shell clams may be cultured. In such instances, the bottom first prepared by removing thick grass, stones and other debris. Pre ators, such as starfish, cockles, conchs and welks are removed by rakir and by towing floor mops over the bottom (to entangle starfish). Th young clams, raised in tanks, are then spread over the area by scatterir them from the side of a boat, using a shovel. Once seeded, the area is le undisturbed to permit the clams to grow to harvesting size. Predato may be removed from time to time. After seeding, the area may see itself naturally.

The Soft-shell Clam

The anatomy of the soft-shell clam is similar to that of the oyster. However, unlike hard-shell clams and oysters, the neck, containing the syphons, is quite large. The shell is oval in shape and is not thick. After the fertilized eggs have hatched, the larvae, free swimming for about 14 days, become young clams which attach themselves to some type of hard material on the bottom. At a size of about 3/4 in. (1.9 cm), they burrow into the mud to a depth of one to several inches (1 in. = 2.54 cm), projecting the neck to syphon water from which food and oxygen are obtained. The many predators of the soft-shell clam include the green crab, the horseshoe crab, gulls, seaducks, rays and fish. No attempt is made to control predators.

The soft-shell clam is found in the Western Atlantic as far north as the Arctic regions and as far south as Virginia, being most plentiful off the coasts of New England, New Jersey and Virginia.

In New England, soft-shell clams are harvested when the tide is low by digging into the mud with the short handled clam hoe and removing them by hand.

In the Chesapeake Bay area, clams are harvested from boats using water-jet dredges and an escalator. They are placed in bags or baskets and brought to the processing plant, where they are washed with sea water and sorted according to size. Specimens 3 in. (7.6 cm) in length or under are usually cooked by steaming. The larger sized clams are removed from the shell by hand and placed in metal containers, after which the containers are refrigerated by being surrounded with crushed ice. In this form, they are shipped to restaurants to be served as a breaded, deep fat-fried product.

Soft-shell clams may be removed from restricted areas and depurated in chlorinated sea water as described for the hard-shell variety. This must be done under supervision by state authorities.

Surf Clams

The surf clam or "skimmer" is large, reaching a length of 8 in. (20.3 cm). It is found just below the surface of sandy bottoms in waters 30 to 100 ft (9 to 30 m) deep off Atlantic Coast states from Massachusetts to and including Virginia. Most of the harvesting of this species is done off New Jersey with water-jet dredges having V-shaped scoops. Aboard the boats, the clams are placed in baskets or jute bags and brought to the processing plant without refrigeration.

Surf clams are used primarily for canning. The viscera are not utilize as food. At the canning plant, the clams are washed, then steamed light to cook the meat partially and cause the shell to open. The meat is the removed by hand, the nectar (liquid left in the shell) being saved. Th lower part of the neck (syphon), the mantle, the adductor muscle and th foot (muscular portion which allows the clam to anchor itself in the mud are then removed with scissors and diced into pieces about 3/8 in. (1 cm wide. The diced portions are then filled into cans together with some hot nectar and salt, after which the cans are sealed and heat processed. Fo 1/2 lb (227 g) flat cans, processing times of 40 min at 240°F (115.6°C) o 60 min at 228°F (108.9°C) are used. The large No. 10 cans of product ar heat processed for 100 min at 240°F (115.6°C).

A number of other species of clams are utilized as food. On the Wes Coast, the butter clam reaches a maximum length of 9 in. (23 cm) an the pismo clam reaches a maximum length of 4.5 in. (11.4 cm).

Some of these clams are preserved by canning and heat processing. I such instances, the meat is removed from the shell after steaming, th syphon is slit and cut off near the end with scissors, the soft visceral part are squeezed out and discarded, and the remaining portions are washe and ground. The ground meats are then filled into cans with some hot nectar, after which the cans are sealed and heat processed.

It should be noted that all clams and mussels, which feed mainly o algae, may at times become toxic to humans (shellfish poisoning). Thi happens when bivalves feed on certain algae (dinoflagellates) whic contain substances that are toxic to humans but not to molluscs. Publi health officials periodically test bivalves for toxin and close the shellfis beds when there is danger of shellfish poisoning outbreaks.

Scallops

There are several types of scallops, of which the sea scallop and the ba scallop are best known. The internal anatomy of the scallop is similar t that of the oyster, but the adductor muscle, the only part of the scallo which is eaten, is much larger than those of oysters and clams. Onc beyond the larval stage, the scallop may attach itself temporarily to som object, but the adult scallop is quite mobile. By closing the opened she with its adductor muscle, thus forcing water through two holes in the to shell, the scallop becomes jet propelled. These bivalves cannot be hel out of water in the live state, as can clams and oysters, since the wate drains from the shell which cannot be tightly closed.

The bay scallop is generally circular in shape with a grooved upper and lower shell and a rectangular projection at the back near the ligament (the bay scallop is the logo that can be seen in any Shell Gasoline sign). The bay scallop reaches several inches (1 in. = 2.54 cm) in diameter, and the adductor muscle may be as large as 1 in. (2.54 cm) in diameter. Bay scallops are harvested with basket rakes in shallow water and with dredges in deeper water.

The sea scallop is much larger than the bay type. It may reach a size of 8 in. (20.3 cm) in diameter, and the adductor muscle may be as large as 3 in. (7.6 cm) or more in diameter. Unlike the bay scallop, the shell of the sea scallop is not grooved. Sea scallops are found in ocean waters 60 (18.3 m) or more in depth. While this bivalve ranges from Labrador to New Jersey, it is most plentiful on Georges Bank off Cape Cod. Sea scallops are harvested with dredges. Aboard the boat, the "eyes" or adductor muscles are removed from the bivalves with the aid of a knife, placed in muslin bags and iced and brought this way to port. The remaining portions are discarded at sea.

Sea scallops are sold in the fresh or frozen form. If frozen for purposes of selling after defrosting, they are placed in freezer rooms in muslin bags and held until shipped to retailers. Some sea scallops are breaded and may be deep fat-fried prior to packaging and freezing. At 0°F (− 17.8°C), the storage life of scallops is longer than one year.

Other species of scallops now used include the calico scallop, found off the coast of Florida, and bay-type scallops, found off the coasts of Alaska and Australia and in the Irish sea.

CRUSTACEANS (CLASS DECAPODA)

Several types of crustaceans are used as food for humans, most of which are prized as delicacies. Included among these are shrimp (several species); lobsters (American, European and Norwegian species); crabs (several species); and crayfish (several marine species and the fresh-water species).

Crustaceans have a hardened external skeleton which is made up of a calcified polymer of glucosamine [a 6 carbon sugar containing an amine (NH_2) group] called chitin.

The external anatomy of crustaceans consists of the mouth parts, the eyes, the antennae, varying greatly in size, the body or cephalothorax to which five pairs of legs are attached, and the abdomen or tail which

consists of a number of jointed segments adjoined to the body. In som
crustaceans, the first pair of legs is chelate or enlarged and develope
into biting and crushing appendages called "claws." The end section c
the tail has several parts, including the fan-shaped telson. In som
species, the tail may be contracted or flexed to provide for movement i
the water. On the under side of the tail there are a number c
attachments called pleopods or "swimmerets," which for certain specie
are the main appendages providing for movement in the water. Som
crustaceans will shed an injured claw, then generate and grow a new one
Crustaceans grow by shedding the old shell (moulting) to become soft
shelled for a short period (the new larger shell soon becomes hard) an
filling up the new larger shell, which allows more room for growth
moulting occurs most frequently in the early years of growth. Matin
takes place when the female is in the soft-shell stage. The fertilized egg
are attached to the swimmerets, are eventually hatched, and afte
several larvae stages, the small crustaceans sink to the bottom an
assume the general habits of the adult.

The internal anatomy of crustaceans includes the gills, which lie ove
folds of body tissue adjacent to the inner shell and over blood vessel
which connect the gills to the heart and the circulatory system. Th
mouth is anterior, inside the body, and is connected to a grindin
apparatus (similar to the gizzard in fowl) called the proventriculus. Thi
is connected to the digestive gland or liver (the tomalley), and th
digestive gland is connected to a straight gut, which runs through the ta
to the anus, located near the telson.

Lobsters

Lobsters (see Fig. 18.16) have either a well developed first pair c
walking legs or biting claws. The European lobster is found aroun
certain parts of the British Isles and mainland Europe. The Norwegia
lobster is found mainly around the coast of Norway and the west coas
of Sweden. The American lobster ranges from Labrador to the coast c
North Carolina in an area which extends seaward for a distance of 50 m
(80 km). However, there are deep sea lobsters which are found more tha
200 mi (322 km) from the coast. The depth of water where lobsters ar
taken is usually 30 to 150 ft (9 to 46 m), but deep sea lobsters live a
depths up to 1200 ft (366 m). The American lobster is most abundant o
the coasts of Maine and the maritime provinces of Canada. The averag
American lobster caught measures 9–10 in. (23–25 cm), weighs 1 to 2 l

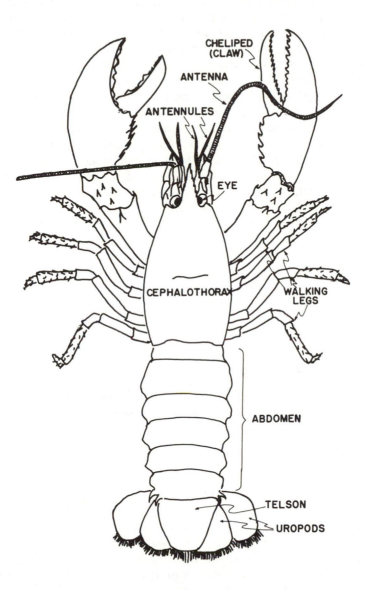

FIG. 18.16. AMERICAN LOBSTER *(HOMARUS AMERICANUS)*

(454 to 908 g) and is 4 to 7 years old. However, deep sea lobsters are larger, and specimens weighing more than 40 lb (18 kg) have been caught which are believed to be more than 50 years of age. The food of lobster includes fish, clams and other molluscs.

Lobsters are caught in pots (see Fig. 18.7) and are held in the live state aboard boats without refrigeration, since they are brought to port shortly after harvesting. Lobsters may be held in the live state, out of water at low temperatures above freezing, for more than one week, given sufficient air space, since they are able to obtain oxygen from what dissolves in the water on their gills (the gills must be kept moist). They may also be held in the live state for a month or more in ocean pounds which allow the free flow of water, or in tanks in which sea water is filtered, aerated and circulated. When held in tanks, the biting claw may be immobilized by the insertion of a wooden plug into the flesh above the thumb or by an elastic band encircling the thumb and claw.

Lobsters are sold mostly to restaurants or to the consumer, in the live state. Lobsters should be cooked from the live state or killed and cooked immediately. The reason for this is that lobsters have a very active proteolytic enzyme system which soon digests part of the tissue of the dead lobster, partially liquefying the meat or causing it to become soft and crumbly (a condition known as "short meated"). Some cooked lobster meat is sold as the canned or frozen product but does not make up a significant part of the catch. The storage life of frozen lobster meat at 0°F (−17.8°C) is at least 8 months. Whole lobsters in the raw, cooked or partially cooked state cannot be frozen successfully since when frozen in the cooked state, the tomalley becomes rancid and affects the flavor of the meat, and, when frozen in the raw or partially cooked state, the flesh undergoes proteolysis.

Shrimp

There are numerous species of shrimp (see Fig. 18.17) which are used as food for humans. The edible types vary in size from very small, about 2 in. (approximately 5 cm), to more than 10 in. (25 cm). The large shrimp are called prawns. The overall size of Gulf shrimp as caught is to 8 in. (17 to 20 cm). Most shrimp caught by United States fishermen are taken from the Gulf of Mexico, and these consist of three main types: white, brown and pink. Some shrimp are taken from Atlantic waters of the Carolinas, Georgia and Florida, and some are taken off Alaska, Maine and Massachusetts. Shrimp are imported from Mexico, India,

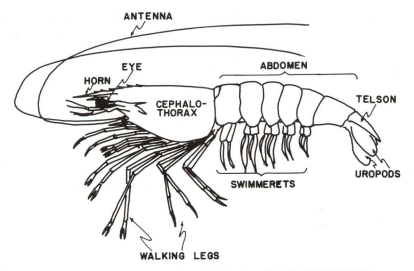

FIG. 18.17. SHRIMP (*PENAEUS, PANDALUS,* AND *XIPHOPENAEUS*)

'anama, Venezuela, Brazil, Guiana, Ecuador, Nicaragua, Colombia, El alvador, Honduras, Thailand, Surinam, Malaysia and other countries.

The shell color of shrimp is brown, green, pink or gray, but, like all rustaceans, when cooked, the shell and surface of the flesh is red in olor. The first pair of walking legs is not enlarged to form biting claws.

In the Gulf of Mexico, young shrimp seek the brackish waters of bays nd estuaries along the coast, but, as they grow, they move to ocean aters of higher salinity. Shrimp spend most of their time on or near the ottom and may bury themselves in the mud. They swim forward by moving the swimmerets but may move rapidly backwards by flexing the ail. It is believed that adult shrimp eat worms, small crustaceans, small olluscs and plant debris.

Shrimp are caught with otter trawls that are somewhat modified from hose used to catch cod and haddock. In some instances, two boats may e used to tow the trawl which is attached to an outrigger. When inshore shing is practiced, the whole shrimp may be brought to port in the iced r uniced condition. In most cases, however, and especially on boats hich may fish for one week or more before landing the catch, the shrimp re dumped from the trawl onto the deck where the head (cephalothorax) removed by hand from the tail, and discarded. The tails are then ashed with sea water and placed in hold pens or boxes, more or less

surrounded with ice. Some boats may now be holding shrimp tails i
tanks of sea water refrigerated with crushed ice. Some fishermen dip th
shrimp in a 1 to 1 1/4% solution of sodium bisulfite prior to icing t
prevent the formation of "black spot," a melanin compound formed b
enzymatic oxidation of the amino acid tyrosine.

Few shrimp reach the consumer in the fresh state. Usually, shrimp ar
frozen prior to distribution, although they are oftentimes defrosted at th
retail outlet and sold to the consumer this way. Shrimp may be frozen i
the shell, in which case they are weighed into cartons and frozen betwee
refrigerated plates or in cold air. This type of product may be glazed afte
freezing by opening the package, spraying the product with water
closing the package and returning it to the freezer storage room. Larg
quantities of shrimp are peeled (shell removed) and deveined (gu
removed) before freezing. This may be done by hand, the vein bein
removed with the aid of a knife, although most processing is done b
machine. The shrimp are then washed in running water. Following th
process, some shrimp may be dipped in a solution of sodiun
tripolyphosphate prior to freezing. This is done to prevent textur
changes and loss of water during frozen storage. Such treatment als
causes the product to pick up water and, if excessive concentrations o
phosphate are used, the texture of the product is affected so that eve
after cooking, the shrimp have an undesirable texture and uncooke
appearance. Peeled and deveined shrimp may be frozen individually i
liquid freon or liquid nitrogen or placed on trays and frozen in cold air
Before packaging, the frozen shrimp are usually glazed by passing then
on a belt through a container of water. When prepared in this manner
the shrimp are usually packaged in plastic bags with or without a
outside carton. Some shrimp are peeled and deveined, with or withou
splitting (butterfly form), then battered and breaded prior to freezing
The telson (end part of the shell) may be left on in some forms of breade
shrimp. In the breaded condition, they are packaged in waxed paper
board cartons with strips of waxed paper separating the various layers. I
such instances, the shrimp are usually frozen between refrigerated
plates. Some shrimp are cooked in boiling water or weak brine or breade
and cooked in hot oil at 375°F (190.6°C) prior to freezing.

Raw shrimp in the shell, when protected against dehydration, have a
high quality storage life of at least two years at 0°F (−17.8°C) or below
Cooked shrimp, especially those cooked in hot oil, have a storage life o
3 to 5 months at 0°F (−17.8°C). Uncooked shrimp, prepared and froze
in the butterfly form, are also subject to storage changes, since there is s

uch space in the package which is not occupied by the product that
ehydration occurs through a continuing two-step process: 1) moisture
om the product vaporizes and fills the voids and 2) moisture from the
oids condenses on the inner surface of those parts of the package that
re adjacent to the voids.

Frozen shrimp imported from other countries must be defrosted before
ey can be processed. This is usually done by tempering the product at
bout 40°F (4.4°C) for 24 hr, then completing defrosting by holding the
npackaged shrimp in running water. A more sanitary defrosting method
mploying microwave heating is now available and, in some cases,
lready in use.

Considerable quantities of shrimp are canned. For this purpose, they
ay be delivered to the cannery with the heads on. The shrimp are first
ashed and separated from the ice. The tails are then removed from the
eads, usually by machine. The shell is then removed and the vein taken
it by machine. Individual specimens are then inspected, broken and
ecomposed shrimp being discarded. The product is then blanched or
eated in boiling saturated salt solution (25%) for a period of 3/4 to 3
in. After blanching, they are graded for size and filled by machine into
ne of several can sizes. Hot, dilute salt solution is added to the product
the cans, and the cans are sealed immediately. Heat processing to
:ovide commercial sterility is carried out at 250°F (121.1°C) for various
mes depending upon the size of the container.

In the Pacific Northwest and Alaska, very small shrimp may be
anned without deveining. Shorter blanching times are used for this
roduct and small amounts of citric acid are added to the brine used to
ver the shrimp. The brine is added cold, and the cans are vacuum-
aled prior to the heat processing.

rabs

Crabs have the general anatomy of other crustaceans, but the body is
al-shaped or disc-like instead of cylindrical, as in lobsters, shrimp and
awfish. Also, the abdomen (tail) is comparatively small, flattened and
:rmanently flexed under the body. Several species of crab are used as
od by humans.

The Blue Crab.—The blue crab (see Fig. 18.18) is found from Nova
:otia to Mexico, including the Gulf of Mexico, and is especially
undant in the Chesapeake Bay region. It is commercially important

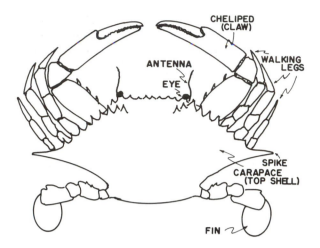

FIG. 18.18. BLUE CRAB *(CALLINECTES SAPIDUS)*

only south of New Jersey. The semi-oval body of the blue crab has spike
peaks near the back end. The first walking legs are well developed in
biting claws and the last pair of legs (called back fins) is flattened an
used to propel the crab in the water. When fully grown, these crustacear
may measure 7 in. (18 cm) or more across the body. The under-she
surfaces are white and the back shell is dark or brownish green. The to
of the biting claws have varying amounts of blue. Blue crabs live i
shallow waters near bays or river estuaries. They eat a variety of livir
and dead plant or animal tissues. They are caught with crab pots
traps, or with trot lines.

Blue crabs are brought to the processing plant in the live state. The
are then cooked in live steam or boiling sea water, or in steam at 240°
(115.6°C) for 10 min, or in steam at 250°F (121.1°C) for 8 to 10 min. Aft
cooling, the back shell, viscera, claws and legs are removed; then th
meat may be removed from the shell with the aid of a small, sharp knif
In hand picking, the body meat adjoining the back fin is separated fro
the finer body meat, since it is considered to be of better quality and
higher value. Certain machines are now available for removing crab me
from the shell. This may be done by impact or a roller process may l
applied to the cooked or partially cooked debacked body, legs and claw
Somewhat better yields of meat are obtained by machine picking an
much less labor is required to do the job, but machine picking does n
provide for the separation of back fin lump meat from the other bod
meat, unless it is done by hand before the crabs are machine processe

After the meat has been picked, it is packed in metal cans which are then closed and heated in boiling water until an internal temperature (at the center) of 185°F (85°C) is reached. This temperature is held for one minute. The product is then cooled and held at 33°–38°F (0.6°–3.3°C) prior to distribution.

Blue crab meat is neither frozen nor heat processed to provide commercial sterility, since either treatment results in a product of poor quality.

If blue crabs, when caught, are nearing the molting stage, they may be held in seawater pounds, until they shed their shells. They are then soft-shell crabs, and they are sold in the live state at a premium price, since soft-shell crabs are considered to be a delicacy. High mortality rates are usually encountered during the holding of crabs for molting.

The Dungeness Crab.—The Dungeness crab is found from the Alaskan peninsula to Southern California but is most abundant in the area between San Francisco and Southeast Alaska. It may attain a size of 9 in. (23 cm) across the back. This crab is reddish-brown on the back with lighter streaks and spots. The underside is whitish to light orange. The biting claws are well developed, but the last pair of legs is not modified to form swimming appendages and is smaller than the other legs.

Dungeness crabs are caught in water which is 12 to 120 ft (3.7 to 36.6 m) deep. The circular pot is used in deep water, the rectangular pot in shallow water. Ring nets may also be used. Dungeness crabs are brought to port in the live state aboard the boat, held in wells of sea water, and may thereafter be held in tanks of sea water until sold to restaurants in the live state.

In processing, to obtain meat from Dungeness crabs, the back shell of the live crab is first removed, the viscera and gills are then torn away, and the body is broken in half with the legs attached. The sections are then cooked in boiling sea water for 10 to 12 min, then the meat is removed by hand by shaking or by impacting against a metal container, or it may be removed by running the body and legs between mechanized rollers. Pieces of shell may then be separated by floating the meat in a salt solution of the appropriate specific gravity for the meat to float and the shell to sink. Fresh meat is packed in cans, the cans are sealed and the product is held at 32°–40°F (0°–4.4°C) for purposes of distribution.

Some whole or eviscerated Dungeness crab is frozen in brine at 5° to 0°F (−15° to −17.8°C), packaged or glazed, and stored at 0°F (−17.8°C) for distribution in the frozen state. A larger amount of this type of crab meat is packed in hermetically sealed cans and frozen in moving air at 0° to −10°F (−17.8° to −23°C). This type of crab meat is not especially stable in frozen storage but may be held for as long as 6 months at $-10°$F ($-23°C$) with fairly good results.

Dungeness crab meat is also canned and heat processed. The meat is packed in cans holding 6.5 oz (185 g) of product. A weak solution of salt and citric acid (pH 6.6 to 6.8) is then added, covering the meat to prevent discoloration. The cans are then sealed and heat processed at 240°F (115.6°C) for 60 min, then cooled in the retort. The quality of this product is inferior to that of the fresh meat.

The King Crab.—This species is not a true crab but is similar in structure and habits to true crabs. It has a heavy rough shell and a tail of relatively small size which is not permanently flexed under the body. The last pair of walking legs is small and inserted under the body. The first pair of walking legs is modified into biting claws which are not large considering the size of this species. The King crab is very large, having an overall spread, including the legs, of up to 5 ft (1.5 m), and weighing as much as 24 lb (10.9 kg). The walking legs are very large.

King crabs are caught off central Alaska to the Aleutian Islands and of the islands of northern Japan. They are harvested with large rectangular pots. Aboard boats, the crabs are held in the live state in wells of circulating sea water.

King crab meat is either canned or frozen. In canning, the whole crab is cooked in boiling water, after which the meat is squeezed out between rubber rollers, but legless bodies may be given a 10 min cook at lower temperatures (160°–165°F or 71.1°–73.9°C) than the legs prior to removing the meat. The meat is washed and packed in cans which hold 7 1/2 oz (212.6 g) of product and are either lined with parchment paper or lined at the ends only. A small amount of weak brine at pH 6.5 is added. The cans are then sealed and heat processed at 240°F (115.6°C) for 50 min, then cooled in the retort.

King crab meat is frozen in large blocks for the restaurant trade. The blocks comprise 250 oz (7.1 kg) of meat and up to 24 oz (680 g) of water —added to fill voids—and packed in plastic-lined, cardboard cartons. The blocks, when frozen, are glazed, and may be sawed into 1 or 2.5 lb (454 or 1135 g) portions and repackaged for the restaurant trade. Cartons of product holding 6 oz (170 g) are packed for the retail trade. Cooked intact, crab legs and claws are also frozen, glazed and packaged for restaurants and retail outlets.

If properly packaged, and frozen to a temperature of 0°F ($-17.8°C$) or below and held at this temperature, King crab meat has a high-quality storage life of at least 12 months. Lower storage temperatures provide for an even longer storage life.

The Snow or Tanner Crab.—This species is relatively large, reaching size of 5–6 in. (12.7–15.2 cm) across the back and 2.5 ft (76.2 cm) between the tips of the outstretched legs. The snow crab is taken in deep water off central and western Alaska and in the Bering Sea and some is taken off Nova Scotia and Newfoundland. It is caught in large, baited pots, as is the King crab.

Snow crabs are handled and processed in much the same manner as re King crabs, but most of the meat is canned and heat processed. The meat of the snow crab is inferior to that of the King crab.

Red Crab.—There is now a new crab industry—the red crab industry. The red crab is found from Nova Scotia to South America but is taken almost entirely in deep waters off southern New England. Red crab meat is removed mechanically from the debacked, partially cooked specimens with machinery employing the roller process. Red crab meat is sold mostly as the fresh refrigerated product but some is sold as frozen.

Jonah Crab.—The Jonah crab is found in waters from Nova Scotia to North Carolina and is caught in lobster pots. The meat of this crab is difficult to remove from the shell, and the product is sold mostly as cooked-refrigerated or frozen whole crabs or claws. This crab, like the rock crab, is not taken in industrial quantities at this writing, largely because of a lack of effective machinery to pick the meat.

Marine Crayfish.—The crayfish or spiny lobster has become a popular food in the United States. There are a number of different species of marine crayfish ranging from Florida and the Gulf of Mexico to Central and South America. They are also found off Australia, New Zealand, South Africa and other countries. These species have the general anatomies of lobsters, but the first pair of walking legs is not developed into biting claws.

Since only the tail portion is eaten, this is removed from the live specimen, packaged, with shell on, in moisture vapor-proof material, and frozen, for sale to restaurants or for the retail trade.

Freshwater Crayfish.—Freshwater crayfish are grown in ponds. Although they have the general anatomy of the true lobster, with well developed biting claws, they are much smaller, the maximum weight being about 8 oz (227 g). There is presently a small industry in which the small crayfish are placed in rice fields after the rice has been harvested. Here, they eat the rice roots and also serve to fertilize the fields. By planting time, the fields can be drained and the crayfish harvested. Generally, these specimens are handled in the fresh, refrigerated state and are only processed by cooking.

19

Cereal Grains

Of all the plants on which man has depended for his food, those that produce the cereal grains are by far the most important, as they have been since earliest recorded time. Cereal grains are the seeds of cultivated grasses that include wheat, corn, oats, barley, rye, rice, sorghum and millet. There are a number of reasons why cereals have been so important in man's diet. They can be grown in a variety of areas, some even in adverse soil and climatic conditions. They give high yields per acre (0.4 ha) as compared to most other crops, and, once harvested, their excellent storage stability combined with their high nutritional value makes them the most desirable of foods for holding in reserve. They are easy to package and transport and they can be used to produce a large variety of highly desirable foods both for humans and animals, as well as beverages for human consumption.

Cereal grains are the most important source of the world's total food. Rice alone is reported to supply the major part of the diet for more than 1/2 of the world's population. Cereal grains are the staple food of the peoples of developing countries, providing them with about 75% of their total caloric intake and about 67% of their total protein intake. The grains are eaten in many ways, sometimes as a paste or other preparation of the seeds, more often milled and further processed into flour, starch, oil, bran, syrup, sugar, dried breakfast forms, etc. They are also used to feed the animals that provide us with meat, eggs, milk, butter, cheese and a host of other foods.

All cereal grains comprise three parts—the bran (a layered protective outer coat), the germ (the embryonic part of the plant) and the endosperm (the large starchy part, containing some protein). Except for two amino acids, lysine and tryptophan, most cereals contain th essential amino acids required by man, as well as well as vitamins an minerals. When they are consumed with other foods that can suppleme the nutritional elements that are low in cereals, the minimum dieta requirements may be met or nearly met. Research in cereal genetics m; be expected to produce hybrid cereals that will be complete or near complete foods, containing more of the nutritional elements required t man. Triticale, a hybrid of wheat and rye, first produced in the la 1800s, combines the high total protein content of wheat with the hig lysine content of rye. It is also more adaptable to unfavorable grow conditions and seems to resist wheat rust (a disease caused by molds The improvement of this hybrid is continuing and can lead to mo; beneficial genetic changes. This cereal is now being grown on more tha 1,000,000 acres (404,000 ha) in 52 different countries. A composi proximate analysis of cereal grains indicates that they have a prote: content of about 11%, fat about 3%, moisture about 12%, carbohydra about 68% and fiber about 6%.

On a world basis, rice is the most important cereal, being produced f human food in the largest amount, while in the United States, corn produced in the largest amount, although it is used for animal food ar other products, as well as for human food. The grain grown in the large quantity for human food use in the United States is wheat.

For most food uses of cereals, the bran and the germ are removed; tl bran, because it is indigestible by man and because of its adverse effe on the appearance and on some functional properties of flour, and tl germ, because of its high oil content which may subsequently becon rancid. The germ is used to produce oil (e.g., corn oil). The bran go mainly to feed for animals.

The first ready-to-eat cereals were produced just before the turn of t century, with flaked and puffed cereals following within a decac Ready-to-eat cereals, made from the endosperm of wheat, corn, rice ar oats, are convenient, nutritious (despite adverse reports to the contrary and they come in a very large variety of forms, textures and tastes. T processing of cereals into breakfast commodities was started in t United States and is still largely carried out in the United States, wi considerable quantities being exported throughout the world. The mc popular of the breakfast cereals are those which are ready-to-eat. The are formed or puffed and oven baked.

In the United States, grain is usually sold by the growers to operate of storage elevators, near the farms, where the grain is cleaned a: stored. It is then sold directly to processors or to operators of stora elevators (near processors). They may sell directly to the processors or their brokers.

A brief description of the handling, processing and use of the more important cereal grains in the United States is found in the following paragraphs.

WHEAT

Whole wheat, consisting of about 13% protein, can contribute considerably to the diet. The flour, made from the whole wheat, is higher in biological value than white flour (made from the endosperm only). Table 19.1 gives some examples of the higher nutritional value of whole wheat flour over white flour.

Wheat is perhaps the most popular cereal grain for the production of bread, and cakes and other pastries. Wheat produces a white flour. In addition, the unique properties of wheat protein alone can produce bread doughs of the strength and elasticity required to produce low density bread and pastries of desirable texture and flavor.

There are many varieties of wheat. They may be classified as hard red winter wheats; hard spring wheats; soft red winter wheats; white wheats and durum wheats. Winter wheats are planted in the fall and harvested in the late spring or early summer. Spring wheats are planted in the spring and harvested in the late summer. Hard wheats are higher in protein content and produce more elastic doughs than soft wheats. Therefore, hard wheats are used for breads, and soft wheats are used for cakes. Durum wheats are most used for alimentary pastes (spaghetti, macaroni, etc.) and for the thickening of canned soups.

Wheat is harvested by combines which cut the stalk, remove and collect the seed and either return the straw to the soil, to be plowed under with the stubble and thus provide humus, or compress and bale it for future use as litter, ensilage, etc.

Wheat may be bagged in jute sacks and stored in warehouses, or it may be stored in bulk in elevators. The latter method provides the best

TABLE 19.1. A COMPARISON OF SOME NUTRIENTS IN WHOLE WHEAT AND WHITE FLOURS

Nutrient	Whole Wheat Flour[1]	White Flour[1]
Protein	13%	11%
Thiamin	2.3 mg/lb	0.3 mg/lb
Riboflavin	0.6 mg/lb	0.2 mg/lb
Niacin	26.0 mg/lb	3.5 mg/lb
Pyridoxine	2.0 mg/lb	1.0 mg/lb

[1] 1 lb = 454 g

protection against rodent and insect infestation. The moisture content of bulk-stored wheat should not be higher than 14.5% and that of sack stored wheat not higher than 16%. Otherwise, microorganisms may grow and cause heating and spoilage. When it is necessary to lower the moisture content of wheat, it may be dried in bins by blowing hot air (not higher in temperature than 175°F (79.4°C) across the bins.

In preparing wheat for milling (see Fig. 19.1), it is blown into hopper scales which record the quantity of uncleaned wheat. Some of the coarser impurities are removed by this process. The grain then passes over a series of coarse and fine sieves which further remove contaminating materials, including chaff and straw. Still in the dry state, stones may be removed by passing wheat over short openings which allow the heavier stones to fall out of the mass and be trapped. The wheat is next passed

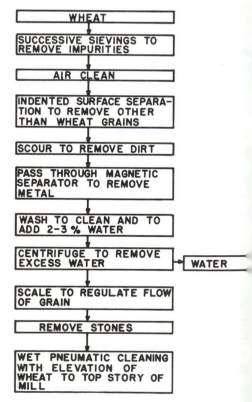

FIG. 19.1. PREPARATION OF
WHEAT FOR MILLING

ver discs or cylinders containing indented surfaces which remove seeds
horter or longer than wheat, following a pass through a magnetic
eparator to remove any metals which may be present. The next cleaning
rocess is dry scouring to remove adhering dirt. The wheat is then
ashed in water, a process which both removes dirt and adds 2–3% water
o the grain. The added water is necessary to provide desirable conditions
or milling. A stone trap is included in the washer. Excess water is
emoved by centrifugation (rotating at high speed). A second wet
leaning with a light brushing action is ordinarily used, followed by
spiration (blowing air through the grain), which is the final cleaning
peration. The grain is then carried into a bin from which it is fed to the
illing operation. This bin is located on the top floor of the flour mill, the
rain having been elevated to this position during the various cleaning
perations.

In milling (see Fig. 19.2), grain is fed automatically through scale
oppers which regulate the flow of the seeds at rates corresponding to
hose of the following operations: Milling may be carried out by passing
he grain through a series of corrugated rolls rotating towards each other,
hich remove chunks of the endosperm from the bran. After each
assage through the break rolls, the material is sifted through cloth, or
ire sieves, and separated according to particle size. The various streams

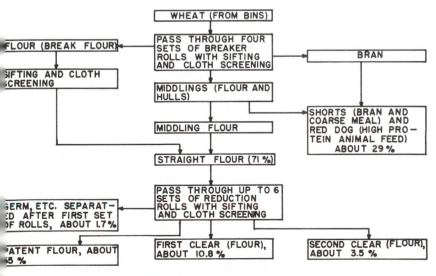

IG. 19.2. MILLING OF WHEAT

of different size flour particles are finally blended to provide the differer grades of flour. The more finely ground flour is nearer to white but le: nutritious than the coarser ground flour. This results from the mo: effective removal of bran and germ from finely ground flour. Impa(milling is now used in some operations. With this method, the seed broken open by impact in a machine called an Entoleter, first develope to control insect infestation. Flour particles of different sizes are sepa rated by air classification or by centrifugation.

High-protein flour is desirable for some types of baked products, flot of moderate protein content for others, and high-starch low-protei content flour is desirable for still other baked goods. The smaller flo\ particles are higher in proteins, the larger flour particles are higher i starch. Through air classification in a turbomill it is possible to separat flour particles into various sizes which can be blended to provid whatever protein or starch content is required by the baker or other use: of flour. Turbomilling, developed in the late 1950's, is considered to be significant milling innovation, because only through this process is th variety of flour blends for different products possible.

In the United States, wheat flour is enriched with the mineral, iron (ε a salt). Enrichment with calcium salts is optional for some types of flou but mandatory for enriched, self-rising flour. Wheat flour is also enriche by the addition of small amounts of the vitamins thiamin, riboflavin an niacin. The levels at which these materials must be present in enriche flour was increased in October 1973.

Wheat flour is used to make leavened products, such as bread, cake: pastries and doughnuts, and unleavened products such as alimentar pastes (macaroni, spaghetti, noodles, etc). Cake mixes are also prepare with flour, and flour is used for thickening canned and homemade stew: soups, gravies and white sauces.

Various breakfast cereal products are made from wheat. Generally, i these products, the wheat is precooked and passed through heated rol\ to form flakes. It may also be shredded, or it may be heated to above th boiling point of water under pressure, with puffed wheat formed whe the pressure is released. Wheat bran may also be produced as flake: High-protein cereals may be produced from wheat together with adde wheat starch, sugar, malt, minerals, such as phosphates, vitamins an other ingredients. Some wheat flakes are coated with very thin layers (sugar. The variety of breakfast cereals produced from wheat and othe cereals seems almost endless, and the rate at which new types are bein developed and marketed seems to be increasing rather than decreasing

CORN

Many types of corn are grown in the United States. Sweet corn is produced as a vegetable and eaten fresh, canned or frozen. Popcorn is also used as a food. However, the type of corn most utilized in the United States and considered as a grain rather than a vegetable is field corn. There are a number of varieties of corn usually classified as starchy or waxy, depending upon the characteristics of the carbohydrate present. The development of hybrid strains has improved the yields of field corn, which is lower in protein than wheat, and, like all vegetable proteins, including wheat, corn is deficient in some amino acids and so does not provide a complete protein for humans. Corn is especially deficient in the amino acid, lysine, but a new variety of high-lysine corn has recently been developed which may eventually have a great impact on human nutrition in some parts of the world.

Ears of field corn are harvested by a machine which strips the matured ears from the stalks. If harvested in wet weather, corn may have to be dried before it is stored. Usually, it is allowed to dry on the stalk in the field, is harvested, and stored in small roofed bins or silos with metal or wire mesh walls. Much of the corn storage is done on the farm, since most of the corn crop is used as feed for animals. Stalks and leaves may be harvested, chopped, and placed in piles or in silos to form ensilage for animal feed. Stalks and leaves may also be chopped and returned to the soil for humus.

Corn milled for flour (corn meal) is cleaned, as is wheat, then moistened to a water content of 21%. The germ is removed mechanically. The endosperm is then dried to a moisture content of 15%, passed through crushing rolls and sifted to remove the bran. With the use of sieves, milled corn is separated into grits (largest sized particles) and meals and flours (smallest sized particles).

Most of the corn crop is used for animal feed, but considerable amounts are used to produce cornstarch, corn syrup and the various sugar derivatives. In producing cornstarch, the corn is cleaned, then placed in vats, where it is steeped in warm water (slightly acidified with sulfur dioxide to prevent fermentation) for about 40 hr. If not previously degerminated, the steeped kernels are passed through mills which separate the germ and loosen the hull. The mass is then passed through tanks of water where the germs (being lighter) float and are skimmed off. The remaining endosperm, containing starch, corn gluten and hulls, is then finely ground in steel mills. The finely ground material is then passed through shakers (nylon sifters) to remove hulls, the starch and gluten passing through. Starch may then be removed from the gluten by flowing the starch-gluten slurry over long, narrow, flat-bottomed, shallow troughs (called starch tables) which are sloped slightly towards the discharge end. Starch particles are heavier than gluten particles and sink to the bottom while the gluten flows over the end of the table. Eventually, the starch is flushed from the bottom of the table, then washed and dried. The method by which starch is usually separated from gluten today is centrifugation. The starch particles, being heavier, are separated at the outer region of the centrifuge, the lighter gluten migrating to the center. Starch may be produced from potatoes, rice, tapioca or wheat by methods similar to that described for corn, except that with potatoes and tapioca, it is not necessary to degerm the product.

Starch may be modified chemically to provide properties which are suitable for various manufactured products. Starch is used as a filler in pies. It is used in biscuits and crackers and as a filler or carrier in baking powder. Starch is used as a thickener in canned cream-style corn, in canned soups and in canned baby foods. Starch is also used in many types of candies and as an ingredient of the white sauce contained in some frozen foods. Some desserts, such as instant puddings, are largely composed of modified starches, and starch is used for sizing several types of cloth and textiles, and in leather, adhesives, pharmaceuticals, paper and tobacco. Various types of sweeteners are made from cornstarch, since starch consists of a long straight or blanched chain of glucose molecules that may be broken down to short chains of glucose molecules (dextrins), to maltose (two molecules of glucose) or to glucose (dextrose). Corn syrup is produced by heating starch in water acidified with hydrochloric acid. The hydrolysis, in this case, is only partially completed so that the mixture contains some glucose, some maltose and some longer chains of glucose. The hydrolysis may be carried out by first heating with acid followed by treating with an enzyme which hydrolyzes starch. The latter method produces a syrup which is higher in maltose than does the straight acid hydrolysis method. After hydrolysis, the syrup is neutralized by adding sodium carbonate, filtered and concentrated to 60% solids, again filtered through bone charcoal, then passed through resins (ion exchange) which take out the salt (sodium chloride formed from the acid and sodium carbonate). The corn syrup may be spray- or drum-dried to about 3% moisture to obtain corn syrup solids, or a more completely hydrolyzed syrup after purification may be concentrated, seeded with fine corn sugar crystals and crystallized to produce crude corn sugar. Dextrose or corn sugar can be produced in a similar manner from a completely hydrolyzed starch. This product is centrifuged, washed and dried, the liquor from the centrifuge (first liquor and washings) being concentrated and recrystallized. Corn syrups and corn sugar are used in bakery products, pharmaceuticals, carbonated beverages, confectioneries, ice cream, jams and jellies, meat products, and dessert powders. The cruder, less refined products are used in tanning, for brewing, to produce vinegar, to produce caramel coloring and in tobacco.

Corn is also used to produce popcorn. The variety used is a specific one. When the dried kernels are heated, internal moisture creates a vapor pressure due to the rise in temperature, and when the pressure is sufficient, the hard outer shell is burst and the pressurized grain is expanded. Essentially, popcorn is a puffed cereal.

OATS

Oats, one of the popular nutritious present-day cereals, was once regarded as useful for feeding only cattle. Oats can grow in colder and wetter climates than can wheat. Oats are harvested much in the same manner as is wheat. The moisture content at the time of harvest should not be higher than 13%.

Milling of oat kernels requires that they first be washed and cleaned and then dried in a rotary kiln or pan drier to a moisture content of about 12%. They are then hulled by impact, the seeds being thrown from a rotating disc against a rubber ring which splits off the hull and leaves the groat mostly intact. After the hulls are removed by passing the product through sieves, the groats are steam heated and passed between rollers to produce rolled oats, or, they are cut into pieces about 1/3 of the original size and then steamed and rolled to produce quick-cooking rolled oats. Small amounts of oatmeal may be produced by grinding the steamed groats. Steaming facilitates cooking and inactivates enzymes which, if not inactivated, may cause bitter flavors to develop. Oat flour may be produced for use as an ingredient of bread or a thickener for soups. If made from unheated groats, there may be a problem with the development of rancidity.

BARLEY

Barley products do not bake as well as wheat products, thus barley, containing little or no gluten, is not as popular as wheat when there is an option. However, barley has the advantage of growing in climates too cold and in soils too poor to grow wheat, and, in addition to being a hardy grain, its growth requires a shorter time than does that of wheat.

Some barley is produced in the United States. Spring and winter varieties are planted as in the case of wheat. In the United States, barley is used as feed for cattle and poultry, for the production of malt used in brewing, and as an ingredient of soups. Small amounts of barley flour are also produced.

For producing malt, the grain is soaked in water for several days or until the moisture content reaches approximately 50%. It is then removed from the steep tank and placed in containers where air at 65°–70°F (18.3°–21.1°C) can be drawn through it over a period of approximately 1 week. This allows the barley to germinate or sprout. The sprouted barley is then kiln dried over a period of 24 hr. Drying is begun at a low temperature, which is gradually raised as drying proceeds. The purpose of malting barley may be to produce enzymes which will hydrolyze starch to maltose, a sugar which can be utilized by yeasts to produce ethyl alcohol and carbon dioxide. Nondiastatic malt (will not hydrolyze starch) may be produced for its flavor components. Therefore, in drying the sprouted barley, the temperature must not be raised to the point where the starch-splitting enzymes, produced during sprouting, will be inactivated. It would appear, however, that temperatures are raised to the point where some of the sugars present in the sprouted barley are caramelized, hence the dark brown color of malt. Malt is used in the brewing industry for converting the starches present in rye, rice, corn or other grains to maltose, which can be utilized by yeasts. It is also used in bread baking for much the same reason, although in this case, the purpose is to have the yeast produce carbon dioxide for reasons of leavening (raising the dough), the alcohol produced being largely dissipated during the heating involved in baking.

RYE

Rye, like oats, can grow in colder, wetter climates than can wheat. It is botanically similar to wheat, and also in appearance.

As in the case of wheat, there are winter and spring varieties of rye. In the United States, rye is used for the production of bread and cracker-like bakery products. It is also used as an ingredient of animal feeds and as the source of carbohydrates in the production of rye whiskey.

Rye flour is produced much in the same manner as wheat flour, although it is more difficult to separate the bran from the endosperm; hence much of the rye flour produced contains some of the bran.

In using rye for baking bread, some wheat protein (gluten) must be used as an ingredient, since the protein in rye is not suitable, by itself, to form and retain the structure of the loaf of bread. Rye is richer in lysine than is wheat.

RICE

Rice is a much more important grain even than wheat, as far as world-wide direct utilization by humans is concerned. By far the greatest consumption of rice is in Asia.

In the United States, rice is grown in Louisiana, Texas, Arkansas, Mississippi and California. Rice is now harvested in much the same manner as wheat. Its varieties are classified by the shape of the grain as round, medium or long. About 60% of the rice kernel constitutes the rice ordinarily obtained by the consumer, the remainder consisting of hulls, bran, polishings and broken kernels. Mechanical driers are used to reduce the moisture content of rice to about 14%, at which level it can be stored without becoming spoiled by the growth of microorganisms.

Prior to milling, rice is ordinarily steeped in warm water (parboiling), then dried to a moisture content which will facilitate milling. This process loosens the hull and carries some of the soluble vitamins and minerals into the kernel. Parboiled rice is often referred to as converted rice. Once the rice has been dried after steeping, the hulls and bran are removed. In rice milling, the grains are not crushed as in wheat. Instead, they are abraded so that only the surface (hull) portion is removed in shelling machines (also called hulling machines) between abrasive discs or rubber belts. The kernels are then polished. Hulls, bran and broken kernels are screened out, the broken kernels being separated from the hulls and bran. The more effective the polishing, the whiter and less nutritious the rice.

Some quick-cooking rice is produced by pre-cooking the kernels and redrying. This process provides for the preparation of rice for human consumption by merely bringing the water used for rehydration to the boiling point and allowing the mixture to stand for short periods.

Some puffed rice cereal is produced by heating the rice to a temperature above the boiling point of water in closed containers and suddenly releasing the pressure, which causes the kernel to increase in size, as the water vaporizes, allowing it to escape from the interior to the outside.

About 1/3 of the rice produced in the United States is used by the brewing industry. This consists mostly of broken kernels, but some whole grain rice is also used for this purpose.

A small amount of rice flour is produced, and most of this is used by those who are allergic to wheat flour. Rice flour may be used, too, for the preparation of white sauces, especially for prepared frozen-food products since certain types of rice flour produce sauces which do not curdle and weep (separation of liquid from the sauce) when frozen and defrosted.

Rice kernels may be enriched, as is wheat flour, by mixing with a powder containing vitamins and minerals. This powder sticks to the surface of the kernels. The enrichment materials may then be coated with a waterproof, edible film to protect them from being washed off. The federal standards for rice enrichment are shown in Table 19.2.

The protein of rice is comparable to that of wheat in composition although rice is lower in total protein than wheat. Neither of these grain contains a complete protein, that is, the proteins do not contain sufficient amounts of certain amino acids to provide for the requirements of the human.

OTHER CEREAL GRAINS

Sorghum

Sorghums, comprising four general classes (sweet sorghum, broom corn, grass sorghum and grain sorghum), are grown in southern section of the Great Plains and in parts of the Southwest. Some varieties of the grain sorghum class yield glutinous starch, similar to that of corn During World War II, sorghum was used as a substitute for tapioca because the importation of tapioca was impeded by the war situation The deterrent to the use of grain sorghum for the production of starch i the pigmentation of the grain's pericarp, which complicates the production of a white starch. However, enough progress has been made in the development of desirable sorghums to warrant the consideration of sorghum for the production of starch in the future.

TABLE 19.2.   FEDERAL STANDARDS FOR RICE ENRICHMENT (MG/LB)

Nutrient	Minimum	Maximum
Calcium	500.0	1000.0
Iron	13.0	26.0
Niacin	16.0	32.0
Riboflavin	1.2	2.4
Thiamin	2.0	4.0
Vitamin D	250.0 U.S.P. Units	1000.0 U.S.P. Units

Buckwheat

Buckwheat is not a true cereal grain. All the cereal grains belong to the botanical family Gramineae, whereas buckwheat belongs to the family Polygonaceae. However, from a use standpoint, it is considered to be a cereal food. While it is a minor crop in the United States, only Russia and France produce more buckwheat than the United States. It is grown mainly in New York, Pennsylvania, Michigan, Maine and Ohio. Of the few varieties used, the Silverhull is used mainly for producing flour because of the higher yield of the endosperm. Buckwheat is dried to about 12% moisture, cleaned, graded by size and milled similarly to wheat. Most of the flour is used for making pancakes.

Cottonseeds

Although cottonseeds come from plants of the family Malvaceae, and soybeans and peanuts from plants belonging to the family Leguminosae, it should be mentioned that they have been used to produce edible flours. However, these starting materials must be heated to lower their moisture contents and to inactivate their enzymes in order to stabilize them during storage. The heating also improves the flavor of all three sources. A very important function in the heating of cottonseeds is to destroy gossypol ($C_{30}H_{30}O_8$), a toxic compound that is decomposed by heat. Neither of these products is important as a substitute for the true cereals for producing flour; however, for use under conditions that might limit availability of the true cereals, the demonstrated potential of these alternative sources makes them worth investigating.

Millet

Millet is used for food in Asia and, to some extent, in Europe. In many parts of Europe, it is used for hay, as it is used in the United States. Some varieties are used as food seeds for caged birds and poultry.

20

Bakery Products

Bakery products include those leavened (raised) by the carbon dioxide produced by the growth of yeasts (i.e., breads, rolls, etc.), items leavened by carbon dioxide produced chemically through the use of baking powder (including cakes, doughnuts, biscuits, etc.), items leavened by the incorporation of air (i.e., batter-whipped breads, angel food cake, etc.) and unleavened products (crackers, pie crusts, etc.).

Bread and other baked food products (cakes, cookies, rolls, pies, doughnuts, etc.) are important items belonging to the class of foods that is sold in ready-to-serve form. Some of these products are partially baked and require a final baking prior to serving. Growing amounts of baked goods are handled and sold in the frozen form, especially since the quality of baked goods is exceedingly well preserved by freezing. Some baked items, such as cookies and biscuits, are canned, and these products, which must be held under refrigeration until used, need to be baked before serving. The dry ingredients used for some baked goods, especially cakes, are premixed industrially and sold as prepared mixes, and while the user must add the fluid ingredients and bake the product, is still a convenient system for the consumer. In self-rising flours, chemical leavening agents are added directly to the flour.

Generally, the high quality of bread and other baked goods goes into a rapid decline soon after the products are removed from the oven. Freezing them is the only method now known to preserve them effectively for long periods.

BREAD

Bread is the oldest and most important baked product. It has been made from many of the grains, including wheat, corn, rye, rice, barley oats and even buckwheat. The development of its popularity has been due to a number of factors, but an important one is that grains of one type or another have been grown in nearly all the inhabited parts of the world. The composition of the ingredients of a loaf of white bread is approximately 57% flour, 36% water, 1.6% sugar, 1.6% fat or shortening 1% milk powder, 1% salt, 0.8% yeast, 0.8% malt and 0.2% mineral salts. The flour used for making bread is usually of the hard wheat type which is higher in protein than that from soft wheat types. The reason for this is that in yeast leavened products, the gluten (protein) in the unbaked loaf must be sufficient in quantity and of adequate elasticity to form a stretched mass which will entrap bubbles of carbon dioxide. This increases the volume and forms the structure of the loaf. It also allows retention of the structure until sufficient heating has occurred. When due to heating which coagulates the gluten, a more rigid structure has been formed, the structure of the loaf of bread is fixed.

Maturing or oxidizing agents and bleaching agents are usually added to flour at the flour mill. Included in these are benzoyl peroxide, chlorine dioxide and potassium bromate. Benzoyl peroxide bleaches the flour which, without treatment tends to have a yellowish color. Chlorine dioxide has a bleaching effect, and it has a "maturing" effect on the protein (improves the elasticity of the gluten). Potassium bromate is maturing agent.

When the ash or mineral content of flour is high, its color is generally darker. This is due to the fact that the minerals in wheat are concentrated in the bran and adjacent layers. Even though the bran is removed the adjacent layers are retained and because of their high mineral content they impart a darker color to the flour.

Flour millers are able to supply bakers with flour which has essentially the same protein content from delivery to delivery, as specified by the baker, by blending a variety of flours that have been classified according to protein composition. Also, there are several kinds of equipment, such as the farinograph and the extensograph, which are used to determine the characteristics of different flours by measuring the physical properties of the doughs made from the flours.

Water is a chief ingredient of the dough in baking. The amount of water added is such that the finished loaf cannot contain more than 38% water, according to federal regulation. If the water available to a baker hard (contains minerals), the amount of yeast food (mineral salts) to e added may be modified. Also, during the mixing of the dough, a ertain amount of heat is generated because of friction encountered uring the forcing of the mixing bars through the dough, and from the otor which runs the mixing apparatus, as well. If the temperature rises bove a certain point [82°−85°F (27.8°−29.4°C)], the yeast may be estroyed and the gluten and starch of the dough adversely affected. 'herefore, the water must be cooled prior to adding it to the ingredients the mixer, or part of the water must be added as ice, which melts uring the mixing and controls the temperature.

Sugar (cane or beet sugar) in small amounts is used in baking bread, nce it serves to supply a source of readily utilizable carbohydrate to the easts and hence provides for a suitable fermentation which produces arbon dioxide to raise the dough.

Some fat or shortening is added to bread mixtures. Ordinarily, this is solid fat (an oil which has been made into a solid material through ydrogenation). This facilitates mixing, tenderizes the crumb of the loaf nd prevents staling of the bread. Today, as a rule, small amounts of onoglycerides are also added, since they are more active antistaling ;ents than fats. A monoglyceride consists of glycerine with which a fatty :id has been combined with only one of the three alcohol groups, the :her two alcohol groups remaining as such.

Milk powder is usually added to bread dough, since it has a desirable 'fect on the texture of the crumb (inside the loaf) of the finished bread. he yeast used in bread baking is *Saccharomyces cerevisiae*. This may e used as a dried material or as a moist compressed cake containing 70% ater (the latter type must be held under refrigeration prior to use). ither type of yeast must be suspended in warm water prior to adding to e material in the mixer, in order to obtain an even distribution roughout the dough.

Small amounts of salt (sodium chloride) are used in making bread, nce it is desirable for the flavor of the finished loaf and trace amounts ay be utilized by the yeast during its growth.

During mixing, proofing and the early part of baking, the yeasts grow d produce ethyl alcohol and carbon dioxide. The latter, a gas, causes e dough to rise, and provides for the volume of the loaf. Leavening tion may result to a degree by the vaporization of water in the dough en the temperature of the mass is raised sufficiently in the oven. The crease in loaf volume is also believed to be effected in the presence of ortening which can entrap air during mixing and release it when the air pands during baking. The ethyl alcohol is largely dissipated during baking, although residual amounts of alcohol, esters and other compo nents may remain and contribute to the flavor of the loaf.

The malt used in bread making is usually of the diastatic type (contains active enzymes which will convert starch to maltose or glucose). As proofing continues, the small amount of sugar present in flour, and the sugar which has been added to the dough mixture, may become mostly used up by the yeast. Therefore, to continue the yeast growth, the action of the malt enzymes on starch may provide a source of sugars during the latter stages of proofing and the early stages of baking Since a small amount of sugar is essential to the browning of the crust of the loaf of bread, the malt may also be important in that it provides the sugar in the development of crust color.

The mineral salts added to dough mixtures are called "yeast foods." Yeasts require small amounts of nitrogen-containing and phosphorus containing salts for growth and the production of carbon dioxide. For this reason, small quantities of ammonium salts and phosphates are added to the ingredients of the dough.

In addition to the above components, many special types of bread are produced which contain one or more additional ingredients (i.e., butter extra milk powder, buttermilk or dried buttermilk solids, dried vegetable powders, honey, etc.).

There are two general methods of handling dough in baking bread— the straight dough method and the sponge method. With either process flour (stored in bulk in bins), water, fat (usually melted), suspended yeast, milk powder, etc., are weighed and added to the mixer auto matically.

In the straight dough method (see Fig. 20.1), all the ingredients are added to the mixer and the material is mixed first at low speed (about 3? rpm), then at high speed (about 70 rpm). The mixed dough is then placed in large metal containers or troughs and held in an insulated room at about 80°F (26.7°C) and in an atmosphere of high humidity to allow fermentation. During fermentation, the mass of dough is kneaded severa times to allow the escape of some carbon dioxide, which is produced continually during fermentation. In addition, working of the dough in this manner assists in stretching and conditioning the gluten, which is the important ingredient responsible for the formation and retention of the structure of the loaf.

In the sponge dough method, 50–75% of the flour, enough water for a moderately stiff dough, all the yeast, the malt, and the yeast foods are added to the mixer and combined. This sponge is fermented for 3–4 hr

hen returned to the mixer and combined with the remainder of the flour
and water, the shortening, the sugar, the milk powder and the salt. The
sponge method of baking bread produces a crumb of finer texture and
with smaller gas holes than that obtained when the straight-dough
method of bread baking is used.

After fermentation, or after fermentation and mixing, the dough is
divided into pieces of a size which will eventually make up the finished
loaf. This is done by a machine which measures the dough by volume and
cuts off pieces of the desired size. When cut, the dough has an irregular
shape with cut ends through which the leavening gas can escape. It is

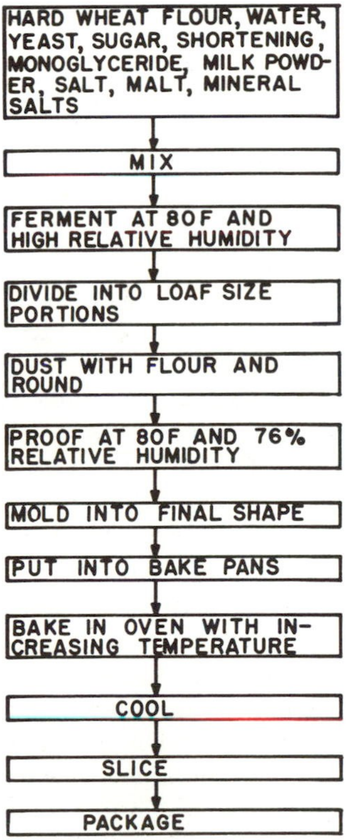

```
┌──────────────────────────────────┐
│ HARD WHEAT FLOUR, WATER,          │
│ YEAST, SUGAR, SHORTENING,         │
│ MONOGLYCERIDE, MILK POWD-         │
│ ER, SALT, MALT, MINERAL           │
│ SALTS                             │
└──────────────────────────────────┘
                 ↓
┌──────────────────────────────────┐
│              MIX                  │
└──────────────────────────────────┘
                 ↓
┌──────────────────────────────────┐
│ FERMENT AT 80F AND                │
│ HIGH RELATIVE HUMIDITY            │
└──────────────────────────────────┘
                 ↓
┌──────────────────────────────────┐
│ DIVIDE INTO LOAF SIZE             │
│ PORTIONS                          │
└──────────────────────────────────┘
                 ↓
┌──────────────────────────────────┐
│ DUST WITH FLOUR AND               │
│ ROUND                             │
└──────────────────────────────────┘
                 ↓
┌──────────────────────────────────┐
│ PROOF AT 80F AND 76%              │
│ RELATIVE HUMIDITY                 │
└──────────────────────────────────┘
                 ↓
┌──────────────────────────────────┐
│ MOLD INTO FINAL SHAPE             │
└──────────────────────────────────┘
                 ↓
┌──────────────────────────────────┐
│ PUT INTO BAKE PANS                │
└──────────────────────────────────┘
                 ↓
┌──────────────────────────────────┐
│ BAKE IN OVEN WITH IN-             │
│ CREASING TEMPERATURE              │
└──────────────────────────────────┘
                 ↓
┌──────────────────────────────────┐
│              COOL                 │
└──────────────────────────────────┘
                 ↓
┌──────────────────────────────────┐
│              SLICE                │
└──────────────────────────────────┘
                 ↓
┌──────────────────────────────────┐
│            PACKAGE                │
└──────────────────────────────────┘
```

FIG. 20.1. MANUFACTURE OF BREAD
(STRAIGHT DOUGH METHOD)
See metric conversion tables in Appendix.

immediately dusted with flour and rounded. This is done by machine. Rounding dries the surface with flour and closes up the cut ends, thus preventing the escape of gas. The rounded dough is then carried on a belt to a proofer, where it is emptied onto another belt and held at 80°F (26.7°C) and 76% rh for a period during which the dough relaxes and increases in volume as more carbon dioxide is produced. The pieces of dough are then molded and shaped by machine. In this process, the floured dough is rolled out into a sheet which is curled into a loose cylinder, again rolled and the ends sealed. In some operations, two cylinders of dough are twisted together in the molding operation. The cylinders of dough then fall into pans which are conveyed to the oven for baking. During the first stages in the oven, the dough continues to ferment and increase in volume. As the dough passes through the oven, the temperature is increased, further expanding the dough (increase in volume of the gas due to temperature), and eventually, the gluten is set by the heat, the starch first gelatinized and then set by the heat, and some water, ethanol, etc., are evaporated. Eventually, the outer layers of the dough become browned to form the crust. Browning is probably due to both the reaction between proteins and sugars, and caramelization of sugars. After baking, the loaves of bread are cooled as they are carried through air-conditioned tunnels. The loaves are cut into slices by machine, and the sliced loaves are packaged automatically by bread wrapping machines.

Bread and other baked goods are subject to spoilage by molds. Therefore, small amounts of mold inhibitors, such as sodium or calcium proprionate, or sodium diacetate are usually added to bread. Sodium diacetate may be added at levels of 0.4 parts per 100 parts of the flour used, and sodium or calcium propionate at levels of 0.32 parts per 100 parts of flour. Since water and other ingredients are used in bread, the actual concentration of these inhibitors in the finished loaf is much lower than in the flour. Mold inhibitors not only delay the growth of molds in bread, but they also inhibit the growth of certain bacteria which produce a slime in the crumb of the loaf, a condition known as "ropiness."

In today's continuous batter-whipped process of making bread (a third method), a liquid mixture of water, sugar, yeast, milk powder, salt, and yeast food, small amounts of flour, and some vitamins is fermented for 2–3 hr. It is then cooled prior to mixing with liquid shortening and the bulk of the flour. The mixture is then agitated at high speed in a developer, which incorporates air. The dough is then extruded directly into baking pans which, after a short proofing period at 80°F (26.7°C) are conveyed directly to the baking oven. Automation in bread manufacure has led to higher production volumes, lowered production costs, shorter production time (to about 1/4 of original time) and to better control over the properties of the finished product, making it easy to produce large volumes of uniform bread that can meet whatever specifications are desired. Bread made by the continuous process is finer in texture, but it is questionable that much flavor is developed in processes of this type.

Standards for bread and flour, shipped in interstate commerce and labeled "enriched," were initiated by the federal government in 1952. These standards require the enrichment of bread with thiamin (vitamin B-1), riboflavin (vitamin B-2), niacin (a member of the complex of B vitamins, also called nicotinic acid) and iron. Calcium and vitamin D may be added as enrichment agents at the option of the producer.

CAKES AND COOKIES

Processes for baking cakes and cookies have not been automated to the extent of those for bread making. Also, these products are usually produced in smaller batches. Leavening in these products is produced chemically, and there is no yeast fermentation. The chemicals used to produce carbon dioxide are sodium bicarbonate (the source of carbon dioxide) and sources of acid to react with the bicarbonate, such as potassium hydrogen tartrate (cream of tartar), sodium hydrogen pyrophosphate, calcium hydrogen phosphate, and alum (sodium aluminum sulfate). The desirable chemical leavener produces small bubbles of gas at a constant rate consistent with the period involved in mixing and baking to temperatures which set the structure of the cake. Cookies are formed with dies or by extrusion, and, in many instances, the process is quite complex, since the dough may have to be extruded around a central component of fig paste or other type of filling.

Many prepared cake mixes are also produced, some of which may be used by bakers, although many are used by the home baker. Soft wheat flour or flour of low or moderate protein content is used for cakes. In premixes, the flour, egg powder, shortening, fruit or flavoring components and leavening agents are combined in the dry state, although, when mixed, the shortening is in the melted or liquid state, and emulsifiers, such as monoglycerides, which improve air incorporation during mixing, may be used. Much of the successful preparation of cake mixes depends upon the kind and quantity of chemical leaveners used. Angel cakes contain egg white and are leavened by the incorporation of air into the mixture. During baking, the air in the cake expands and acts as the leavening agent.

DOUGHNUTS

The ingredients for dough used in the manufacture of doughnuts are similar to those of cake, especially pound cake. The dough, after mixing is extruded, cut into doughnut form, and cooked in hot oil (370°–380°F [187.8°–193.3°C]). Fat absorption is reported to be about 15%. Fat absorption may be higher when processing parameters (e.g., temperature of the dough) are not controlled, resulting in "greasy" doughnuts. When fat absorption is insufficient, the keeping quality of the doughnuts is diminished. When doughnuts are to be sugared, the temperature and relative humidity must be controlled (70°–75°F [21.1°–23.9°C] and 85%, respectively) for optimum sugar pick-up. Generally, doughnuts are prepared from materials already premixed elsewhere, and seldom do the manufacturers of doughnuts mix their own formula.

CRACKERS

Crackers are unleavened or only slightly leavened. The flour used for these products consists of wheat flour, although some rye flour is often used. For producing crackers, flour, liquid shortening, salt, and small amounts of a chemical leavening agent, with or without sugar, or a flavoring agent, such as onion powder, are mixed into a dough. The dough is then extruded into the desirable shape and baked without proofing. Milk powder, whey powder, and emulsifiers, such as monoglycerides, may be used with some combination of the ingredients listed.

PIE CRUSTS

Pie crusts, which are also unleavened bakery products, can be made from all-purpose flour, but highest quality pie crusts are obtained from unbleached soft wheat pastry flour of low protein content. When the protein content is too high, the desirable flakiness characteristic is minimized, and, to compensate, a larger amount of shortening has to be used. The shortening should be a solid or hydrogenated type (e.g., lard) and it should be medium firm. Milk powder or milk may be used in small quantities to enhance color. Eggs have the same effect. Salt and sugar may be added. The composition of a high quality pie crust is approximately 47.5% flour, 1.0% salt, 3.3% nonfat milk powder, 2.0% cerelose dextrose, a 6-carbon sugar), 32.4% shortening and 13.8% water. To ensure a high-quality crust, all ingredients should be mixed at a temperature of 60°–65°F (15.6°–18.3°C), and the ingredients combined with a minimum of mixing and handling.

21

Vegetables

Vegetables are plant foods that include various edible parts such as leaves, shoots, roots, tubers, flowers, and stems. They normally do not include fruit, which is the subject of the next chapter. However, tomatoes and olives, which are technically fruit, are included in this chapter, because their culinary role is related more to vegetables than to fruit, e.g., a garden salad almost always includes tomatoes and sometimes olives, but it never includes any of the other fruit. Consistent with this, fruit salads never contain tomatoes or olives. Vegetables belong to an important class of foods that supply man with many of his nutritive requirements, including proteins, starches, fats, minerals, sugars and vitamins. Vegetables also supply bulk to the diet, as well as a large variety of flavors and odors that provide the knowledgeable chef with a repertoire of culinary tricks. On a world basis, vegetables make up a considerable part of man's diet, with the largest proportion of the total vegetable crop being consumed in the fresh state (not preserved). In the United States, however, the proportion of processed vegetables consumed may be quite high. In the fresh state, vegetables continue to carry on life processes, and they are susceptible to various forms of deterioration, hence requiring sanitary handling under controlled temperatures.

The detention of vegetables after harvesting is detrimental to their quality. They undergo microbial spoilage, they lose water, in many cases they lose sugar and they give up considerable energy in the form of heat value reported is more than 100,000 Btu per ton per day [27,784 Cal kg)/MT/day]). Of course, by the heat they produce, vegetables hasten their own deterioration by microbial action which is accelerated as the temperature is increased (within limits). Enzymic deterioration, especially at sites where bruises occur, is also accelerated by higher tem peratures (again within limits).

Various procedures are involved in the production, preservation an distribution of vegetables. The seeds may have to be treated, the soil ha to be fertilized, the crops have to be sprayed or dusted with insecticide and/or fungicides, and the crops have to be harvested and subjected t one of the various methods of preservation. The finished product is the distributed to the retailer or held in a temperature-controlled warehous until it is distributed. Vegetables are sold at retail as fresh produce, a canned and heat processed, as frozen, and occasionally as dried products

Seeds are treated with fungicides or insecticides to prevent loss c decay to either insects or fungus prior to germination and growth of th plant. Arsenic or mercury-containing compounds are ordinarily used fc seed treatment, the seeds being dusted or treated with a slurry (wate suspension) of the compound, then dried. Bean and pea seeds may als be inoculated with bacterial cultures which take nitrogen from the a and make it available to the plant, which requires the nitrogen fc growth.

Various fertilizers may be applied to the soil prior to planting Generally, fertilizers consist of some combination of nitrogen compound (ammonium salts, nitrates or urea), phosphates, and potassium con pounds. These chemicals, required for plant growth, are apt to b deficient in the soil. Soil is frequently treated with calcium-containir compounds to neutralize its acidity and provide a pH suitable for plar growth. Liquefied fish wastes are also sometimes used as fertilizers, an chemical fertilizers are sometimes applied in liquid form.

After crops have started to grow, they may require spraying or dustin with insecticides or fungicides to destroy insects, such as aphids, worm grasshoppers, etc., and to destroy molds, viruses or bacteria which ma invade the plants and cause them to rot or otherwise deteriorat Insecticides consist of chlorinated hydrocarbons, chlorinated aromat organic compounds or organic phosphorus-containing compounds. Plar extracts, such as pyrethrum, rotenone, or extracts of pepper may be use with other insecticides to intensify the effect of the former. Variou antibiotics (materials extracted from mold or bacterial cultures whic inhibit the growth of other molds or bacteria) are sometimes applied plants to prevent rotting due to the growth of microorganisms.

Vegetable crops may be produced by the organization which wi eventually handle and process them, or often the processor will contra farmers to grow the crops. Large processors have a field departmer which employs a number of horticulturists. The field department su plies the seed, specifies the kind and extent of fertilization and sc treatment required, and supplies the ingredients. Soil fertilization an

eatment, and planting are done by the farmer. Weeding may be done
actual removal of weeds or by chemical treatment, and the latter may
done by the farmer or the processor. Spraying or dusting with
ecticides or fungicides may be done by the farmer or the processor.
rsonnel from the processor's field department decide when the crop is
the right maturity for harvesting and arrange for harvesting of the
op. If the crop cannot be handled quickly enough, the vegetables may
ss their optimum maturity, resulting in a loss of product quality.

With some vegetable crops (especially beans, corn, and peas), the heat
it system is used to accommodate postharvest handling capacities.
e heat unit is a measure of the temperature (above the minimum
owth temperature) and the time in days (or hours). For example, the
inimum growth temperature of peas is 40°F (4.4°C). If the average
mperature on two successive days is 60°F (15.6°C), then the number of
at units accumulated for the vegetable on those two days is determined
multiplying the number of degrees exceeding the minimum (60°F −
)°F = 20°F; 15.6°C − 4.4°C = 11.2°C) by the number of days (2); 20°F
2 = 40°F days (11.2°C × 2 = 22.4°C days) heat units (degree days),
nverted to degree hours the number of heat units would be 40°F × 24
/day = 960°F hr (22.4°C days × 24 hr/day = 537.6°C hr). In cases
ere the heat unit is used, the number of heat units required for specific
getables to reach maturity is known. By keeping careful records,
erefore, the precise time when these vegetables mature can be de-
rmined accurately. Such a system assists the planting schedule, since
enables planting to be done so that only the quantity of the product
ich can be handled adequately by the processing plant will be planted
any one day. The heat unit system provides for a better quality
getable since it prevents the maturing of crops beyond their best
ality, as well as the losses that occur when the crop harvested is
eater than the processing capacity.

All vegetables which are sealed in cans or glass jars must be heat-
ocessed (for typical production sequence see Fig. 21.1) (usually at
0° or 250°F [115.6° or 121.1°C] and sometimes at higher tempera-
res), to make them commercially sterile. Commercially sterile means
at all disease bacteria have been killed, and all bacteria and bacterial
res, which might grow out and cause spoilage under the conditions
which the product will be handled after processing, have been de-
oyed. The time during which a heat preserved vegetable must be
ocessed to attain commercial sterility depends upon the size of the
ntainer, the temperature at which the product is processed, the type
container (glass, metal or plastic), whether or not the product is ag-
ted during heating, whether the vegetable product heats by convec-
n or conduction, and other factors.

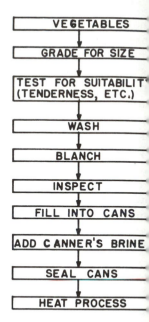

FIG. 21.1. PRODUCTION OF CANNED VEGETABLES

When a vegetable or vegetable product is preserved by freezing, should be brought to a temperature of 0°F (–17.8°C) or below in all par and held at this temperature or below until sold to the consume Vegetables are packaged and frozen on moving chain mesh belts in co air tunnels. Packages of vegetables are sometimes placed on trays, th trays placed on racks, and the product frozen as the racks are move through cold air tunnels. In other instances, vegetables are packaged, th packages placed on trays and the trays placed between refrigerated met plates for purposes of freezing the product. Some vegetables are loos frozen on chain mesh belts in cold air tunnels prior to packaging.

Following are descriptions of planting, harvesting, processing an general handling of the more important vegetables used in the Unite States. No attempt is made to cover all vegetables.

ASPARAGUS

Asparagus may be planted from the seed or as roots. If planted fro the seed, 700–1000 days will be required before the first harvest. planted as roots, a crop may be obtained the first year, depending up he age of the roots. Asparagus may be attacked by fungus growth or by nsects and may, therefore, have to be sprayed or dusted with chemicals uring the growing season. While mechanical methods of harvesting sparagus may be available, it is now usually harvested by hand, with a pecial knife, when the stalks are about 8 in. (20.32 cm) long. Once arvested, asparagus soon loses its sugar and also becomes tough and tringy. When cut, it is collected in boxes and should be promptly cooled n chlorinated ice or refrigerated water. It should thereafter be held at 2°–40°F (0°–4.4°C) until sold at retail or processed.

In processing for freezing, the asparagus stalks are arranged on a belt ith the butt end against a guard rail. As the belt moves along, the stalks ass under rotary knives which cut off the butt end and further cut the pears to a length of 5 in. (12.70 cm). The middle portions are packaged s cut asparagus. Spears and cut portions are then washed (separately) 1 a soaking-type washer, after which the spears are sorted for size diameter of butt end). Spears and cut portions are then blanched heated in free flowing steam to inactivate enzymes) and cooled. 3lanching times are 3.5–5 min for spears, depending on size, while cut ortions are heated for about 3 min. Both spears and cut portions are ackaged in cartons by hand, after which they may be overwrapped prior o freezing in plate or air-blast freezers.

Asparagus which is canned (heat processed) is handled in a similar 1anner to that which is to be frozen up to the point of blanching, lthough the spears are usually cut to longer lengths (6–7 in. [15.24–17.78 m]) for canning. Canned asparagus is not blanched, due to the difficulty f getting it into the container after heating. In order to fill the container, he stalks are arranged within a metal band, and the band is then ightened to allow the insertion of the spears into the container. The and is released when the spears are part way into the can or glass jar.

Glass jars containing asparagus spears are filled with a hot, weak salt olution containing small amounts of stannous chloride (chloride of tin), hen vacuum sealed. The stannous salt is added to help retain the green olor which otherwise would fade during heating. Asparagus in cans is overed with a hot weak brine. It is not necessary to add stannous hloride, in this case, since during processing, this compound will be ormed from the tin lining of the container. Asparagus in cans may or 1ay not be vacuum sealed, since the hot brine is sufficient to provide ome vacuum after processing and cooling. Canned asparagus heats by onvection, the brine solution within the container forming convection urrents which speed up the heating of the spears within the container. ue to the shape of the product, in order to obtain convection heating, the containers must be placed in the upright position during processing since if placed in the horizontal position, the spears will impede the convection currents, slowing down the heating process.

GREEN AND WAX BEANS

Green and wax beans are planted from seed in the spring or when temperatures will allow. A period of 50–70 days is required from planting to harvesting. Mechanical methods are now available for harvesting bush-type green or wax beans. Beans of this type may be invaded by molds, bacteria, viruses or insects and may have to be sprayed or dusted with chemical inhibitors during the growing season. Beans, when harvested, are collected in bags. If they are to be held, they should be promptly cooled to 40°F (4.4°C) and held at this temperature until processed or sold to the consumer.

During processing, the beans are first size-graded (since the large beans are cut in a different manner from the smaller specimens). They are then washed, and the ends are snipped off by passing them through a perforated cylinder. Here, their ends project through the perforations and are cut off by moving knives which pass over the perforations of the cylinder. The beans are then cross-cut (smaller types) or frenched (cut along their length into narrow strips) by machine. They are next blanched in a free-flowing steam (the cross-cut types may be blanched in water at 210°F (98.9°C) for 2–4 min), then cooled. Cross-cut beans may be filled into packages mechanically (french style by hand) for a product which is to be frozen. The packages may then be overwrapped. Freezing is accomplished by one of the methods previously indicated. Some cross cut beans are loose frozen, then packaged in plastic bags or held in containers for use in mixed vegetable products, when the mixing is done after freezing. Canned, cross-cut, green beans are filled into the container by machine, and the container is then filled with a hot, weak brine, sealed (with or without vacuum) and heat-processed.

LIMA BEANS

There are two types of lima beans, baby limas and the Fordhook-type limas. Lima beans are planted as seed. A period of 70–90 days is required between planting and harvest. Lima beans are subject to the same insect, mold, bacterial and virus diseases as are green beans and may have to be treated in a similar manner to prevent these diseases. Lima beans are harvested by mowing down the whole plant which is then passed through a viner. Here the pods are beaten with rotating paddles as the vines pass

:om one end of the viner to the other. This releases the beans from the ods, after which the beans fall through perforations in the viner shell nd are collected in boxes or in bins. The beans are carried to the rocessing plant where they should be promptly processed, since in this ondition they tend to heat. The vines and empty pods are cut up and lowed under the soil as mulch or placed in ditches or stacks, and ermitted to ferment to form ensilage, which is eventually fed to cattle.

At the processing plant, lima beans are washed and blanched in water t 210°F (98.9°C) for 2.5 min for baby limas or 3–4 min for Fordhook ypes. The latter type of lima beans is sometimes blanched in steam. fter blanching, the beans are cooled in water, then passed through a uality grader. The quality grader consists of a circular trough of brine of 1ch density that the lighter beans of the desired maturity float in the rine and can be separated from the top brine layer, while the heavier, 1ore mature beans, sink through the brine and can be separated from 1e bottom brine layer. Those which sink may be packed as a lower grade ozen product or canned. After passing through the quality grader, the eans are washed to remove brine and passed over an inspection belt here skins, broken beans, weeds, etc., are removed and discarded. In 1e freezing plant, lima beans are loose frozen on mesh chain belts in air last systems, then filled automatically into cartons or plastic bags hich are then sealed, or they are filled automatically into cartons and ozen. The cartons may or may not be overwrapped with cellophane. ome baby lima beans are canned, and they are handled in much the 1me manner as the frozen product except that they may or may not be assed through the quality grader. The beans are automatically metered 1to cans which are filled with hot, weak, salt solution. The cans are :aled and heat-processed. Some lima beans to be canned are allowed to 1ature and dry out on the vine before harvesting. In such cases, the :ans must be soaked in water for purposes of rehydrating prior to irther processing.

EETS

Beets require 50–70 days from planting to harvesting. They may quire dusting or spraying with chemicals to prevent bacterial, virus or sect infestation. They are harvested by machine and brought to the ocessing plant in hoppers or trucks. Generally, they are not cooled prior processing. Beets are canned but are not frozen. At the canning plant, e tops are cut off by machine, after which the roots are held for several 1ys to allow the skin to wilt, thereby loosening it. The beets are then aded for size by machine. After sizing, the beets are washed with sprays of water or in a soaking tank, and are then peeled. Beets are peeled by steaming at 220°F (104.4°C) for about 20 min, after which the skin i removed manually or in a centrifugal abrasion peeler after cooling. Th peeled beets are then trimmed by hand. Small beets are canned whol while the larger beets are sliced prior to filling into cans. The cans, t which beets have been added to a point about 3/8 in. (0.956 cm) from th tops, are then filled with a weak salt solution. The cans are then seale and heat-processed. Beets sold in the fresh state are washed, but usuall not topped. They should be cooled to 32°–35°F (0°–1.7°C) and held i this manner until sold to the consumer. In this condition, they have storage life of 10–14 days. When topped and cooled to 32°F (0°C) the may be held for 3–5 months.

BROCCOLI

Broccoli requires 60–70 days from planting to harvesting. The plant usually require spraying or dusting with chemicals to prevent inse infestation, as well as bacterial or viral diseases. Broccoli is harvested k hand or by semi-automatic machine methods and transported to th plant in baskets or hampers. Essentially no broccoli is canned. If brocc is to be held prior to processing, it should be precooled to 35°–40° (1.7°–4.4°C). Much broccoli is handled in the fresh state. In suc instances, it is sorted to remove blossomed and insect-infested head after which it is thoroughly washed in a combination spray and so tank. Then the stalks are bound together in small retail-size bunch with a paper or plastic band. Fresh broccoli should be held at 32°–35° (0°–1.7°C) at all times after inspection and washing until sold to t consumer. In this condition, it may be held for 7–10 days.

In general, fresh produce, such as broccoli, Brussels sprouts, cabbag etc., should be precooled prior to shipment. This may be done, aft placing in trucks or railroad cars, by blasts of cold air from refrigerati generators, or after placing in rooms in which blasts of cold air can l circulated. In the latter case, the vegetables are placed in refrigerat railroad cars or trucks after cooling. Sometimes, cars or trucks filled wi fresh produce are placed in a large metal chamber. The chamber is clos and subjected to a vacuum. This causes water to evaporate from t produce, cooling it. About 5% of the water is lost from the product cooling in this manner. Broccoli to be frozen is handled in much the sar way as fresh broccoli during sorting and inspection. The center heads a split and trimmed so that each head is 1.5–2 in. (3.81–5.08 cm) diameter. The heads are then placed, facing one direction, on a convey

here they pass under a rotary cutter which trims the stalks to a length
5 in. (12.7 cm). After washing, the trimmed stalks are blanched in
ee-flowing steam for 2.5–7 min (usually 3–5 min). The blanched product
cooled in water, then packed into cartons by hand to provide a certain
eight. The filled cartons, with or without overwrapping, are placed on
ays and racks, and frozen.

RUSSELS SPROUTS

This vegetable is handled as the fresh or frozen product, and essential-
none is canned. A period of 90–100 days is required from planting to
arvesting. Dusting or spraying with chemicals may be necessary, as
ith broccoli. Brussels sprouts are harvested by hand and brought to the
ocessing plant in crates. At the plant, the stalks are removed with a
tary trimmer, after which the heads are passed through a dry rotary rod
eleafer and are washed by spraying with water. Brussels sprouts to be
ndled as fresh are weighed into small, retail-size baskets and placed
der refrigeration at 35°–40°F (1.7°–4.4°C) until sold to the consumer.
. this condition, they have a storage life of 3–4 wks. The precooling may
similar to that of broccoli. If they are to be frozen, Brussels sprouts are
anched in free-flowing steam for 3–9 min (usually 5–6 min), then
oled. They are next weighed into cartons by hand after which the
rtons are closed and may be overwrapped by machine. The product is
en frozen.

ABBAGE

Essentially no cabbage is canned or frozen as such. Most of it is
ndled in the fresh state or processed to produce sauerkraut. Cabbage
quires 60–100 days from planting to harvesting and may need chemical
eatment as does broccoli, during the growing season. Cabbages are
rvested by hand and placed on a mechanical conveyor belt which loads
em into hoppers or trucks. Cabbage is washed, trimmed of loose leaves
hand, and precooled, as is broccoli. This vegetable should be held at
°–35°F (0°–1.7°C) until sold to the consumer. In this condition, it may
held for 3–4 months.
Cabbage to be used for the manufacture of sauerkraut may be held out
refrigeration until the outer leaves wilt, which facilitates cutting and
redding. The cabbages are then cored with a power-driven cylindrical
tter. After coring, the outer coarse and green leaves are removed with

a knife. The trimmed and cored heads are then washed with sprays water. The washed cabbages are then sliced or shredded by machine in strips which have a width of about 1/32 in. (0.08 cm). The cabbage salted during or after shredding and then placed in large vats (about 1 ft [3.7 m] in diameter and about 8 ft [2.4 m] deep). It may be salted it is placed in the vats, by alternating layers of shredded cabbage wi salt. The amount of salt added is 2–3% of the total weight. In th condition, the cabbage is then allowed to ferment for 40–90 day depending upon ambient temperatures. During fermentation, bacter grow and produce lactic acid. At the completion of fermentation, t lactic acid content should not be less than 1.5%.

Most sauerkraut produced today is canned. It may or may not heated to about 110°F (43.3°C) before it is filled into the cans by han In any case, after it has been added to the can, a weak salt solution 165°F (73.9°C) or higher is added to complete the filling of the can. Aft adding the brine, the lactic acid content of the mixture should not be le than 1%. The cans are then sealed and heat processed. Due to the low p of this product (below 4.0), sauerkraut can be rendered commercial sterile by heating it in boiling water to the point where the temperatu of all parts reaches 200°–210°F (93.3°–98.9°C). It may be heat-processe at higher temperatures.

CARROTS

Carrots require 70–85 days from planting to harvesting. Since worm bacteria or fungus may invade carrots, the soil may have to be treate with chemicals prior to planting, and the plants may have to be spraye or dusted with chemicals to protect the leaves against the invasion microorganisms or insects. Carrots are harvested by special machin and brought to the processing plant in trucks. If carrots are to be handle as fresh produce, they are trimmed of the stems and leaves (topping washed with water sprays and then cooled in refrigerated water to temperature of about 35°F (1.7°C). They are then weighed and package in plastic bags. For maximum shelf-life, carrots should be held temperatures of 32°–35°F (0°–1.7°C) until sold to the consumer. In th condition, they may be held for 4–5 months.

Carrots may be preserved by freezing or canning. Carrots to be froze are passed through a high-pressure steamer for about 10 sec aft topping, then washed with water sprays to remove the thin outer ski They are then trimmed by hand and inspected. The carrots are ne: diced into cubes of about 1/4 in. (0.6 cm), blanched in free-flowing stea or 4–6 min and cooled in a water flume. The flume empties into a circular channel separator filled with water, where chips of carrots can be separated from the cubes, which sink to the bottom. The chips are discarded. Diced carrots are frozen on trays which are placed on racks that move through a cold air blast system, or on mesh chain belts in a cold air tunnel. The frozen product may be packaged in plastic bags or held frozen in metal containers for mixing with frozen peas or other frozen vegetables, in which case, the mixed vegetables are packaged in cartons which may or may not be overwrapped with plastic. The cartons are filled automatically using a volumetric fill method. Carrots may be in diced, whole, cut or sliced forms for canning. Carrots for canning may be peeled by water sprays after steaming or by immersion in a lye solution. In the latter case, the trimmed and sized carrots are passed through a hot solution of sodium hydroxide (1–3% NaOH) at 200°–212°F (93.3°–100°C) for 18–25 sec. They are then washed in water sprays. Since the larger sizes are packed as the sliced or diced product and only the very small specimens canned as whole carrots, size grading of this product is done prior to peeling. Carrots are blanched in free-flowing steam for 5–15 min, then cooled prior to placing in cans, the blanching time depending upon the size of the product. Whole small carrots are placed in cans or glass jars by hand. Diced or cubed carrots may be added to containers by hand or by volumetric automatic methods. The containers are filled with hot (about 180°F [82.2°C]) or cold canner's brine (4% sugar, 1 1/2% salt). If cold canner's brine is used, the open containers are exhausted or heated in steam until all the product reaches a temperature of 140°–150°F (60°–65.6°C). The containers are then sealed and heat-processed.

CAULIFLOWER

Cauliflower is not often canned since it becomes mushy and discolors during heat processing. It is handled mostly as the fresh product, although it may be preserved by freezing. A period of 55–60 days is required from planting to harvesting. Since cauliflower is subject to infection or infestation by the same type of microorganisms and insects that deteriorate broccoli, it may also require treatment with chemicals during the growing season.

Cauliflower is harvested by hand and brought to the processing plant in baskets or crates. If it is to be handled as a fresh or frozen product, it should soon be cooled to 31° to 34°F (−0.6° to 1.1°C) and held in this state until further processed or sold to the consumer. Handled as the fresh product, the outer leaves are first trimmed off. The heads are then washed in a soaker-spray combination washer. The vegetable is then packed in crates and precooled in cold air or by a vacuum system. Cauliflower has a short storage life (2–4 weeks) and, handled in the fresh state, should be held at 31° to 34°F (–0.6° to 1.1°C) until sold to the consumer. When it is to be frozen, the leaves and base of the stem are cut off with a knife, the core is removed and the head is broken and cut into individual flowerettes or curds. The curds are then passed through a cylindrical rod cleaner which eliminates loose leaves, small pieces, etc. and flumed to a washer where they are subjected to heavy sprays of water. After washing, the curds are blanched in free-flowing steam for 3–10 min (usually 4–5 min) and then cooled. Cauliflower to be frozen is placed in cartons by hand. The cartons may or may not be overwrapped with cellophane. The product in filled cartons is then frozen.

CELERY

Celery is handled almost entirely as a fresh product, since it softens during canning or freezing. Celery requires 112–125 days from planting to maturity. It is subject to insect and worm infestation and may require spraying with chemicals to control such pests.

Celery is harvested by hand, placed in crates and brought to the processing plant. At the plant, the heads are washed in a combination spray-soaker type washer, the outer leaves are trimmed off by hand, and the product is repacked into crates and promptly cooled to 31° to 34°F (–0.6° to 1.1°C). At this temperature, it may have a storage life of 2–3 months. Pre-cooling may be done with refrigerated air or by a vacuum system. Celery should be held at 31° to 34°F (–0.6° to 1.1°C) until sold to the consumer.

SWEET CORN

Sweet corn is handled as fresh, frozen or canned on the cob, or as canned or frozen kernels. It may also be canned as a mixture of partially macerated corn kernels, starch and sugar. Corn requires 60–85 days from planting to harvesting. It is subject to deterioration by bacteria, more so by various molds, and several types of worms or borers. Therefore, the plants must be sprayed or dusted with inhibitory chemicals during the growing season. Corn is harvested by a machine which may also husk or remove the leaves from the ears and is brought to the processing plant in trucks.

CORN ON THE COB

If corn on the cob is to be shipped in the fresh state, it is handled as picked, or the base of the cob is cut off and the leaves are left on, or the base of the cob is cut off and the leaves and silk are removed between rollers. It is then packed in crates and precooled in refrigerated air, or it may be precooled and packaged in cardboard trays covered with plastic. Corn should be promptly precooled after harvesting, since at ambient temperatures, it soon loses its sweetness and takes on off-flavors. Promptly cooled and held at 32°–34°F (0°–1.1°C), it has a storage life of about 2 weeks. Fresh corn should, therefore, be held at 32°–34°F (0°–1.1°C) until sold to the consumer. It should be realized that corn loses its sweetness rapidly with some sugar being converted to starch and some consumed in respiration.

Corn to be canned on the cob is cut at the base, husked and washed. The ears are then cut to can-sized lengths (from the point) and placed in the can by hand. A hot, weak, salt and sugar solution is then added to cover the ears, and the cans are sealed and heat-processed.

Corn to be frozen on the cob is handled in the same manner as canned cob corn as far as husking and cleaning are concerned. It is then blanched in free-flowing steam for about 9 min and cooled in water. The blanched and cooled corn may be frozen prior to, or after, packaging.

Corn as Kernels

Corn, canned or frozen as kernels, is handled like corn on the cob as far as husking and washing are concerned. To cut the kernels from the cob, the ears are passed through a machine in which knives, which encircle the cob, follow the shape of the cob so that the cut is made only to a depth sufficient to remove most of the kernels. The cut kernels are passed through strainers made of parallel wires that are set wide apart, in order to remove strands of silk which may be present. The kernels may also be floated in water to remove detritus.

Canned, whole corn kernels are packed in two ways. The kernels may be added to the can and entirely covered with a hot, weak solution of salt and sugar, after which the can is sealed, or the kernels may be added to the can, after which only a small amount of weak solution of salt and sugar is added. In the former case, the product heats by convection, currents of hot brine moving up along the wall of the can, across the top and down the central axis. In the latter case, the can is sealed under a very high vacuum drawn on the can by mechanical means during sealing.

The high vacuum product heats both through steam, generated from the small amount of liquid present in the container (the vacuum allowing the steam to spread through the product), and through conduction. The advantage of the latter process is that soluble solids are not, to an extent, leached out from the kernels into the brine. The more mature ears of corn are used for cream-style canned corn, which is a mixture of corn kernels, sugar, water, starch and salt. The mixture is prepared and cooked to 190°F (87.8°C) prior to filling into the cans. The cans are sealed and heat-processed. Cream-style corn takes longer to process than do the whole corn kernels, because heat transfer through it is by conduction only.

Corn kernels to be frozen are blanched in free-flowing steam for 2–min on a wire mesh belt, then cooled with water sprays. During cooling some sugar and starch present in the kernels are leached out, resulting in some loss of quality. Whole corn kernels to be frozen are filled into cartons automatically by a volumetric filler, and the cartons may or may not be overwrapped.

CUCUMBERS

Cucumbers are used as the fresh product or to make pickles or relishes. A period of 55–60 days is required from planting to harvesting. Cucumbers are subject to virus, mold, and bacterial infections and to insect beetle and worm infestation. They may, therefore, require treatment with chemical sprays or dusts to control the invasion of these deteriorative agents. Cucumbers may be harvested by hand into baskets which are picked up by a boom attached to a trailer, or they may be harvested by automatic methods and placed in lug boxes on a truck or trailer. Larger cucumbers are used for sale as the fresh product, while smaller and medium sized cucumbers are used for pickling.

At the processing plant, cucumbers to be used as fresh produce are washed, and may be covered with a thin layer of wax and polished by machine. They should be promptly cooled to 45°–50°F (7.2°–10°C) and held at this temperature until sold to the consumer. In this condition they have a high-quality storage life of 10–14 days. At temperatures lower than 45°F (7.2°C), they may develop pitting or dark-colored, watery areas.

Pickles are manufactured as sweet, sour, dill, kosher-dill, etc. In the manufacture of pickles, small or medium-sized cucumbers are submerged in a solution of 10% salt (sodium chloride) and allowed to ferment. During fermentation, which takes several weeks, the sugar in he cucumbers is gradually utilized by bacteria, and salt penetrates the ucumber. The salt in the brine is increased to 15%. (Dill-type pickles are ermented in a less concentrated salt solution.) After fermenting, the ucumbers are soaked in warm water, then packed in glass jars with some ombination of vinegar, sugar, spices, garlic, etc. Pickles in jars should e pasteurized in hot water by bringing their temperature up to 60°–180°F(71.1°–82.2°C). Some cucumbers are sliced vertically or orizontally prior to packing them. Sliced pickles are covered with a veak solution of salt, vinegar and spices, with or without sugar, and heat-rocessed as indicated.

Relishes may be made from fermented cucumbers which are freshened, hopped, and mixed with vinegar and spices. Since they are used with ther foods, some type of gum, such as locust bean gum, may be added o relishes for liquid retention.

LETTUCE

Lettuce is used entirely as a fresh product, since when heated or rozen, the leaves wilt. This is unacceptable, because lettuce is expected o be crisp and firm when eaten. There are a number of varieties of ettuce—the open head or loose leaf, the Romaine, the iceberg and others. ceberg lettuce, used in this country, requires a period of 75–80 days from lanting to maturity. Lettuce plants may be sprayed or dusted with hemicals to prevent the invasion of microorganisms and insects during he growing season. Lettuces are harvested by hand, but may be ollected on conveyors which place them in hampers or baskets on a ruck which takes them to the processing plant. At the processing plant, ettuces are trimmed, washed and placed in crates; the crates are placed n refrigerated freight cars or trucks. The cars or trucks are then moved nto large metal chambers. The chambers are closed and subjected to acuum, which accelerates the evaporation of moisture from the lettuce, owering the temperature of the product to about 33°F (0.56°C) through vaporative cooling. During shipment and until sold to the consumer, ettuce should be held at 32°–34°F (0°–1.1°C). At this temperature, it las a storage life of 2–3 weeks.

MUSHROOMS

Mushrooms are sold fresh and canned, and a small number are frozen ıs an ingredient of various cooked foods. There are a number of types of edible mushrooms, but the variety commonly used for food is the champignon. Mushrooms are grown in darkened humid rooms (e.g. cellars) in a mixture of horse manure, straw, and loam which has been composted. The period from planting to harvesting is 50–85 days. Harvesting may continue for about 2 1/2 months. High humidity and temperature of about 70°F (21.1°C) are required for growing. Mushrooms are harvested by hand and placed in baskets. Those to be sold as fresh are cooled to 32°–34°F (0°–1.1°C). Fresh mushrooms have a high-quality storage life of about 5 days at 32°F (0°C), 2 days at 40°F (4.4°C) and day at 50°F (10°C). The freshly handled product is usually sold within 4 hr after harvesting. Mushrooms to be canned (some mushrooms may be sliced) are soaked in a tank of water for 10–15 min and washed with sprays of water. They are then blanched in water at 175°–180°F (79.4°–82.2°C) for 8–10 min in order to shrink the product, which allows proper fill of the container, and to prevent excessive darkening. The blanched and cooled mushrooms are placed in cans by hand and covered with a hot solution of 1.5% salt and 0.2% citric acid. The acid is used to prevent excessive darkening. In some instances, the solution is not preheated, in which case, open cans are heated in steam until the product reaches a temperature of 150°F (65.6°C). After sealing the cans the product is heat-processed. The growth medium for mushrooms may contain the organism which causes botulism. Therefore, great care should be taken to make sure that, during heat processing, all the product in the container is subjected to that degree of heat necessary to provide at least a minimum botulinum kill (the equivalent of heating all parts to 250°F (121.1°C) for a period of 2.5 min). Many wild varieties are poisonous, so only commercial varieties should be consumed.

OLIVES

Olive trees are started in nurseries as rooted cuttings and allowed to grow for several years before being set out in orchards. The desired varieties are grafted onto the seedling stock as 2- or 3-year stock. Trees are pruned and thinned yearly, and bear fruit after about 5 years. Scale insects affect olive trees; hence they may be sprayed with chemicals during the growing season. The trees may bear fruit for many years. Olives are used for the production of olive oil (described in Chapter 24) for the production of canned black or ripe olives, and green or green stuffed olives, usually packed in glass jars. The fruit is harvested when still green, just before it would turn pink. Large and unblemished olives are used for canning as black or green olives, small and blemished fruit or oil or for chopped olives. Olives must be treated with lye solution (sodium hydroxide) to destroy or decrease the content of a bitter substance.

In the production of black olives, the sized, washed and inspected fruit is placed in tanks and covered with a 1–1.5% lye (sodium hydroxide) solution at 60°F (15.6°C) for 4–8 hr, the mixture being stirred occasionally. The lye solution is then drained off, and the fruit is exposed to air for 3–6 hr with occasional stirring, to oxidize and set the color (to black). The fruit is then covered with water, agitated by compressed air and held for 3–4 days. It is next drained and again treated with a 0.5–0.75% lye solution for 3–5 hr, then drained and washed once more. Black olives may or may not be pitted prior to canning. The prepared olives are sized, inspected, placed in cans volumetrically and covered with a boiling 2.5–3% salt solution. The cans are then sealed and processed at high temperatures.

Green olives are placed in a brine (sodium chloride) solution (2.5–5%), the salt content being increased daily until it reaches 7.5–10%. The fruit is held in this manner for 30–45 days, during which time it undergoes a fermentation, due to the growth of lactic acid bacteria. After washing without agitation in tanks of water for several days, the fruit is treated with lye solution, as in the case of black olives, then rewashed. The fruit is inspected, graded for size, pitted or not pitted, and canned as in the manner of black olives. Olives in large containers (No. 10 cans) must be exhausted (heated) to bring the temperature in all parts to 180°F (82.8°C) prior to sealing the cans, then heat-processed at high temperatures. Green olives may be pitted and stuffed with pimientos or nuts prior to canning. In some instances, olives (usually pitted) preserved in a 10% brine (or higher concentrations) are shipped in barrels to repackers who freshen the pickled product, stuff the individual fruit with pimientos or nuts, and repack it in glass jars. The fruit is then covered with a hot 2.5–3% brine, the jars sealed and the product heat-processed at high temperatures.

ONIONS

Onions are utilized mostly as the fresh product, but small quantities are canned, pickled or frozen. Onions require 100–110 days from planting to harvest. The soil may require chemical treatment to prevent rotting, and the plants may be sprayed or dusted to prevent insect infestation. Onions may be harvested by mechanical diggers and placed in bags, baskets or lug boxes which are brought to the processing plant on trucks.

At the processing plant, onions are topped (stems are removed), dry cleaned, and placed in mesh sacks or bags holding 50 lb (22.5 kg) of product if they are to be stored. Stored onions should be held at 32°–35°F (0°–1.7°C) and at low relative humidity (no ice). Higher temperatures promote sprouting and high humidities promote decay. When distributed to the consumer, they are weighed into mesh sacks in quantities of 2 or 3 lb (0.9 or 1.4 kg). Some onions, especially small ones, are canned. The onions to be canned are topped by hand, dry cleaned, and peeled in a flame peeler. The heat from open flames loosens the skin which may then be removed with sprays of water. After washing, the onions are trimmed and inspected. They are then filled into cans volumetrically and covered with hot canner's brine (weak solution of sugar and salt). The cans are then sealed, and the product is heat-processed. Frozen onions are used primarily with other vegetables. They are precooked in steam before freezing because they are used as a cooked vegetable. Onions are loose frozen on chain mesh belts in a cold air tunnel, then stored in metal containers prior to mixing with other frozen vegetables, such as peas. Freezing and storage requirements are the same as for other vegetables. Some onions are dried, usually in the form of minced onions. They are topped, peeled, washed and inspected in the usual manner, then diced by machine. The raw, diced product is then dried to a moisture content of about 5% on metal mesh belts, where temperature and humidity can be controlled. They are then packaged in glass containers.

PEANUTS

Peanuts are legumes, and botanically classified as vegetables. They require more than 100 days from planting to maturity, producing a low-growing, vine-like plant, the pods (shells) containing the nuts which are formed underground. The shoots bearing the pods are first formed on the plant above the ground but eventually insert themselves into the soil. The plants are subject to invasion by molds, bacteria, viruses, beetles and weevils; therefore, they may require dusting or spraying with chemicals during the growing season. Peanuts are harvested by pulling the vines and arranging them in rows to dry for several days. During this drying, the moisture content of the nuts is reduced from about 50 to about 25%. The nuts, in the shell, are then removed from the vines by machine and are further dried to prevent heating and spoilage. This is usually done in bins through which warm air is blown, the peanuts being piled in the bins to depths of 5–6 ft (1.5–1.8 m). The air used for drying should not exceed 10°–15°F (5.56°–8.34°C) above ambient (outside) temperatures. Dried to a moisture content of 5–8%, peanuts have an almost indefinite storage life at 48°–50°F (8.9°–10°C). Peanuts in the shell are roasted in hot air to a moisture content of about 4%. The temperatures used are sufficiently high to eliminate the raw bean-like flavor of the uncooked nut. The nuts may be removed from the pods by machine after which the thin skins covering each nut may be removed by machine, if desired. In general, peanuts with or without skins, which are to be sold in the shelled state, are cooked in hot vegetable oil at 300°–380°F (148.9°–193.3°C) until slightly browned or until the moisture content has been lowered to about 2%; then they are salted. Some peanuts are dry roasted (without using oil for blanching or cooking).

Peanut butter is prepared by grinding the nuts to a fine consistency, mixing with about 3% salt (sodium chloride) and small quantities of emulsifiers, such as monoglycerides. The emulsifiers are used to prevent separation of peanut oil from the other components. In preparing peanut butter, the nuts are first blanched or fried in oil, to develop color and flavor, and to lower the moisture content to about 2%. Some types of peanut butter are prepared this way, but with chunks of broken blanched peanuts mixed with the ground nuts to provide the chunky texture.

Peanut oil may be obtained from peanuts by first subjecting the nuts to pressure, the nuts being enclosed by pressing cloths, then further extracting the cake with hydrocarbon solvents. The presscake is generally used as cattle fodder.

GREEN PEAS

Green peas are mostly canned or frozen. Only small quantities are sold in the pod as fresh. A period of 55–75 days is required from planting to harvesting. Past the optimum maturity, peas have a lower quality. Peas are subject to bacterial diseases and mold and to infestation with insects. Therefore, the plants may be sprayed or dusted with chemicals for protection. Peas are harvested as are lima beans. The vines are mowed and passed through a viner which has rotating paddles that strike the pods and liberate the peas. The peas fall through perforations in the shell of the viner and are collected in lug boxes. The vines pass through the viner and are cut up and used for mulch in the growing area or are stacked in piles and allowed to ferment to form ensilage for cattle feed. Peas should be processed within 3–4 hr after vining, if they are to be frozen, since they soon take on off-flavors and may ferment and become sour, especially if held at high ambient temperatures. Peas may be cooled in cold water prior to processing, especially if they are to be canned, but freezing varieties develop tough skins under such condition and off-flavor development may not be entirely prevented by precooling

At the canning plant, peas may be size-graded or subjected to tenderometer tests using a machine which determines tenderness by the pressure required to crush a certain volume of peas. They are then washed, sometimes in detergent solutions, then in clear water. The washed peas are next blanched in water at about 210°F (98.9°C) for 1– min, the time depending upon the maturity and size of the peas. After cooling, the peas are inspected to remove detritus, filled into can volumetrically, and then covered with a hot weak solution (about 170° [76.7°C]) of sugar (2–4%) and salt (1–2%). The cans are then sealed and heat-processed.

Peas which are to be frozen are handled in much the same manner a those which are canned, except that different varieties are used fo freezing. Due to the fact that off-flavors are more easily detected in frozen peas, they should be processed within 4 hr after vining, even when ambient (outside) temperatures are not higher than 70°F (21.1°C). A higher ambient temperatures, off-flavors develop in shorter periods After blanching for about 1 min in water at 210°F (98.9°C), the peas are cooled in flumes of water or in cylindrical reels fitted with sprays. The are then passed through a quality grader—a cylindrical trough of brine o such density that peas of the desired maturity float and can be separated at the top. The over-mature peas are more dense, and sink through the brine; they can be separated at the bottom layer. The more mature pea may be packed as a grade of lower quality; they may be canned or, if too mature to make a good quality canned product, they may be discarded After passing through the quality grader, the peas are washed in wate and passed over an inspection belt where weeds, like thistle buds, as wel as stalks and other detritus, are removed. The vegetables are then filled into cartons volumetrically; the cartons are frozen, and may or may no be overwrapped with plastic. Some peas are individually frozen on meta mesh chain belts that pass through a tunnel through which cold air i blown. The frozen peas are then packed in cartons or plastic bags as loose frozen, free-flowing product, or they may be placed in large meta containers and held in frozen storage until mixed with other vegetable and packaged in cartons or plastic bags to be sold as a mixed vegetabl product.

POTATOES

White potatoes, of which there are many varieties, are used mainly a a fresh product. Some are preserved by freezing or drying. Other varietie

re best suited to use as boiled or mashed potatoes. Some varieties make
a superior baked product, while still others have good all-purpose
qualities. Potato plants are grown from sprouted potatoes which have
been cut so that each piece includes a sprout and some of the tissue. A
period of 60–70 days is required from planting to harvesting. Since potato
plants are subject to diseases caused by bacteria, molds and viruses and
to infestation by beetles, aphids, worms and weevils, they may need
spraying or dusting with chemicals. Potatoes are dug by machine and
placed in barrels by hand or placed in trucks automatically. They are
trucked in bulk or in barrels to the processing plant. At the processing
plant, potatoes to be sold as fresh are passed through a dry reel to remove
soil. They are then washed, dried, and packed in paper or plastic bags of
5- or 10-lb (2.37 or 4.5 kg) capacity. White potatoes, held at temperatures
above 50°F (10°C), lose sweetness as their sugar is converted to starch.
At 40°F (4.4°C) or below they become sweeter, as their starch is
converted to sugar. For ordinary purposes, when sold in the fresh state,
high sugar content is not desirable. Also, if potatoes are to be canned as
such or used as a component of corned beef hash or fish cakes, high sugar
content is not desirable, since during heat-processing, off-flavors and off-
colors may develop as a result of nonenzymatic browning. On the other
hand, if potatoes are to be processed as a frozen product, such as French
fried, hash-browned, etc., the sugar content should be high enough to
provide some color without prolonged heating. The sugar content in
potatoes that are used in large quantities for the manufacture of potato
chips must be controlled. If the sugar content is too low, the desirable
light brown color will not be attained during frying. If the sugar content
is too high, the potato chips will burn or become black during frying.
Because of these changes in sugar content, white potatoes may have to be
conditioned or held at a particular temperature for 1–3 weeks prior to
processing. Potatoes handled as fresh may be held for 2–4 months at 40°F
(4.4°C) without sprouting. Generally, they are held at 40°–45°F
(4.4°–7.2°C) prior to shipment, since during transportation and subsequent handling, they will be held at higher temperatures which will
cause the conversion of sugar to starch.

Frozen white potatoes are processed as French fried, baked stuffed,
hash-browned, or as some other prepared product. Frozen French fried
potatoes are prepared from the fresh product. The potatoes are either
washed, peeled in lye, and washed in a neutralizing weak acid solution or
they are heated in steam at 80 psig (5.6 kg/cm^2) for about 10 sec and
subjected to water sprays to remove the peel. They are then inspected to
remove eyes, etc., after which they are treated in a weak solution of citric
acid and sodium bisulfite. This treatment prevents discoloration (brown-

ing) due to enzyme action. The potatoes are then cut into French-styl
pieces. If the potatoes have not been conditioned by holding at th
temperature that regulates the sugar content to the desirable concentra
tion, they may have to be blanched in water at about 180°F (82.2°C) fc
sufficient time to remove excess sugars or, if the sugar content is too low
they may be heated in weak sugar (glucose) solutions. Blanching c
precooking is done in water or free-flowing steam for about 2−4 mi
after which the potatoes are drained, then cooked for short periods in hc
vegetable oil at 350°–375°F (176.7°–190.6°C). Deep-fat frying is done fo
only sufficient time to give the French fried potatoes a light brown colo
After cooling, the potatoes are frozen.

Baked stuffed frozen potatoes are baked whole with the skin on. The
are then cooled and cut in half lengthwise. The inner material i
removed, leaving two half shells of the skin and outer layers of the potatc
The inner material is then mashed and mixed with cooked onions
margarine and various flavoring materials. The half shells are then fille
with the prepared mashed product, placed in cartons, and frozen.

Hash-browned potatoes are prepared in the manner of French frie
potatoes except that they are frenched to a smaller size and cooked in oi
for somewhat longer periods to provide a darker color. Hash-browne
potatoes are packaged in cartons and frozen.

Some small-sized potatoes are canned. These potatoes are peeled b
heating in lye and washed in a weak acid solution. The peeled potatoe
are then filled into cans by hand, covered with a 1–2% salt solution, th
cans exhausted by heating in free-flowing steam for 10 min, closed anc
heat processed.

Potatoes preserved by drying are mostly used as mashed potatoes b
rehydrating them just prior to use. Some dried potatoes are produced a
specialty products, such as scalloped potatoes.

SWEET POTATOES

There are several varieties of sweet potatoes, most of which are sold i
the fresh state, small quantities being either canned or frozen. A perio
of 70–90 days is required from planting to harvesting. Since swee
potatoes are subject to diseases caused by bacteria and molds and ma
be infested by beetles, worms, weevils, flies or moths, the soil may hav
to be treated with chemicals and the plants sprayed or dusted wit
chemicals. Sweet potatoes may be harvested mechanically or manually
Generally, they are brought to the processing plant in bulk. Those to b
handled in the fresh state are dry cleaned to remove soil, washed, place

bags holding 25 lb (11.3 kg) or more, and cooled. Since sweet potatoes ave a relatively slow respiration rate, they can be brought to storage mperature over a period of several days. Fresh sweet potatoes should be eld at 50°–55°F (10°–12.8°C) at which temperature they have a storage fe of 4–6 months. Sweet potatoes to be canned or frozen are dry cleaned, ashed, and then peeled in steam or in a hot lye solution, as are white otatoes. However, prior to peeling, they may be graded for size. Canned veet potatoes may be handled as a solid pack or as individual potatoes acked in canner's brine (2–4% sugar and 1–2% salt). Solid packs are repared by cooking the potatoes, mashing by passing them through a ulper, then filling the product into cans. The filled cans are then heated free-flowing steam so that all parts of the product have a minimum mperature of 160°–180°F (71.1°–82.2°C). The cans are then sealed and eat-processed. Sweet potatoes may also be precooked and canned whole. this case, they are covered with canner's brine and the open cans are eated in free-flowing steam until the internal temperature of the otatoes reaches 160°–180°F (71.1°–82.2°C). The cans are then sealed, nd the product is heat-processed. Sweet potatoes to be preserved by eezing are peeled as is the canned product. They are then precooked in eam at 240°F (115.6°C) for 5–25 min, depending on the mass of the roduct, and then cooled. The cooled potatoes are then treated with tric acid (about 0.3%), then passed through a pulper. The pulped or ashed product is filled into cartons volumetrically, the cartons sealed ith or without overwrap, and the product is frozen.

OYBEANS

Soybeans are legumes which are planted from seed and require 0–130 days from planting to harvesting. At the end of that period, the aves are dry, and the moisture content of the beans is about 9–10%. In le United States, soybeans are grown mainly in the north central states, pecially in Illinois. They are also grown in Arkansas and in the lississippi delta. The plants are subject to infection by molds, viruses or acteria and to infestation with insects; therefore, dusting or spraying ith chemicals may be necessary during the growing season. The plants re reaped mechanically by a combine which cuts and gathers the plant nd threshes the beans from the pod. The beans are collected in trucks nd brought to processing plants in bulk. Soybeans are generally not aten in the fresh state or as a canned or frozen product, but rather as rocessed products.

Soybeans contain about 20% oil and 40% protein in the freshly harvested state. The oil from the ground beans is usually extracted by pressing or by extraction with solvents, after cracking the beans to loosen the hulls which are then removed with suction. Much of the oil obtained from soybeans is used for the manufacture of margarine. Soybeans contain an anti-growth factor and this must be destroyed by heating before the press cake can be used for animal feed or human consumption. Much of the press cake from which the oil has been extracted is used for animal feed supplements.

Soybean protein may be used after grinding the beans, extracting various proportions of the oil and adding some lecithin as an emulsifier. In such form, it may be used as an extender in sausage products, in blended foods, in baby foods and in pet foods. The defatted soybean, extruded after heating, may be extracted with water, aqueous alcohol or dilute acid to dissolve and remove carbohydrates and other ingredients resulting in a product of higher protein content (about 66–70%). As such, this material may be used in processed meat products, baked goods and breakfast cereals.

A high-protein soy concentrate (about 90–97% protein) may be obtained by extracting soyflakes (oil removed) with water in which the pH is controlled to optimize the removal of carbohydrates and other materials. This product is used in sausages and canned meats, in coffee whiteners, in whipped toppings, in frozen desserts and in cheese spreads. Thermoplastic extrusion of fat-extracted soybean flour together with water (flavoring and coloring ingredients may be used), produces a product containing 50–53% protein. This is used as an extender for meat products, such as hamburgers. A soybean base is made from soybean flour by first extracting carbohydrates, etc., with aqueous alcohol and dilute acid, dissolving the remaining material in an alkaline solution and precipitating it in a coagulating bath. This material (about 90% protein) is fibrous, and when mixed with fats, flavoring, coloring, etc., and then texturized, can be made to simulate cooked beef, chicken, bacon, crab meat, scallops, nuts, fruits and vegetables.

In some Asian countries, soybean milk is prepared by soaking the diced beans for a few hours in water, mashing the beans and boiling in water for 30 min (3 parts of water to 1 part of mash), then straining out the solid particles. This soybean milk may be made from whole or defatted beans. The milk may be used as such, or it may be treated, while hot, with calcium or magnesium salts, with rennet (an enzyme) or with lactic acid to precipitate a curd which is then drained and pressed. The curd is eaten in different forms.

It should be noted that soybean protein is not a complete protein as far s human requirements are concerned, being deficient in the amino cids, methionine and tryptophan. It is however, high in lysine, an amino cid deficient in many vegetable proteins. When it is used as an extender ith animal proteins, the resulting mixture is generally adequate in rotein requirements.

PINACH

There are a number of varieties of spinach which may be harvested in he late spring or late fall. Periods of 35–50 days are required from lanting to harvesting. Spinach is subject to invasion by molds, viruses, phids or worms. It may, therefore, be dusted or sprayed with chemicals uring the growing season to control such pests. Mechanical cutters and aders are used to harvest spinach, which is brought to the processing lant in bulk. At the processing plant, the spinach leaves are trimmed nd inspected to remove dead leaves, passed through a dry reel to remove etritus, then washed in a soaker-type water tank in which the product subjected to heavy sprays of water. Handled in the fresh state, spinach usually packed in plastic bags of 1-lb (454- g) capacity. Spinach eteriorates rapidly and should be quickly cooled to 32°–33°F)°–0.56°C) and held at this temperature until sold to the consumer. pinach to be canned may be handled as whole or as chopped spinach the latter type is cut or chopped by machine prior to blanching). It must e blanched in order to fill the cans with the desired amount of product. lanching is carried out in free-flowing steam or in water at 180°–185°F 2.8°–85°C) for about 4 min. After blanching, the spinach is subjected the pressure of a metal roller as it is carried on a conveyor belt to queeze out some of the water picked up during blanching. The hot pinach is placed in cans manually by workers wearing rubber gloves. he chopped product may be filled volumetrically by machine. After lling, the spinach is covered with a 2–3% salt solution at 190°–200°F 7.8°–93.3°C). The open cans are then heated in steam until all parts ave a minimum temperature of 180°F (82.2°C) (about 10 min). Next, he cans are sealed and heat-processed. Spinach to be frozen is steam lanched for 2–4 min and then cooled in a tank of water equipped with ater sprays. It is then subjected to the pressure of a metal roller to queeze out some of the water picked up during blanching and cooling. he cooled product is filled into cartons by hand, the cartons sealed, with r without overwrap, and the product is frozen.

SQUASH

There are a number of types of squash or pumpkins. Summer and zucchini types require 50–60 days from planting to harvesting. Hubbard type squash (commonly called pumpkins) and others require 90–100 day from planting to maturity. Since squash is subject to diseases caused by molds, bacteria and viruses, and to infestation by bugs and borers, i may need spraying or dusting with chemicals during the growing season Squash is harvested by hand and brought to the processing plant in bulk At the processing plant, the summer-type is washed and may be packaged in units of 2–4 small specimens in cardboard container overwrapped with plastic. The larger winter-type varieties are usuall washed, drained, then handled in the unwrapped state as individua specimens. Summer-type squash has a storage life of 1–2 weeks at 40°F (4.4°C) and should be rapidly cooled to, and held at, this temperatur until sold to the consumer. On the other hand, the other varieties ar more resistant to deterioration. At 50°F (10°C) hard shell Hubbard-typ squash may be held for 6 months or longer, acorn squash for 5–8 week and butternut squash for 2–3 months. These types, therefore, do no require quick cooling. Only the hard shell or Hubbard-type squash i canned. It is washed, cut into chunks by machine and then cooked in free-flowing steam for 25–30 min. It is then passed through a pulpe which macerates the product and removes the outer shell. About 1/2% o salt is then added and mixed. Spices such as cinnamon, ginger and mac may be added and mixed with the product. Some molasses may also b added. The hot pumpkin mixture is added to cans volumetrically, an the cans are sealed and heat-processed. Hubbard-type squash to b frozen is handled much the same way as canned squash, except that it i prepared as a vegetable product rather than as a pie filling, as is the cas with the canned product. Generally, therefore, the frozen product ha only salt added. It is also cooled after cooking and before filling into packages volumetrically to be frozen. Summer-type squash to be frozer is washed, mechanically cut into slices 1/2 in. (1.3 cm) thick, blanched for 3 1/2 min in water at 210°F (98.9°C), cooled in running water and packaged manually. The packages are closed, with or without overwrap and the product is frozen.

TOMATOES

Tomatoes are handled as the fresh product, as well as processed products such as canned whole tomatoes, tomato juice, tomato purée omato paste, ketchup and chili sauce. Tomatoes are frozen only as an ngredient of prepared products such as pizza, cooked lasagna, etc. Tomatoes require 70–85 days from planting to harvesting. Usually, the .eeds are planted in soil enclosed by glass or hot frames, and when the)lants have reached a height of about 6 in. (15.3 cm) they are trans-)lanted to the growing area. Tomatoes may be infected by molds,)acteria and viruses. They are also subject to infestation with worms and)eetles. The plants may, therefore, be sprayed or dusted with chemicals luring the growing season. Tomatoes are handled in the fresh state nostly as the mature partially ripened product and sometimes, as the irm ripe tomato. Tomatoes may be harvested by hand and placed in lug)oxes for transportation to the processing plant, or the vines may be cut .nd the tomatoes shaken off by machine. With the latter type of 1arvesting, the product is brought to the processing plant in bins.

At the processing plant the tomatoes are washed and sorted. Green pecimens are separated from the firm ripe type, and diseased specimens .re discarded. Mature partially ripened tomatoes are packaged by hand n cardboard, cellophane-topped cartons. They should be cooled to i5°–60°F (12.8°–15.6°C) and held at this temperature until sold to the :onsumer. In this condition, they have a storage life of 2–3 weeks. Firm ipe tomatoes generally are held at 45°–50°F (7.2°–10°C), at which .emperature they have a storage life of about five days. At 32°F (0°C) .omatoes may have a shelf-life of about two weeks. However, at this .emperature, they are subject to chilling injury and may lose quality. Tomatoes to be canned should be in the firm ripe condition. They are vashed and inspected, with green and rotten specimens culled out and nold and rot trimmed out from infected but otherwise good specimens. Trimming is very important since enforcement authorities have methods)f detecting mold and rot in finished tomato preparations and can :ondemn tomato products on this basis. After trimming, the tomatoes .re cored by hand or by a rotating burr, that is, the pistil and stamen section is cut out. Tomatoes to be canned as whole are scalded with sprays of hot water and cooled with sprays of cold water. This wilts the skin which can then be peeled off by hand with the aid of a knife. Some varieties have a skin which does not require peeling. The whole peeled :omatoes are then placed in cans by hand, after which a brine (2% sugar .nd 1% salt) is added to cover the tomatoes. In some instances 1% salt .nd juice from whole tomatoes or from peels and cores are added to fill the cans. The open cans are then heated in water at 175°–180°F (79.4°–82.2°C) until all parts of the product have reached 145°–150°F (62.8°–65.6°C). The cans are then sealed and heated in boiling water until all parts of the product reach a temperature of 180°–200°F (82.2°–93.3°C) to provide commercial sterility.

Tomato juice is obtained by passing cored and quartered tomatoe through a pulper and finisher to remove seeds and residual skin. Thi product is best prepared from quartered tomatoes which, directly afte cutting, are heated in steam to 180°–185°F (82.2°–85°C). This inac tivates pectinase enzymes which would otherwise break down pectin an cause separation of the liquid and solids after processing. Heating in thi manner to conserve vitamin C (the hot break method) also inactivate other enzymes and eliminates oxygen which would otherwise destro vitamin C. Juice obtained from unheated tomatoes may be passe through special machines which break up the cellulose and prevent th settling out of solids in the finished product. Juice treated i this manner is thicker than that obtained with the "hot break" method and the consistency may be somewhat less desirable than that obtaine by the latter method. Tomato juice may be heated to 160°F (71.1°C) placed in cans or bottles and the containers sealed and heated in boilin water to a temperature, in all parts, of 205°–210°F (96.1°–98.9°C). Thi process does not always provide commercial sterility, since certai bacterial spores may survive and grow out, causing spoilage. A superio process is to flash heat juice in a heat exchanger to 250°F (121.1°C), hol at this temperature for 0.7 min, cool to 200°–210°F (93.3°–98.9°C), fil into pre-sterilized cans and seal and invert the cans to sterilize th covers, holding in this manner for several minutes before cooling. In orde to increase the acidity slightly and add nutrients to the juice, smal amounts of citric acid and vitamin C may be added.

Tomato purée and tomato paste are prepared by vacuum concentrat ing tomato juice. Tomato purée must contain at least 8.37% tomat solids, tomato paste must contain at least 22% tomato solids, and heav tomato paste must contain at least 33% tomato solids. Salt may be adde to these products and baking soda (sodium bicarbonate) may be added t heavy tomato paste to neutralize some of the acid. Generally, thes products are added to the container at a temperature near 180°–200°I (82.2°–93.3°C). The containers are then sealed and inverted, allowed t stand for several minutes and cooled. Due to the low pH (high acidity) o concentrated tomato products, additional heating is not required t attain commercial sterility.

Tomato ketchup or catsup is made by concentrating tomato juice adding sugar and salt and then a vinegar (10% acetic acid) extract o spices (headless cloves, black pepper, red pepper, cinnamon, mace onions, garlic, etc.). The finished product should have a salt content o 3% and a total solids content of about 30%. This product is bottled ho and capped under vacuum to eliminate air. Removal of air prevent darkening from tannins which are extracted from the spice mixture annins turn dark in the presence of oxygen. Further heating to attain ommercial sterility is not required.

Chili sauce is made from vacuum-concentrated, finely-chopped, eeled, cored tomatoes to which a vinegar extract of spices and red eppers has been added, together with onions and garlic. This product is ottled and handled much in the same manner as ketchup.

Tomato soup, Italian tomato sauce, with or without meat, and other omato products are canned and bottled to some extent. These products ay require various heat treatments to attain commercial sterility, epending upon the pH (acidity) of the finished product.

URNIPS

Turnips require 60–100 days from planting to maturity. They may be fected by molds and viruses, or infested by aphids, bugs and worms; erefore, dusting or spraying with chemicals may be required, although isease or infestation is usually not serious. Turnips are often harvested anually and topped (stems and leaves removed) in the field. They are rought to the processing plant in bulk. Essentially all turnips are andled in the fresh state. At the processing plant, they are dry cleaned, ashed, dried, covered with a thin layer of wax, and machine polished. urnips may be held for 4–5 months at 32°F (0°C), so they may be slowly ooled to 32°–35°F (0°–1.7°C) and held at that temperature until ipped out in large bags or crates. Turnips are sold as unpackaged, dividual vegetables. They should be held at 32°–35°F (0°–1.7°C) until ld to the consumer.

22

Fruits

Fruits are botanically classified as those plant parts that house seeds; in other words they are mature plant ovaries. Fruit includes tomatoes and a few others that are considered as vegetables in the supermarket. Since the popular definition of fruit applies only to what is naturally sweet and what is normally used in desserts, it is understandable that, for example, tomatoes and olives are often treated as vegetables.

Berries belong to a class of fruit that is usually small and is very delicate. On the other hand, melons are of a class that is usually large, often with a tough, and sometimes thick, outer skin.

Fruit is often picked prior to maturity, and allowed to ripen in the distribution chain, reaching the consumer when about ready to eat. Fruit is considered to be ripe when it reaches the optimum succulence and texture and there is a desirable balance between sugars and acidity, as well as the subtle elements that contribute to aroma. Fruit that goes past its optimum ripeness enters senescence, a stage of over-ripeness and breakdown. In this stage, the texture loses its firmness, succulence is diminished, and sugars, acids and aroma elements generally all decline in concentration. Some fruit, like the banana, has an early senescence, and deterioration, once begun, is rapid. On the other hand, some fruit, like apples, resists the onset of senescence as well as its progression. The onset of senescence can be controlled by keeping the fruit at the lowest temperature that it can tolerate and by increasing the amount of atmospheric carbon dioxide to a controlled level. Too much carbon dioxide can be harmful.

All fruit preserved by canning should be heat-processed to attain commercial sterility. Whereas vegetables and certain other foods require the application of high temperatures [240°F, 250°F (115.6°C, 121.1°C) or higher] for significant lengths of time to attain commercial sterility, most fruit is sufficiently acid (pH usually below 4.5) that commercial sterility can be attained by heating the containers in boiling water to the point where all parts of the product reach a temperature of 180°–200° (82.2°–93.3°C).

All fruit preserved by freezing should be brought to a temperature of 0°F (−17.8°C) or below during freezing and thereafter held at 0° (−17.8°C) or below until sold to the consumer. Fruit may be packed in retail-sized containers and frozen in one of three ways: (1) The containers are placed on trays, the trays are placed on racks and frozen as the racks are moved through a tunnel in which blasts of cold air are circulated; (2) the cartons are placed on chain mesh belts which move slowly through a tunnel in which blasts of cold air are circulated; or (3) the cartons are placed on trays, the filled trays are placed between refrigerated metal plates, and the product is frozen with the containers in contact with the cold plates. Therefore, except in those cases where the method of freezing is different from those indicated in the foregoing, no description of freezing methods will be given.

There is a large variety of fruit used in the United States with varying degrees of popularity. Following are descriptions of planting, harvesting processing and general handling of the more important classes of fruits.

BLACKBERRIES

Blackberries grow on 2 year old brambles or canes. They are subject to diseases caused by molds (rust) or bacteria (crown gall) and to infestation by insects (leaf-miners). The plants may, therefore, require spraying with chemicals during the growing season. Blackberries are harvested by hand picking. At the processing plant, they are washed with sprays of water, then passed over an inspection belt where leaves, twigs, etc., are removed. Blackberries handled in the fresh state are placed in baskets holding 1 pt or 1 qt (473 ml or 946 ml), overwrapped with cellophane and cooled to refrigerator temperatures above freezing. They should be cooled quickly to 31° to 32°F (−0.56° to 0°C) and held at this temperature until sold to the consumer. Even at this temperature, the storage life of fresh blackberries is only 3–4 days. Blackberries are frozen for the bakery trade or for the manufacturing of jams and jellies. For the bakery trade blackberries are individually frozen on metal mesh belts in a cold air tunnel, and then placed in slip-cover metal cans holding 20–30 lb (9.1–13.6 kg). The filled cans are covered and placed in frozen storage. Blackberries for jam and jelly manufacturing are cleaned, washed and inspected, mixed with sugar (5 parts berries to 1 part sugar), placed in slip-cover metal containers holding 30 lb (13.6 kg) of product and frozen in bulk in cold-air rooms.

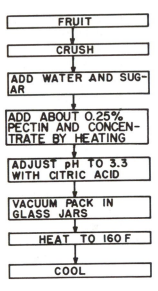

FRUIT

CRUSH

ADD WATER AND SUG-AR

ADD ABOUT 0.25% PECTIN AND CONCEN-TRATE BY HEATING

ADJUST pH TO 3.3 WITH CITRIC ACID

VACUUM PACK IN GLASS JARS

HEAT TO 160 F

COOL

FIG. 22.1. MANUFACTURE OF FRUIT JAM
See metric conversion tables in Appendix.

In the manufacturing of blackberry jam (see Fig. 22.1 for general sequence of jam production), the fruit is defrosted and sugar is added to attain a 1:1 proportion of fruit to sugar. Some water, and about 0.25% pectin are also added. The pectin must be dispersed in the water prior to adding to the mixture. The mixture is then heated in an open steam kettle until sufficient water has been evaporated to provide a soluble solids content of about 68%. The pH is then adjusted to 3.3 with citric acid. The hot preserve mixture is added to glass jars and the jars are capped under vacuum and passed through sprays of hot water to bring the temperature of the product to about 160°F (71.1°C) in all parts. The jars are then cooled. Blackberry jelly is manufactured in a similar manner to blackberry jam, excepting that the fruit is passed through a finisher to obtain juice free from seeds and cellulosic material. About twice as much pectin is used as for jam, and the mixture is evaporated to a soluble solids content of 65%. At the finishing point, citric acid solution is added to adjust the pH to 2.9–3.2.

BLUEBERRIES

Blueberries are harvested from both wild plants and cultivated varieties. In the United States, blueberry cultivation is mainly in New Jersey,

Michigan and North Carolina; elsewhere, cultivation is negligible. Blu
berry bushes require an acid soil and yield fruit the first year aft
planting or the first year after the plants have established growth. The
are many varieties of both high- and low-bush blueberries. The fruit
the blueberry ripens 50–65 days after the blooms occur. Blueberry plan
are subject to mild infections and may have to be sprayed or dusted wit
chemicals or pruned to eliminate infected parts. Low-bush blueberri
are harvested mostly by pronged hand scoops which strip the berri
from the twigs. High-bush blueberries are harvested by shaking devic
which drop the fruit onto catching frames. The harvested berries a
brought to the plant in picking boxes holding about 20 lb (9.1 kg).

Blueberries are chiefly frozen, but some are sold as fresh, and sma
quantities are canned. At the processing plant, the berries are passe
through a fanning mill where small twigs and leaves are separated by ai
They are then graded for size, washed in a flotation-type washer, an
passed over an inspection belt where green or partially-ripe specimen
are picked out by hand. Blueberries handled as the fresh product ar
usually hand-poured into small baskets, holding 1 pt or 1 qt (473 ml
946 ml). The baskets are then overwrapped with cellophane, and th
product is held at refrigerator temperatures above freezing. Fresh blue
berries should be cooled to 32°–35°F (0°–1.7°C) and held at thi
temperature until sold to the consumer. In this condition, they have
storage life of 4–8 weeks.

Blueberries to be canned are cleaned and inspected in the usua
manner, placed in cans by hand or with scoops, and the cans then fille
with water or with a 10–30% sugar solution. The open cans are the
exhausted or heated in free-flowing steam for 10 min, and finally seale
and heated for short periods in boiling water.

Frozen blueberries are generally used for baking pies or other pastrie
The usual procedure is to freeze the cleaned and inspected berries o
trays which are placed on racks and moved through a cold air tunnel. Th
frozen berries are then run through a machine which breaks up cluster
of the frozen fruit and the frozen product is packed volumetrically int
metal cans holding about 20 lb (9.1 kg) of berries. The cans are close
with slip covers and placed in frozen storage. Some frozen berries may b
mixed with dry sugar (4 parts berries to 1 part sugar) or with 50% suga
solution, placed in large metal slip cover containers and frozen in bulk i
cold air rooms. Some blueberries are used for making blueberry jams.

CRANBERRIES

In the United States, commercial cranberry production is carried ou
mostly in Massachusetts, Wisconsin, New Jersey, Washington and

egon. There are a number of commercial varieties of cranberries grown
swamp lands or under similar conditions. A period of about 4 years is
quired from planting to the first harvesting period. Since the blossoms
velop in the spring and are susceptible to frost damage, and the plants
e susceptible to freezing damage, bogs where the berries are grown may
ve to be flooded with water, as a protection against cold damage.
esel oil or chemicals may be used to control weeds and moss in
anberry bogs. Chemicals may be used to prevent infestation with
rms and other insects, and nets may be spread over bogs to prevent
sect infestation. Cranberries can be harvested by hand, using a
onged cranberry scoop, but generally are harvested by machine. When
achines are used, the bogs may be flooded with water to float the vines
om which the berries can then be shaken off by machine and collected
om the water. In other instances, the bogs are not flooded; the berries
e stripped from the vines mechanically and collected on catching
ames.

At the processing plant, cranberries are cleaned in fanning mills, then
opped some distance to eliminate soft or rotten specimens (the
fective berries do not bounce, those which are suitable for food bounce
over a barrier), then washed, first in acid or alkaline solutions to
move spray residues, then in water. Destemming is carried out in a
tating vegetable peeler fitted with a smooth bottom plate. Cranberries
ndled as fresh are packed in paper-lined wooden boxes, and the
oduct is slowly cooled to 36°–40°F (2.2°–4.4°C). Cranberries should be
ld at 36°–40°F (2.2°–4.4°C) until sold to the consumer. At this
mperature, cranberries have a storage life of several months. Cranber-
es may be held frozen prior to the manufacture of jelly or sauce. They
e placed in large metal containers, frozen in bulk in cold air rooms, and
ld in this condition until defrosted for purposes of preparing cooked
oducts.

Large amounts of cranberries are used to produce cranberry jelly and
anberry sauce, both retailed as canned products. In the manufacture of
anberry sauce, water, about twice as much sugar as fruit, and about
3% of dispersed pectin are mixed together. The mixture is then cooked
d evaporated in an open steam kettle to a soluble solids content of
%. Citric acid solution is then added to bring the pH to 3.0–3.4. In
anufacturing cranberry jelly, the clear juice, from the boiled fruit
hich has been passed through a pulper, is mixed with sugar, about
35% of dispersed pectin is added and no water is used. The mixture is
eated briefly to bring the soluble solids content to 65%, and citric acid
lution is added to regulate the pH to 3.0–3.2. Both cranberry sauce and
anberry jelly are canned in metal containers. If filled into the container
about 190°F (87.8°C), further heating is not necessary to attain

commercial sterility after sealing the containers. If packed at low
temperatures, the containers should be sealed and heated in a water bat
or with sprays of water at 185°–200°F (85°–93.3°C) for 6–10 min prior t
cooling.

GRAPES

In the United States, grapes are grown mainly in California, New Yor
Pennsylvania, Michigan and Ohio. Washington, Missouri and Arkansa
also produce some grapes. There are many varieties of grapes, but thre
are predominant in this country. Grapes are utilized to produce u
fermented grape juice, vinegar, wine, raisins, jams and jellies, and as th
fresh product for the table. Grapes are planted as vines or cuttings fro
older plants. The cuttings produce arms bearing fruit, the greatest yiel
coming after 3 years of growth. Properly pruned and cared for, vineyar
produce fruit for many years. Since grapes are subject to infestation b
insects and especially to infection by molds, the vines may need sprayir
with chemicals during the growing season.

Grapes are picked by hand as bunches, and brought to the processir
plant in baskets or hampers. Grapes to be shipped as fresh are packed i
wooden crates, then precooled to about 40°F (4.4°C) in railroad cars o
refrigerated rooms. Generally, the grapes will be fumigated with sulfu
dioxide prior to or during cooling to prevent mold growth. Grapes tha
are to be stored for future shipment should be packed in crates, precoole
to 36°–40°F (2.2°–4.4°C), placed in refrigerated storage (29° to 32°
[−1.67° to 0°C]), and fumigated with sulfur dioxide. They should be hel
in this manner until shipped. Periodic refumigation with sulfur dioxid
may be required to prevent spoilage by molds. Under these condition
grapes have a storage life of 1–7 months depending mainly upon th
variety.

The Concord variety is chiefly used for the manufacture of grape juic
The grapes are washed in acid or alkaline solutions, then in water t
remove spray residues, then destemmed and crushed by mechanic
means. The crushed grapes are heated to about 180°F (82.2°C) to extrac
pigment from the skins, after which the heated material is subjected t
mechanical pressure while enclosed in cotton press cloths. The juice
then filtered, pasteurized by heating to 170°F (76.7°C), and stored i
bulk in covered tanks at about 40°F (4.4°C). This provides for th
separation of tartaric acid salts (cream of tartar or potassium hydroge
tartrate). The juice is then syphoned off from the tartrate and treate
with enzymes, which break down pectins, or with casein for purposes o

arification. It is then filtered and bottled. The bottles are capped, and
en pasteurized by heating in water at 170°F (76.7°C) for 30 min.

Considerable quantities of wine are manufactured in the United
ates. The European varieties of grapes are mainly used for making
ne. The various procedures used in the production of wines will not be
scribed, but the pressed juice, usually after treatment with sulfur
oxide or compounds which liberate sulfur dioxide (to destroy un-
sirable types of yeast), is allowed to ferment by natural yeasts
rviving the sulfur dioxide treatment. During fermentation, sugars are
nverted to ethyl alcohol until a level of 12–14% alcohol is reached.
andies may be made by distilling wine containing ethyl alcohol, and
ese may be used as such for fortifying wines to obtain a higher alcohol
ntent.

In the manufacture of wine vinegar, fermented grape juice (containing
cohol) is allowed to drip over wood shavings in an enclosed cylindrical
ntainer. The shavings have been previously soaked in a high-quality
negar. Air may be introduced into the generator under pressure.
acteria of the *Acetobacter* group present (from the vinegar) on the
avings, convert the ethyl alcohol in the wine to acetic acid.

$$CH_3CH_2OH + O_2 \xrightarrow{\text{bacteria}} CH_3COOH + H_2O$$

Ethyl alcohol — Oxygen — Acetic acid — Water

he effluent from the vinegar generator may be collected and recycled to
tain a complete conversion of the ethyl alcohol. The finished vinegar
ay be stored for several months at 40°–50°F (4.4°–10°C), then filtered,
ttled and pasteurized.

Raisins are produced from grapes by sun drying and artificial drying.
sun drying, the grapes are picked as bunches and placed in a single
yer on wooden trays between the rows of grape vines. The trays are
ted to face the sun. After they are partially dried, the grape bunches
e turned and allowed to dry to the point where no juice can be pressed
it. The trays are then stacked, and the air drying is continued in the
ade until a moisture content of about 17% is reached. After drying, the
isins are placed in sweat boxes to equilibrate, or to even out, the
oisture which is present, and eventually they are packaged for retail in
axed cardboard cartons of 1 lb (0.45 kg) capacity, or in larger contain-
s to be sold to the bakery trade.

In artifical drying, grapes are first dipped in 0.25–1% lye (sodium
ydroxide) solution at 200°–212°F (93.3°–100°C) for 2–5 sec to remove a
atural wax which impedes drying and to check or crack the skin of the
ape to facilitate drying. They are then washed, placed on trays, and
bjected to the fumes of burning sulfur in an enclosed shed over a

sulfur-burning pit. This causes the grapes to absorb sulfur dioxide whic prevents nonenzymatic and enzymatic browning during drying. Henc instead of having the brown or black color of raisins produced fror grapes without sulfuring, they will have a light yellow color when driec They are then dried in artificial dryers at a temperature not exceedin 165°F (73.9°C) and at low relative humidity, about 25%. After th moisture content has been lowered to about 16–18%, the grapes ar packaged as previously indicated.

Some grapes, e.g., the Concord variety, are used for producing jell and jam, with jelly making up the larger part. In jelly making, sugar an about 0.25–0.3% of dispersed pectin are mixed with the clarified grap juice, and the mixture is concentrated in open kettles to a soluble solic content of about 65%. Citric acid solution is added to adjust the pH t 3.0–3.2. The product is then poured into glass jars, vacuum capped, an sprayed with hot water to bring the temperature of all parts to abou 160°F (71.1°C), after which the product is cooled.

Only small quantities of grapes are canned or frozen, and then only a ingredients of mixtures, such as fruit salad.

RASPBERRIES

There are many varieties of raspberries that are either red, black c purple in color. The purple varieties have been produced by cros breeding the red and black varieties. Raspberries grow on canes th second year after planting, and since the canes produce fruit only once they must be pruned each year. Raspberries are subject to molc bacterial and virus diseases, and to infestation with beetles, borers an flies, so the plants should be sprayed or dusted with chemicals. Raspber ries are grown over most parts of the United States, but they are sensitiv to both extreme heat and extreme cold. Therefore, in certain areas, th plants may need some type of protection from extremes of weather. I harvesting, raspberries are picked by hand and placed in shallow tray for delivery to the processing plant, trays being used because the fruit i soft and easily crushed.

At the processing plant, the fruit is washed with gentle sprays of wate and drained on a metal mesh belt. If the product is to be handled in th fresh state, it is placed in baskets of 1 pt or 1 qt (473 ml or 946 ml capacity, with or without cellophane overwrapping and quickly cooled t 31° to 32°F (−0.56° to 0°C). In this condition, the fruit has a storage lif of only 5–7 days.

Some raspberries are frozen for the bakery trade and many for the
anufacture of raspberry jam and jelly. To prepare them for freezing,
e washed and drained berries may be placed in wooden barrels without
gar, then placed in a room at 0°F (−17.8°C) or below and allowed to
eze slowly. Usually, however, they are mixed with sugar (three parts
uit to one part sugar) then packed in slip cover cans of 50 lb (22.7 kg)
pacity. The covered fruit is allowed to freeze in cold rooms at 0°F
17.8°C) or below. In the manufacture of jam, the defrosted fruit is
ssed through a pulper which allows essentially all the berries with the
eds to pass through the strainer. Sugar and water are then added, the
nount of sugar depending upon whether the frozen product was packed
th or without sugar. About 0.1–0.15% dispersed pectin is then added
d the product is heated in open kettles to concentrate it to a soluble
lids content of about 68%. Citric acid solution is then added to adjust
e pH to 3.3, and the fruit is packed in glass jars and vacuum capped.
ie jars of product are then sprayed with hot water to bring the
mperature of all parts to about 160°F (71.1°C), after which the product
cooled. Few raspberries are canned.

TRAWBERRIES

Different varieties of strawberries are grown in many areas of the
iited States, harvested in the winter and spring in southern states and
late spring and summer in northern states. Strawberries require fertile
il, hence soil treatment with fertilizers is usually required. Straw-
rries are planted as 1-year-old plants which produce the next year. The
awberry plant is a perennial which, if properly cared for, will produce
several years. Strawberries are subject to diseases caused by molds
d viruses and to infestation with weevils, nematodes, slugs, bugs,
rms and mites. Therefore, both the soil and the plants may need
atment with chemicals to control such pests.
The berries are harvested by hand. For handling in the fresh state,
ey are usually picked with the calyx or cap intact, since this enhances
e keeping quality of the fruit. For processing, the berries are picked
thout the cap. The berries are placed in baskets of 1-lb (454 g)
pacity, and the baskets are placed in crates for transportation to the
ocessing plant. At the processing plant, the berries are washed,
spected on belts to remove green and rotten specimens, then repacked
baskets and crates for shipment as the fresh product. They are then
oled to 31° to 32°F (−0.56° to 0°C) and should be held at this

temperature until sold to the consumer. In this condition, they have storage life of about 10 days. In some cases, strawberries for the fre trade are precooled in refrigerated water.

Large amounts of strawberries are frozen to be sold to the consumer such or to be used for the manufacture of jams and jellies or for use in t bakery trade. Few strawberries are canned. Frozen strawberries for ret are sliced, mixed with sugar (four parts fruit to one part sugar), a packed in 12-oz or 1-lb (341 g or 454 g) cardboard, metal end, package Some strawberries are frozen individually (whole) on wire mesh belts a packaged in cartons, the cartons overwrapped and placed in shippi cases for holding in frozen storage.

Strawberries used for the manufacture of other foods are usually froz whole, in wooden barrels, mixed with sugar. A proportion of fruit to sug of 3:1 or 2:1 may be used. The barrels of fruit are placed in cold rooms $-10°F$ ($-23.3°C$) or below. In order to mix the sugar with the fruit, t barrels are rocked during filling and rolled periodically during the sever days required for freezing. The frozen fruit is sometimes stored at 10 ($-12.2°C$). Strawberry and sugar mixtures may also be packed in s cover cans holding 30 lb (13.6 kg) of product.

In the manufacture of strawberry jam, water is added to the defrost fruit mixture, as well as more sugar, the amount of sugar depending up how much sugar was added to the frozen product. About 0.25–0.3% dispersed pectin is added, and the product is heated in open kettles t soluble solids content of 65%. The pH is then adjusted to 3.3 with cit acid solution, and the preserve is packed in glass jars and vacuu capped. The sealed jars are then heated in water sprays to a temperatu of about 160°F ($71.1°C$) (all parts), then cooled. Strawberry jelly manufactured in a similar manner to that of strawberry jam, except th the fruit is first put through a finisher, after which the juice is clarifi or filtered. In manufacturing strawberry jelly, about 0.3–0.35% dispers pectin is used, and the pH is regulated to 3.0–3.2.

APPLES

In this country, apples are grown in practically every state. They grown commercially in 35 states with the heaviest production taki place in Washington, New York, Virginia, Michigan, California a Pennsylvania. There are hundreds of varieties of apples which may grown on trees produced from seedlings that were grown in nurseries, from grafts on existing apple trees. The fruit is developed on spurs whi are formed by branchlets of three or more years of growth, the t

elding fruit for many years thereafter. Fertilization of the soil and
eriodic pruning and thinning of apple trees are considered necessary for
ood apple crops. Apples and apple trees are subject to infection with
molds and bacteria and to infestation with many types of insects,
including worms, moths, mites, borers, aphids, beetles, caterpillars and
ugs. Generally, therefore, the trees must be sprayed with chemicals
nce or even several times during the growing season.

Apples are harvested by hand and brought to the processing or storage
lant in boxes or baskets. They are used in many ways—as the fresh
roduct, to produce apple juice, apple cider, and vinegar, and so on.
hey are also canned as apple slices or as apple sauce, and they are
ozen as apple slices. Some jelly is manufactured from apple juice.
ectin, used in jam and jelly making, is extracted from the peels and
ores of apples. Since apples are generally sprayed during the growing
eason, spray residues, which may be toxic to humans, must be removed
t the processing or storage plant. This is done by washing the fruit in
ilute hydrochloric acid or sodium hydroxide solutions, then thoroughly
nsing to remove all traces of acid or alkali. The tissues of apples
ontinue to respire after the fruit is picked; hence, during storage, apples
re subject to many deteriorative physiological changes, including bitter
it, brown core, internal breakdown, internal browning, scald, and soggy
reakdown. At the storage plant, apples are packed in boxes, cooled to
bout 32°F (0°C), and stored at temperatures not below 30°F (−1.1°C) or
bove 38°F (3.3°C), until shipped, the temperature of storage depending
pon the variety. Increased storage life of apples can be attained by
•wering the oxygen content and increasing the carbon dioxide content of
1e atmosphere in which they are stored. Recommended carbon dioxide
ontents vary between 1.5 and 8%, and oxygen contents between 2.5 and
%, depending upon the variety of fruit. Under the best conditions of
olding, the storage life of apples is 2–8 months, depending mainly upon
1e variety of fruit.

Apple juice, apple cider and vinegar are made from fruit not suitable
•r being marketed as the fresh product (see Fig. 22.2 for diagram of
roduction of apple juice). The fruit is decontaminated and washed,
aspected to remove decayed specimens, and ground or crushed. The
1acerated apple pulp is then covered with cotton press cloths and
abjected to pressure in racks in a hydraulic press. The juice which is
eparated from the pulp runs by gravity from the press to collecting
anks. The press cake may be broken up and pressed a second time. The
uice is cloudy and may be marketed as such or clarified by heating to
65°F (73.9°C), adding gelatin, and tannin or pectinase (to destroy the
ectin which releases particles to the bottom), and then filtering through

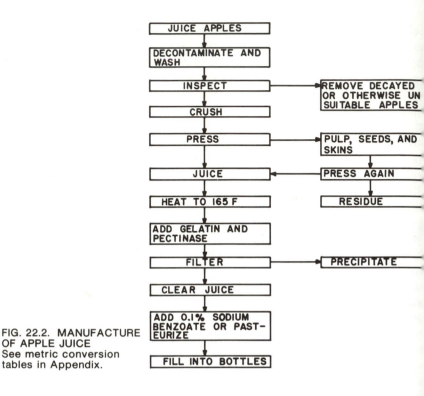

FIG. 22.2. MANUFACTURE OF APPLE JUICE
See metric conversion tables in Appendix.

diatomaceous earth. Juice sold as fresh cider is generally bottled aft 0.1% of sodium benzoate has been added to protect the product frc bacterial or yeast spoilage.

When vinegar is produced from apple cider, the juice in tanks is seed with cultures of yeast and allowed to ferment for up to several weel depending upon ambient temperatures. The fermented juice, containi ethyl alcohol, is then mixed with some vinegar and allowed to drip ov wood shavings which have been soaked in high-quality vinegar a enclosed in a wooden cylinder (closed but not air-tight). Bacteria on t wood shavings convert the ethyl alcohol to acetic acid. This is describ above in the production of vinegar from wine. The effluent from t generator may be run through a second time. When the alcohol has be converted to acetic acid, the vinegar is filtered (if necessary), filled ir bottles, and the bottles are capped and heated in water until a tel perature (in all parts) of about 165°F (73.9°C) is reached.

In canning apple slices or quarters, the apples are washed and inspected to remove rotten specimens. The apples are then size-graded, peeled, and cored by machine, then sliced and quartered by machine, cut portions being discharged into and held in a 3% salt (sodium chloride) solution. Before adding to cans, the cut fruit is washed to remove salt. The cans containing cut apple sections are then filled with a 40% sugar solution, after which the open containers are heated in water at 180°–185°F (82.2°–85°C) for 6 min or longer, depending upon the size of the containers. The heat exhausted containers are then sealed and heat-processed in boiling water.

Apple sauce is made from peeled, cored and sliced apples which have been cooked in steam and run through a pulper while hot. Steam may also be injected during the pulping. The cooked, macerated apples are then mixed with sugar and heated in a steam kettle to about 190°F (87.8°C). The hot mixture of apples and sugar is then filled into glass jars volumetrically, and the jars are capped under vacuum. The capped jars containing the apple sauce are then heated in water to bring the temperature in all parts of the product to about 200°F (93.3°C). After processing, the filled containers are cooled in water or under sprays of cold water.

Some apples are dried, although the volume of this type of product has decreased in recent years. For drying, the peeled, cored and sliced apples are first treated in a weak solution of citric acid and a bisulfite dip. The bisulfite provides sulfur dioxide which inhibits enzymatic browning. The sulfured slices should be held in refrigerated storage for at least 24 hr to allow the sulfur dioxide to penetrate the apple slices. The apple slices are eventually spread on the slatted floors of natural draft, loft-type kilns. In the kilns, heated air rises through the apple slices and removes moisture. After the moisture content has reached about 10%, the apples are packaged in moisture-proof containers to be used in the bakery trade.

Considerable quantities of sliced apples are frozen. There are several methods of preparing apples for freezing. The washed, peeled and cored apples may be sliced, immersed in 3% salt solution, removed from the brine, and subjected to a vacuum (to remove air which would otherwise facilitate enzymatic browning), reimmersed in brine and then washed and packed in 30–50 lb (13.6–22.7 kg) slip cover cans as a mixture of 4 parts fruit to 1 part sugar. Another procedure for preparing apple slices for freezing is to subject the slices to a high vacuum, immerse in salt solution, blanch the brined apple slices in free-flowing steam, cool in water, then pack in 30 to 50 lb (13.6–22.7 kg) slip cover cans. A third procedure is immersion in a solution of sulfur dioxide (0.2–0.25%) or in a bisulfite solution containing citric acid (the acid is used to regulate the pH to 2.8–3.0) for 1 min. The slices are then removed and placed under refrigeration for several hours to allow the sulfur dioxide to penetrate to the center of the apple slices. They are then packed into slip cover cans of 30–50 lb (13.6–22.7 kg) capacity as a mixture of 4 or 5 parts fruit to 1 part sugar. It should be noted that when treatment with sulfur dioxide is used to inhibit enzymatic browning, this chemical must be allowed to penetrate to the center of the apple slice before freezing. Otherwise enzymatic browning will occur at the center of the slices. Apple slices in cans are frozen by placing them in a cold room held at −10°F (−23.3°C) or below.

Pectin or pectin solutions may be manufactured or obtained from dried apple peels and cores. Fruit contains pectin substances in a form which is not water dispersable, and in this form, cannot be used for the manufacture of jams, jelly and other foods. The pectin substances in fruits are in the form of protopectin, a more complex, longer chain of the carbohydrate, galacturonic acid (a 7-carbon compound). To obtain pectin which is dispersable in water, the dried apple substances are heated in boiling water for about 40 min. The heat, together with the acid present in the fruit, hydrolyzes the protopectin to pectin, a substance which contains fewer galacturonic acid units than does protopectin. In the presence of large proportions of sugar (about 65% soluble solids) and acid, to provide a pH around 3.0, pectin will form a gel like that which provides the semi-solid form for jams and jellies. After extracting the pectin from apple peels and cores, the solution is separated, and the residue is pressed between cloths to obtain more of the extract. The extract is then filtered, bottled and pasteurized as a pectin solution to be used as such, or it may be spray dried to a moisture content of about 5% and handled in this form. When used, the dried material (pectin) must be dispersed by fast agitation in water, or mixed with granulated sugar to disperse in water, before adding as an ingredient of foods. If added to foods as such, pectin forms lumps which become caked with gelatinous material on the outside, preventing the pectin from dispersing.

Low methoxy pectin may be produced from pectin by treating solutions of this material with an enzyme (pectin methyl esterase). This treatment removes methyl groups from the ester group of the galacturonic acid unit. Low methoxy pectin has the property of forming gels in the presence of comparatively low concentrations of soluble solids (sugar) and at comparatively high pH (as high as pH 6.5), provided that a source of calcium is present. Low methoxy pectin solutions are dried in the usual manner.

Some apple jelly is produced. While the unclarified juice is high in pectin, after clarifying, most of the pectin has been removed; therefore, sugar, some water, and about 0.4% pectin are added and mixed. The mixture is then concentrated, by heating in open kettles, to a soluble solids content of about 65%. Citric acid solution is added to regulate the pH to 3.0–3.2, and the jelly is placed in glass jars, capped, pasteurized and cooled in the usual manner.

APRICOTS

In the United States, apricots are grown mainly in California, Washington and Utah. Only 2 or 3 varieties are grown commercially. Apricots are budded or grafted on seedling stocks of apricot, plum or peach trees that are 1 year old. The spurs rarely produce well for more than 3 years, so pruning and renewal are required. The trees and fruit are subject to infection with molds and bacteria and to infestation by scale insects. Spraying with chemical solutions may, therefore, be necessary during the growing season.

Some apricots are distributed as the fresh product, but the bulk of apricots are preserved mainly by drying for use in the bakery trade or in the home. Some apricots are frozen for use in the bakery trade. When harvested, the firm ripe (not soft) fruit is picked by hand or by machines which shake the fruit onto catching aprons. The fruit is brought to the processing plant in lug boxes. Handled in the fresh state, the fruit is washed in dilute acid or alkaline solutions to remove spray residues, then in water. It is then packed in shallow boxes and cooled to 31° to 34°F (−0.56° to 1.1°C). The fresh fruit should be held at 31° to 32°F (−0.56° to 0°C) until sold to the consumer. In this condition, it has a storage life of about 1–2 weeks.

In the drying of apricots, the washed fruit is halved and pitted (without peeling) by machine. It is then placed on wire mesh trays and subjected to the fumes of burning sulfur for a period of 3 hr. This causes the fruit to pick up sulfur dioxide which tends to inhibit both enzymatic and nonenzymatic browning. Next, the fruit is placed in single layers on wooden trays and dried in the sun over a period of 1–4 days to provide a moisture content of 26–30%. The trays are stacked for sufficient time to equilibrate moisture, after which the fruit is packaged in retail sized cartons or in paper lined boxes for commercial use. Apricots may also be dried in artificial driers after the fruit has been sulfured.

Frozen apricots are usually packed as the halved fruit in syrup, but may be packed with sugar. The washed and pitted fruit may or may not

be peeled by spraying the halves (cup down) with a hot, weak solution
sodium hydroxide, then washing with dilute citric or hydrochloric ac
solution prior to preparing for freezing. Eventually, the fruit is mix
with a 40% syrup containing 0.1% ascorbic acid (to prevent browning)
proportions of 1 part fruit to 1 part syrup. It is then placed in slip cov
cans holding 30 lb (13.6 kg) of product and frozen in cold rooms at −10
(−23.3°C) or below. Some frozen apricots are mixed with granulat
sugar containing ascorbic acid. Other packs of apricots are prepared fro
blanched apricots (the heat treatment destroying enzymes which mig
cause discoloration). Still others are prepared from fruit which has be
treated to inhibit enzyme action by immersing for 3–4 min in solutions
0.4–0.5% sulfur dioxide. The sugared, blanched, or sulfured fruit is froz
in slip cover cans holding 30 lb (13.6 kg) of product.

Some apricot jam is produced. The fruit is ground and mixed wi
sugar, water and 0.34–0.42% dispersed pectin. The mixture is th
heated in open kettles to provide a soluble solids content of about 65
Citric acid solution is then added to regulate the pH to 3.3, after whi
the mixture is bottled, capped and pasteurized.

BANANAS

Bananas are not grown commercially in the continental United State
but some are grown in Hawaii and shipped to the mainland. In t
western hemisphere, the chief production of bananas occurs in Mexi
and the Central American countries, in Cuba, Jamaica, Haiti, t
Dominican Republic, Honduras, Colombia and Brazil. Bananas are al
grown in some Asian and Middle Eastern countries.

Banana trees are started from young plants which bud from t
underground stem or bulb of older plants. The trees bear mature fru
13–15 months after planting, depending upon climate, and each pla
requires an area of 100–400 ft² (9.3–37.2 m²), depending on soil and wat
conditions. The trees develop flowering stalks with male and fema
flowers, and the female flower eventually becomes the fingers (sing
bananas) of the hand bunch. Only one stem (bunch) of bananas
produced per tree. Banana plants are subject to infection with bacter
and viruses, but particularly with molds. Banana plants are also subje
to infestation with borers, scale insects, thrips and flies. The fruit
subject to infection with molds. Because of their susceptibility to diseas
therefore, spraying or dusting with chemicals during the growing seas
may be indicated.

The banana stem or bunch contains 6–14 hands (clusters of single bananas on the stem) and weighs 30–130 lb (13.6–59 kg). The stem is harvested when the single bananas are mature but green. In harvesting, the tree trunk is cut or nicked a few feet (1 ft = 30.5 cm) below the bunch. This causes the trunk to bend so that the bunch can be caught on the shoulders of a worker who, after the bunch is cut free, will carry it to the packing plant. The tree is then cut down. Bananas are handled mainly as the fresh fruit and shipped from the growing area while still green. At one time, bananas were shipped as bunches, on the stem, but today, they are handled mostly as hands or groups of single bananas, cut from the stem and packed in plastic-lined boxes. They may be treated with fumigants prior to boxing and should be precooled to 57°–62°F (13.9°–16.7°C). Shipment from one country to another by boat may require 2–10 days, and shipment throughout this country may require up to an additional 7 days. Bananas are subject to a chilling injury if held below 55°F (12.8°C); low temperatures kill certain surface cells and prevent normal ripening. Nor should bananas be held for extended periods at temperatures above 70°F (21.1°C); they must, therefore, be shipped under controlled temperatures. Bananas may be ripened in rooms where the temperature is held at 58°–64°F (14.4°–17.8°C) in atmospheres to which some ethylene gas may be added, prior to shipment to the retailer. Ethylene gas hastens ripening. A period of 4–10 days is required for ripening at these temperatures without ethylene, and slightly shorter periods are required when ethylene is used. After ripening, bananas have a storage life of only a few days at room temperature.

While bananas are handled mostly in the fresh state, some are peeled, puréed and canned for the bakery and soda fountain trade. In such cases, the puréed banana pulp is quickly heated to about 280°F (137.8°C) in a heat exchanger, held at this temperature for a few seconds, cooled in a heat exchanger, filled into sterile, large-sized cans and sealed under aseptic conditions. In such an operation, all parts of the process must be accomplished so that the cooling, filling and sealing operations are done with the equipment presterilized and the containers sterilized just prior to filling. The cans must also be sealed under high temperature steam or inert gas.

CHERRIES

Cherries are grown in essentially all states of this country, but commercial production is limited to about 12 states—Michigan, New

York, Wisconsin and Pennsylvania being the chief producing areas. There are many varieties of cherries that come under two general types, sweet and sour. Sweet cherries are used chiefly for sale in the fresh condition and for the production of maraschino cherries. About equal amounts of sour cherries are canned and frozen. Cherry trees, set out as 1- or 2-year-old stock, bear fruit the year after planting. They should be pruned yearly. Cherry trees and fruit are subject to infection by molds and to infestation by aphids, and may, therefore, need spraying with chemicals during the growing season.

Sweet cherries to be shipped as fresh, or manufactured as some types of maraschino cherries, are picked by hand, with stem on, when bright red or black in color. Sour cherries are picked without the stem when deep red. Cherries are brought to the processing plant in lug boxes. At the processing plant, sweet cherries are washed in dilute acid or alkali solution to remove spray residues, then in clear water. They are next packed in baskets holding 1 pt or 1 qt (473 ml or 946 ml), with or without cellophane overwrap, and cooled. Fresh cherries should be cooled to 31° to 33°F (−0.56° to 0.56°C) and held at this temperature until sold to the consumer. Under such conditions, sweet cherries have a storage life of 10–14 days.

Maraschino cherries are produced with or without stems. The fruit is graded for size, then washed and inspected to remove defective specimens. The cherries are next pitted by machine (the stem left on for some types of product). Then the cherries are placed in barrels, covered with a 2% sodium chloride solution containing 0.6% sulfur dioxide (as such or as a bisulfite salt) and about 0.36–0.48% calcium carbonate, after which the barrels are covered. The sulfur dioxide bleaches out the color of the fruit and the calcium in the calcium carbonate tends to firm it. The barrels of fruit are then placed in cold storage (above freezing temperature) and rolled several times daily over a period of about 60 days. After curing, the cherries are leached in running water to remove sulfur dioxide, then cooked in several changes of boiling water. They are then dyed with a hot solution of red dye (approved by the FDA) in water or in a weak soda (sodium bicarbonate) solution. The cherries are set aside in the dye solution for several days. Eventually, the dye solution is poured off and the cherries are covered with a 0.5% solution of citric acid to set the dye. After holding for several days the acid solution is poured off, and the fruit is packed in jars holding 3–28 oz (85–793 g) or in 1/2- or 1-gal. (1.9–3.8 liter) jars and covered with a hot, 12–20% sugar solution, after which the jars are capped. The bottled cherries are pasteurized in water at 170° or 180°F (76.7° or 82.2°C) to bring the temperature of all parts to bout 165°F (73.9°C). Some maraschino cherries are packed in barrels nd covered with a 12–20% sugar solution containing 0.1% sodium enzoate for purposes of preservation. Some maraschino cherries with yrup, in barrels, may be heated to about 165°F (73.9°C) in all parts by oils of metal, through which steam is allowed to flow, prior to heading he barrels.

In the canning of cherries, the fruit is washed and inspected, green and ecayed specimens being floated out in water or removed by hand from n inspection belt. The cherries are then sized to facilitate pitting, pitted y machine and, if stems are present, the stems are pulled out by nachine prior to pitting. Cherries are added to the can by hand, after which the cans are filled with water or syrup. If syrup is used, it may vary n concentration from 10–40%. The open cans are then exhausted or eated in free-flowing steam or water at about 200°–210°F 93.3°–98.9°C) for 10–20 min to remove air, after which they are sealed nd heat-processed.

In the freezing of cherries, the fruit is soaked in cold water and, if verripe, some calcium chloride is added to the water to firm the fruit. It s then graded for size automatically and passed over inspection belts where green and rotten specimens are picked out. After the pits are emoved by machine, the cherries are packed in 30- or 50-lb (13.6 or 22.7 g) slip cover cans or in 50-gal. (189.3 liter) barrels as a mixture of 3 parts ruit to 1 part sugar. The closed cans or headed barrels are placed in ooms at 0°F (−17.8°C) or below for purposes of freezing the product.

GRAPEFRUIT

Essentially, all the grapefruit grown in this country is produced in Florida, Texas, Arizona and California. There are a number of varieties of grapefruit. The trees are planted from nursery stock, and the tree olossoms and fruit must be protected against freezing temperatures. Grapefruit trees require some pruning. They may be infested by scale nsects, flies, mites and aphids, which affect the trees and/or fruit. Hence, during the growing season, spraying with chemicals may be necessary. Grapefruit is harvested by hand picking, with or without the aid of clippers, when the fruit has a soluble solids content of 6.5 parts to 1 part acid. The soluble solids are mainly sugar, and the acid is calculated as citric. The grapefruit is placed in bags which are emptied into pallets or boxes which, in turn, are emptied into trucks. Tree shakers with catching frames may be used for harvesting the fruit, although such methods have not been perfected.

At the processing plant where fresh fruit is handled, the product i inspected to remove damaged or decayed specimens and floated in a soa tank of water containing a detergent and one or more chemical com pounds which inhibit the growth of microorganisms, particularly molds The washer is also equipped with brushes to clean the fruit. Afte washing, the fruit is dried, a thin layer of wax is applied, and the fruit i polished mechanically. It is then sized and packed into boxes or crates fo shipment. If it is not fully colored, it may be held in bins at 65°-70° (18.3°-21.1°C) in an atmosphere containing ethylene in low concentra tion (1 part in 50,000 parts) before processing. The ethylene speeds u coloration. After packing, the fruit is sometimes cooled in refrigerated ai which is blown through the refrigerator cars containing the fruit to b shipped. The fruit is water cooled prior to packing, in some areas. Sinc the fruit is wet after cooling, an inhibitor of microorganisms in concentration of about 0.1% may be used in the cooling water. Grapefrui may be packed in perforated plastic bags or in cardboard cartons fo purposes of shipment. During cooling, the temperature of the fruit i lowered to about 50°F (10°C), and this temperature should be main tained until the fruit is sold to the consumer. Under such conditions grapefruit has a storage life of 4-8 weeks.

Considerable quantities of grapefruit juice are canned, the smalle sized fruit being used. After washing and sizing, the fruit is cut in hal and reamed or pressed by machine between a cup and a cone. This removes the juice and separates the peel. The juice is then screened t remove seeds and coarse tissues, and sugar is added to raise the suga content to about 15% (from about 6.5%). The juice is then heated t about 185°F (85°C) in a heat exchanger and filled automatically into cans which are sealed and inverted. Since grapefruit juice is so acidic, no further heat processing is required. The heating of the juice serves, not only to destroy microorganisms, but also to inactivate enzymes which might act on the pectin components of the juice and cause precipitates to form with the separation of a clear serum.

Some grapefruit is canned as segments. The washed and graded fruit is immersed in boiling water for 4-6 min to loosen the skin. The skin is then scored and removed by hand. This leaves some of the white portion (the albedo) on the fruit. To remove the albedo, the fruit is immersed or sprayed for 25-35 sec with a 2-5% solution of sodium hydroxide at 170°-180°F (76.7°-82.2°C), then washed in cold water. The segments of peeled grapefruit are then separated by hand. After separating, the segments of fruit are added to cans by hand, covered with a 35-65% sugar solution, and the open cans are heated in water at 175°F (79.4°C) for

5–18 min. The cans are then sealed and heated in water at 180°F (82.2°C) for periods depending upon the size of the container.

Most of the pectin produced in this country is manufactured from grapefruit peel, the albedo of which contains a large amount of pectic substances. In the manufacture of pectin from citrus fruits, peel from the decontaminated and washed fruit is first treated with hydrocarbon solvents to remove the flavedo (the colored portion), then ground, and heated in water to which some hydrochloric or citric acid has been added. After heating, the water extract is filtered and spray dried. Low methoxy pectin may be prepared from the extract by treating with the enzyme, pectin methyl esterase, and then spray drying.

LEMONS

Lemon trees may be grown from nursery root stock, but mostly the fruit is grown on grafts made on stock of the sour orange. Once established, the trees may bear fruit for many years. Since the trees, and particularly the fruit, are damaged by freezing temperatures, they may have to be protected in cold weather. Lemons are subjected to infestation as in grapefruit; therefore, the fruit may require treatment with chemical sprays during the growing season. Lemons are produced in this country mainly in California and Arizona. They are harvested by hand when the skin is green or silver in color. Lemons are brought to the processing plant in bulk and are generally ripened to a yellow color (as is grapefruit) in bins, the air of which contains some ethylene. Most lemons are marketed in the fresh state, and as such are handled in much the same manner as are grapefruit. Cooled to about 32°F (0°C) and held at this temperature, fresh lemons have a storage life of 1–4 months.

Some lemon juice is prepared and frozen as a lemonade concentrate. The fruit is washed, sized and inspected, after which the juice is extracted, as with grapefruit. The pulp or juice sacks are then screened out and set aside. Sugar is added to the screened juice to provide a soluble solids acid ratio of 14:1–18:1. The mixture is then evaporated at low temperatures under vacuum to provide a 5:1 concentrate. The screened pulp (which contains some liquid) is then added to the concentrate to provide 2.4–4% pulp, by volume. The concentrated mixture is frozen to a slush having a temperature of 25°F (−3.9°C) in a refrigerated heat exchanger. The slush is filled into cans, the cans sealed and conveyed through a cold air tunnel until the product is cooled to a temperature of −18°F (−27.8°C) or lower. Some lemon juice, evaporated

to a concentration of about 2:1, is packaged in lemon-shaped plast
containers after 0.1% of sodium benzoate has been added. Pulp or jui
sacs and sugar are not added to this product which is handled at ambien
temperatures (no refrigeration).

Pectin is manufactured from lemon peel as in the case of grapefruit

MELONS

Melons belong, as do squash, to the cucumber family. There are tv
general types, the *cucumis* species, which includes muskmelon, ca
taloupe and honeydew melons, and the *citrullus* species, including th
watermelon and the Chinese watermelon. There are numerous varieti
of melon. Although they may be grown in almost any of the 50 state
except possibly Alaska, *cucumis* species require warm weather for goc
growth and 75–130 days from planting to harvesting; therefore, most a
grown commercially in the southern states. The *citrullus* species a
grown in the South, but can also be grown in those northern states whe
130–140 days of growing weather prevail. In warmer climates, about
days are required from planting to maturity.

Melons are planted as seed but may be started in greenhouses or h
frames and set out as plants. Melon plants are subject to infection wit
molds and to infestation with beetles and aphids; hence, they ma
require dusting or spraying with chemicals during the growing seaso
Melons do not improve in flavor after harvesting, hence, are picked whe
fully ripe. They are harvested by hand and placed in trucks with
without the aid of a conveyor. Care must be taken to prevent bruisir
during harvesting and handling. Melons are mostly handled as the fres
product.

At the packing plant, melons are washed, drained, dried and shippe
to retail markets in wooden crates. Some melon types are subject
chilling injury at low temperatures, so are held at 45°–50°F (7.2°–10°C
The storage life of some melons is indicated in Table 22.1.

TABLE 22.1.   STORAGE LIFE OF SOME MELONS

Type	Temperature of Storage (°F)	(°C)	Approximate Storage Life (Days)
Cantaloupe	32–40	0–4.4	5–15
Persian	45–50	7.2–10	14
Honeydew	45–50	7.2–10	21–28
Casaba	45–50	7.2–10	28–42
Watermelon	40–50	4.4–10	14–21

Melons are not preserved by canning or drying, and only minor quantities are frozen as melon balls. In preparation for freezing, the melons are halved, the seeds are removed, and ball-shaped pieces are removed with a hand scoop. The melon balls are washed with sprays of water, drained and filled into liquid-tight cartons and covered with syrup (25–30% sugar), and the cartons are sealed. This product is frozen and stored at 0°F (−17.8°C) or below until shipped to the retailer. Frozen melons should be held at 0°F (−17.8°C) or below until sold to the consumer.

ORANGES

The orange is utilized as a food to a greater extent than any other citrus fruit. The trees are set out from nursery stock and must be protected from freezing weather. As with other citrus trees, some pruning has to be done each year. The five states which produce oranges commercially are Florida, California, Arizona, Texas and Louisiana, with Florida being, by far, the greatest producer. About three-quarters of all oranges, in this country, are used for the production of frozen juice concentrate and for the so-called "fresh" orange juice. Fresh oranges are picked and handled much in the same manner as are grapefruit. When picked, the solids to acid ratio should be 12:1–18:1. Oranges may be dyed by immersing in a solution of certified food dye at 120°F (48.9°C) for about 3 min prior to waxing, polishing and cooling, since the color of the skin is often green when the fruit is picked. Some oranges are cooled to 32°–40°F (0°–4.4°C) and others to 40°–44°F (4.4°–6.7°C), depending upon variety. They should be held at these temperatures until sold to the consumer. Under these conditions, they have a storage life of 1–3 months, depending upon variety.

Much orange juice is frozen as a concentrate (see Fig. 22.3). The juice is extracted after washing, inspecting and sizing the fruit, in the same manner as with grapefruit juice. The juice is screened to remove pulp and fruit sacs, then concentrated at low temperature in vacuum evaporators to a solids content of 50–55%. The concentrate is then cut back with fresh juice to 42% soluble solids (this provides most of the flavor, since flavor components are lost during evaporation), at which time some of the pulp and juice sacs may be included. The concentrated juice is then frozen to a slush in a refrigerated heat exchanger and filled automatically into metal or cardboard metal-ended containers. The containers are sealed and further subjected to cold air in a freezing tunnel. In some cases, the slush-filled and sealed cans are placed in cartons and the cartons held in

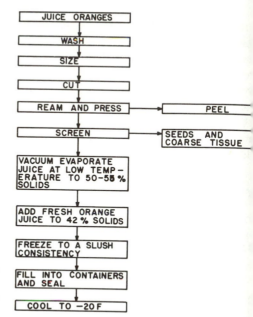

FIG. 22.3. MANUFACTURE OF CONCENTRATED ORANGE JUICE See metric conversion tables in Appendix.

storage at temperatures of $-20°F$ $(-28.9°C)$ or below for purposes of further freezing the concentrated product. Orange juice may be frozen in slabs and eventually defrosted and shipped as a slush to be pasteurized and sold as the single strength product. Some orange juice is canned. The juice, filtered from pulp, is heated to $180°–200°F$ $(82.2°–93.3°C)$, added to the can at this temperature and the cans are sealed, inverted and air cooled.

Pectin and low methoxy pectin may be manufactured from orange peel, as in the case of grapefruit.

PEACHES

Peaches are grown commercially in about 35 states in this country, California producing by far the largest amount. There are a number of varieties which are classified as clingstone or freestone varieties. The flesh of clingstone varieties, used primarily for canning, adheres tightly to the pit. In freestone varieties, used both as fresh and for canning, the flesh is easily separated from the pit. Peach trees, not long lived, are set ut from nursery stock and produce significant quantities of fruit for 7–16 years. Both trees and fruit are subject to infection with molds and viruses and to infestation with scale insects, beetles, borers and nematode worms. Spraying with chemicals may, therefore, be necessary during the growing season. Yearly pruning and thinning of peach trees is practiced. Peaches are harvested or picked by hand when in the firm, ripe condition, but before they become soft. The fruit is picked into buckets or bags which are emptied into lug boxes for transportation to processing plants.

At the processing plant, peaches to be handled as fresh are washed in dilute acid or alkaline solutions to remove spray residues, then rinsed in water and allowed to drain and dry. Decayed specimens are removed as the fruit passes over an inspection belt, and the unblemished fruit is packed in boxes. The fruit is air-cooled to 31° to 32°F (−0.56° to 0°C) and should be held at this temperature until sold to the consumer. Handled in this manner, the fresh fruit has a storage life of 2–4 weeks.

Peaches to be canned are inspected to remove decayed and unripe specimens, graded for size, then passed through a machine which cuts each fruit in half and removes the pit. The peach halves, pit cup down, are then immersed in, or sprayed with, a hot lye solution (usually a solution of 6–8% sodium hydroxide at about 140°F [60°C]) for 45–60 sec. The peach halves are then discharged into a shaker-washer where they are subjected to heavy sprays of water which remove the loosened skin and residual alkali. Clingstone peaches are usually canned as halves, while freestone types are canned as slices; therefore, the latter types will be sliced before placing in cans. Peaches may be peeled by immersing in alkaline solutions and then subjected to water sprays prior to halving and pitting, or they may be halved and pitted and then subjected to free-flowing steam for 1 1/4 min. They are then sprayed with water to remove the peel. The halves or slices are placed in cans by hand, and the cans are filled with syrup, having a sugar content of 20–55%. The open cans containing the product are then exhausted by heating in free-flowing steam or water at 180°–200°F (82.2°–93.3°C) for 10 min. Finally, the cans are sealed and heat-processed, with or without agitation, in water at 212°F (100°C).

Peaches to be frozen are usually prepared as the sliced product as indicated in the foregoing. Packed for the retail trade, the sliced peaches are placed by hand in metal-ended cardboard cartons and covered either with a 50% sugar syrup or a 60% sugar syrup containing 0.2% of ascorbic acid (vitamin C) to prevent browning. The cartons are closed and the product is frozen. Considerable quantities of peaches for freezing are filled into 30-lb (13.6 kg) slip cover cans and covered with a 50% sugar syrup containing ascorbic acid. The cans are closed and placed in col rooms at $-10°F$ ($-23.3°C$) or below in order to freeze the product.

Peaches to be dried, usually the freestone varieties, are washed t remove spray residues, inspected, halved and pitted, placed on wir mesh trays, and subjected to the fumes of burning sulfur (to absor sulfur dioxide which prevents enzymatic and nonenzymatic browning for 4–6 hr. They are then exposed to the sun on wooden trays for 1–4 day to lower the moisture content to 26–30%. The trays are stacked for som time to equilibrate moisture in the product, after which the product i packaged in retail cartons or in paper-lined boxes for the bakery trade Peaches are sometimes dried in artificially-heated drying tunnels.

PEARS

There are many varieties of pears. The trees are set out as 1-year-ol stock, and once they start to bear fruit, they may continue to do so fo many years. Ordinarily, the trees are lightly pruned each year. Pears ar grown in essentially all states of this country, but California, Oregon anc Washington account for 90% of the commercial production. The Bartlet pear is the most important variety, both for consumption as fresh and fo preservation by canning. Trees and fruit alike are subject to infection by molds and to infestation by scale insects, psylla insects, mites, thrips anc moths, and spraying with chemicals during the growing season may therefore, be indicated. Pears do not ripen successfully on the tree anc are harvested by hand while still green. They are transported tc processing plants in lug boxes or pallet bins, the latter holding about 1000 lb (453.6 kg) of fruit.

At the processing plant, pears are washed in weak acid or alkaline solutions to remove spray residues. They are then washed in water, drained, and inspected to remove defective specimens. They are also usually graded for size, especially if they are to be canned, before placing in storage. Pears to be sold as fresh are cooled in cold water and packed in boxes, or they are packed in boxes and cooled in refrigerated air. Cooled to 30° to 31°F ($-1.1°$ to $-0.56°C$), and stored at this temperature, pears have a storage life of 2–7 months, depending upon variety. The storage life of pears may be extended by about 3 months by regulating the oxygen content of the storage atmosphere to 2.5% and the carbon dioxide content to 5%. Pears which are overripe when picked are subject to scald and core breakdown during storage. Pears to be used as fresh

uit are ripened at 60°–70°F (15.6°–21.1°C) prior to or during shipment.
fter they are ripened, they should be held at 32°–35°F (0°–1.7°C) until
rocessed or sold to the consumer.

Pears to be canned are handled in much the same manner as the fresh
roduct, until removed from storage for processing. They are usually held
ι cold storage prior to canning. The pears are peeled by hand with knives
r by machine or by immersion in dilute lye (sodium hydroxide) solu-
ons (about 5%) at about 140°F (60°C) for 40–60 sec, then washed
ith heavy sprays of water. They are then split in half lengthwise, and
ιe core is removed by machine. The pear halves are placed in cans, by
and, and covered with a 10–40% sugar syrup, depending upon the type
f pack. The filled, open cans are then heat exhausted in water at 175°F
79.4°C) for 10–12 min, after which they are sealed. They are then heat
rocessed for 10–35 min in boiling water, the time depending upon the
ize of the container.

PINEAPPLES

Pineapples grow on plants developed from crowns (top leafy portion of
ιe fruit) or from slips and shoots of the plant. The fruit matures in about
.5 years, but in climates where pineapple is harvested for purposes of
anning, planting is arranged in such a manner that some fruit reaches
naturity and is harvested each month of the year. Both plants and fruit
ιre subject to infection with molds and viruses and to infestation with
piders and bugs, and so the plants may have to be sprayed with
hemicals during the growing season.

There are two main varieties of pineapple. The red Spanish variety is
grown mainly in Florida and the West Indies, as well as some other
countries, and is used in the United States as the fresh product. The
mooth cayenne variety is grown in Mexico, although Hawaii is a more
mportant source. This variety is used as the fresh fruit but more often
or canning. The red Spanish variety, when ripe, is more acid and
contains less sugar than does the smooth cayenne type. When ready for
ιarvesting for purposes of canning, the smooth cayenne variety has about
).5–0.6% acid (calculated as citric acid) and 10–12% sugar. The fruit is
roken from the stalk by hand and placed in canvas bags which are
mptied at the end of the plant rows. If the fruit is to be handled as fresh,
t is placed in crates or trucks for transportation to the processing plant.
f the fruit is to be canned, the crowns are broken off prior to placing in
rucks or in crates and trucks.

Pineapples which are to be used as fresh may be harvested in the mature green or ripe condition. They are packed in crates and air-cooled to 45°–55°F (7.2°–12.8°C). In this condition, the mature green fruit has a storage life of 3–4 weeks, while the ripe fruit has a storage life of 2– weeks. Spoilage of fresh fruit is usually caused by fermentation due to the growth of yeasts or molds.

Pineapple is canned in slices, chunks, crushed or as pineapple juice. In canning, the fruit is first graded by machine for size. It is cut by machines (Ginaca machines) which cut off the top and bottom, remove a cylindrical section within the peel, and remove the cylindrical section of core from the center. The cut portions are then hand trimmed with knives by workers wearing rubber gloves. Rubber gloves are required since raw pineapple contains a very active proteolytic enzyme which will attack and eat away the flesh of hands and fingers exposed to the juice of the fruit. The fruit is next sliced by machine and placed in cans by hand by workers wearing rubber gloves. The pineapple in cans is then usually covered with a 30–50% sugar syrup or with pineapple juice containing some added sugar, after which the filled open cans are heat exhausted in water at 170°F (76.7°C) for 5–12 min, then sealed. Alternately, the syrup is subjected to vacuum and added to the fruit in machines in which the fruit has been subjected to vacuum. In such cases the cans are sealed under vacuum. The sealed cans are heat-processed in boiling water, in open cookers which agitate the cans, for a period of 30–35 min. Pineapple pieces cut from whole slices, or crushed pineapple obtained from broken slices and from the flesh adhering to the shell remaining after the cylinder of fruit is cut out, may be canned in the same manner. In preparing crushed pineapple, sufficient sugar is mixed with the fruit to provide a total sugar content of 20–24%. The mixture is then cooked in open steam kettles for 10–11 min and filled into cans volumetrically while hot. The cans are sealed and heat processed in boiling water.

Pineapple juice may be obtained from the cylinders of fruit which were cut out by machines by grating and pressing them between cotton press cloths. Pineapple juice may also be prepared from the pressed grated pulp, obtained from the shell left after the cylinder of fruit has been cut out. Pineapples for juice should be harvested in the soft ripe stage in order to obtain the best flavored juice. After pressing, the juice is filtered to remove coarse particles. Next, it is heated to 180°–185°F (82.2°–85°C) in heat exchangers and filled into cans at this temperature. The cans are sealed, inverted, held in this manner for 20 min, then cooled.

PLUMS

Plum trees are set out when they are 1 year old and bear fruit the next year. There are many varieties of plums which are grown in many areas of the United States. The prune plum, used for the production of dried prunes, is grown mainly in California. Purple plum types, used for canning, are grown mainly in Oregon and Washington. Plum trees and fruit are subject to infection with bacteria and molds and to infestation with scale insects, beetles, borers and mites. Spraying with chemical solutions may, therefore, be necessary during the growing season.

Plums are hand picked and transported to the processing plant in lug boxes. At the processing plant, plums are washed in dilute acid or alkaline solutions to remove spray residues, then rinsed with water. They are next inspected to remove defective specimens, and graded for size. Those to be sold as fresh are packed in boxes and air cooled to 31° to 32°F (−0.56° to 0°C). In this condition, they have a storage life of 2–4 weeks. Plums to be canned are cleaned, inspected and graded. Then they are placed in cans by hand and covered with a 25–30% sugar syrup. The filled, open cans are then heat exhausted in water at 180°–190°F (82.2°–87.8°C) for 12–15 min, after which the cans are sealed and heated in boiling water for 20–25 min, depending upon the size of the container.

Plums which are dried to produce prunes are washed, dipped in boiling lye solution (0.25–1.0% sodium hydroxide) for 5–30 sec, washed, and placed on wooden trays for drying by exposure to the sun. The lye treatment removes wax which inhibits drying, and it cracks or checks the skin. A drying time of 7–14 days is required to lower the moisture content to 22–25%. After sun drying for 4–5 days, the trays are stacked for further drying and to equilibrate moisture. Plums may be dried in artificial driers at 145°–160°F (62.8°–71.1°C) after treating with boiling lye and washing.

ugar

Sugar, the common name for sucrose (also saccharose), is extracted
ıd refined from sugar cane and sugar beets. There are many substances
ıemically classified as sugars, and when these are referred to, they are
ways used with a qualifier such as in milk sugar (lactose), corn sugar
extrose), and malt sugar (maltose). When the word "sugar" is used
ıthout a qualifier, it generally refers to the common sweetener
ucrose). Other sugars have varying degrees of sweetness relative to
ıcrose, and some sugars differ from sucrose in that they lend varying
egrees of bitterness whereas sucrose imparts only a sweet taste. Other
ıportant sources of sucrose include palm and maple trees and fruits.
ıemically, and in every other way, cane sugar and beet sugar are the
me. In addition to providing energy for the body and sweetness to
ods, sugar performs numerous other roles in the food industry. It is used
baked products where it contributes to the desirable texture of baked
ods, and it stabilizes the foam of beaten egg whites. When it
ıramelizes, it imparts a unique but desirable color and flavor to surfaces
pastries, cakes, etc. It is used in ice creams and dairy products, in
everages, in the home, in institutions and in restaurants for foods and
everages, in canned and frozen fruits, in canned vegetables, in jams and
llies, and in other types of food. Sugar is also used in some nonfood
oducts. Sucrose is the most important of three naturally occurring
saccharides. It has the formula $C_{12}H_{22}O_{11}$. It may be hydrolyzed,
elding glucose and fructose (levulose), both 6-carbon sugars.

$$C_{12}H_{22}O_{11} + H_2O \rightarrow C_6H_{12}O_6 + C_6H_{12}O_6$$

Sucrose $\qquad$ Glucose $\quad$ Fructose

SUGAR FROM CANE

Sugar cane is a giant grass belonging to the genus *Saccharum*. Whi nearly all sugar canes are of the same species, differences in growir conditions (climate, etc.) affect the characteristics of the juices. Fe example, the sugar content of the juices from sugar canes grown in th tropics is higher than in sugar canes grown in cooler climates. In th United States, sugar cane is grown primarily in Louisiana, and some grown in Florida and Hawaii. Cuba, Puerto Rico, the Virgin Islands, th Philippines and other countries also produce sugar cane. Sugar cane grown by planting cuttings from the stalk, each containing a bud. Th length of time that the cane is allowed to grow before harvesting varies i different countries and may be from 7 months to 2 years. The yield e sugar from cane juice is about 14–17%. Sugar cane is harvested by han or by mechanical cane harvesters, the stalks being cut off just above th ground. At this time, the tops of the stalks are cut off, since they contai high concentrations of an enzyme which hydrolyzes cane sugar (sucrose a 12-carbon sugar) into dextrose and levulose or fructose (two 6-carbo sugars) and greatly reduces the yield of cane sugar. Also, at the time e harvesting, the leaves are stripped from the canes, although they may b burned off prior to harvesting. Parts of the sugar cane other than the toj contain some of the enzyme which converts sucrose, and so the cane mus be processed shortly after harvesting in order to obtain maximum yield: The sugar cane is usually transported to the processing plant in trucks e carts.

At plants producing raw sugar (see Fig. 23.1), the cane first passe through shredders and then through 3–7 roller mills which press out th juice. After the first pressing, the bagasse (the pressed cane) may b mixed with hot water or dilute hot cane juice and again pressed to extrac more sugar. Following the final pressing, the bagasse is usually brough directly to the boilers, where it is used as fuel. The wet bagasse : reported to have about ¼ the fuel value of ordinary fuel oil. Howeve bagasse is a wood-like fiber, and it also is used to manufacture wallboarc It has been the object of study for use in other applications.

The juice, which is dark green in color, has a pH of about 5.2. Afte extraction, the cane juice is strained to remove pieces of stalk and othe detritus, and a mixture of lime and water (source of calcium hydroxide $Ca(OH)_2$) is added to raise the pH. The juice is heated to precipitate an remove impurities. The limed mixture is then held in tanks, where th lime-impurities mixture is allowed to settle out, and the clear juice i separated from the sediment. The clear juice is then heated (at tem

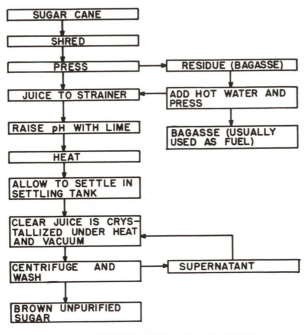

FIG. 23.1. PRODUCTION OF RAW SUGAR FROM
SUGAR CANE

eratures below the boiling point of water) under vacuum to evaporate
water and concentrate the sugar to the point where there is a mixture of
sugar crystals and molasses. More syrup may be added to the evaporation pans as the syrup is concentrated. The mixture (called massecuite)
is centrifuged to obtain the brown unpurified sugar. The liquid centrifuged from the sugar crystals still contains dissolved sugar and is
returned to the evaporation pans. When the molasses from the centrifuge
treatment reaches a low enough concentration of sugar that removal of
sucrose is uneconomical, it is called blackstrap. This is not discarded,
but is not returned to the evaporation pans. Generally, blackstrap is sold
to the fermentation industries for the production of rum. Sugar may be
purified or refined at the plant which manufactures the raw sugar, but
usually the raw sugar is packed in jute bags and shipped to sugar
refineries. Higher grades of molasses for domestic use may be made from
juice obtained from cane which is not pressurized sufficiently to extract
all the liquid. The molasses is heated to 160°F (71.1°C) and canned or
bottled hot. The containers are sealed, heated in water at 185°–200°F
(85°–93.3°C) for 10 min and immediately cooled.

Raw sugar, as delivered to the refinery, contains 97–98% sucrose. T
first step in refining (see Fig. 23.2) consists of mixing the raw sugar wi
a hot saturated sugar syrup. This softens the film of impurities whic
envelop the sugar crystals. This mixture is then centrifuged, and durir
centrifugation, the sugar crystals are sprayed with water to remove son
of the impurities. The washed sugar crystals are then dissolved in h
water, treated with lime to bring the pH to 7.3–7.6, and the temperatu
is raised to 180°F (82.2°C). The hot mixture is then filtered throug
diatomaceous earth or paper pulp. The coloring material in the suga
solution is removed by filtering the hot liquid through bone charcoa
after which the sugar is crystallized out in vacuum pans and centrifuge
to separate it from the liquid. During centrifugation, the crystals a
washed with water. They are then dried, screened for crystal size an
packaged. The finished dried sugar, which has an indefinite shelf-life,
made available in different grades. Large grained sugar is used for th
manufacture of candy and other prepared sweet products. Ordinary tab.
sugar is made up of fine-sized grains. Ultrafine sugar grains (confe

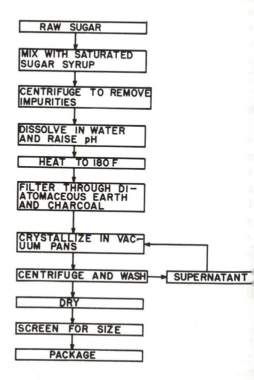

FIG. 23.2. REFINING OF SUGAR

oner's sugar) are produced by grinding the crystals in pulverizing
ammer mills. To prevent caking in confectioner's sugar, about 3%
ornstarch is used. There are other intermediate grades. Sugar is also
repared as cubes or tablets by forming a mixture of sugar crystals and
hite sugar syrup under pressure, followed by drying. A variety of "soft
ugars," ranging in colors from white through various shades of brown,
re produced. These sugars are allowed to retain some of the molasses
hich provides their unique flavor.

Invert sugar is made by heating sucrose in the presence of an enzyme
invertase) and some acid, whereupon it combines chemically with
pproximately $5\frac{1}{4}\%$ of its weight of water. The temperature must not be
igh enough to inactivate the enzyme.

$$C_{12}H_{22}O_{11} + H_2O \xrightarrow[\text{Enzyme}]{\text{Invertase}} C_6H_{12}O_6 + C_6H_{12}O_6$$
$$\text{Sucrose} \quad\quad \text{Water} \quad\quad\quad\quad \text{Dextrose} \quad \text{Levulose or fructose}$$

he dextrose is not as sweet as sucrose but the levulose is sweeter than
ucrose with the result that the final syrup, invert sugar, is slightly
weeter than the syrups having the same concentration of sucrose. Invert
ugar has special uses in the food industry. Mixtures of invert sugar and
ugar are more soluble than sugar alone. In a proportion of 1:1, a mixture
f invert sugar and sugar is more soluble than any other combination of
heir mixtures.

It should be noted that raw sugar contains thermophilic bacteria
spore-forming bacteria which grow at high temperatures—as high as
70°F [76.7°C]). If such bacteria are allowed to grow to high concentra-
ions, they may serve as a source of contamination for such products as
anned foods to which sugar is added. The spores of thermophilic
acteria are very difficult to destroy by heat, and canned foods can
ndergo some deterioration during heat processing. For this reason,
uring the various filtrations in which sugar solutions are held at high
emperatures for comparatively long periods, they should be held at
emperatures high enough to prevent the growth of thermophilic bacteria
about 185°F [85°C]).

BEET SUGAR

The sugar beet *(Beta vulgaris)* stores its sugar in the root, unlike the
ugar cane which stores its sugar in the stalk. Another difference between
ane sugar production and beet sugar production is that the latter is a
ontinuous operation and does not produce the intermediate raw sugar.
Whereas sugar beets contain 16–20% of sugar as sucrose, the yield of

sugar per acre (0.4 ha) is considerably less than that obtained from suga cane, due to the quantity of beets or cane harvested per acre. Sugar bee are planted as seed and require 70 days or more from planting to ha vesting. Since the plants are subject to bacterial or mold infection and t infestation with aphids, maggots and other insects, the plants may nee spraying during the growing season. In the United States, beets for suga are grown mainly in Colorado, California, Michigan, Utah, Idah Nebraska and Montana. Sugar beets are harvested and topped mechan cally and brought to the processing plant in bulk by freight cars trucks. They may be stored outside the processing plant in large pile until treated to extract sugar.

In extracting sugar from beets, the beets are first thoroughly washed t remove mud, stones, etc., and then passed through mechanical slice that slice them into thin shreds called "cossettes." They are then covere with hot water to extract the sugar. Extraction is done even more quickl and more effectively by a continuous counter current extraction wit water. The juice from the extraction process is treated with lime calcium hydroxide and then with carbon dioxide. The lime remove impurities as a precipitate, and the carbon dioxide is used to precipitat the calcium hydroxide as calcium carbonate ($CaCO_3$). The juice is the filtered and again treated with carbon dioxide to precipitate residu calcium hydroxide. After a second filtration, the extract is treated wit sulfur dioxide to bleach out colored components in the liquid. The liqui extract is then vacuum concentrated to 60–70% soluble solids an filtered through bone charcoal. Next, the filtered liquid is concentrate in vacuum pans to form crystalline sugar. The sugar is washed as it centrifuged, dried, screened and packaged, as in the case of cane suga In some manufacturing processes for beet sugar, the liquid from the thir centrifuging is used for the manufacture of monosodium glutamate, flavor enhancer used for many food preparations, including soup gravies, Chinese foods and meat dishes. Extracted beet pulp and the top are dried and used as cattle feed.

OTHER SUCROSE SOURCES

Sucrose can be obtained from the sap of a variety of palm trees, one the most important being the date palm *(Phoenix sylvestris)*. Much this sucrose is obtained in the Middle East by primitive methods whic involve boiling in open kettles, after which there is separation of crystal

om molasses, or the unseparated mass may be allowed to set into a
hole sugar.

In the northern part of North America, sucrose is obtained from the
p of the hard maple tree *(Acer saccharinum)*. Although the maple sap
largely sucrose, it contains unique impurities that impart to it (when
ncentrated) a special delicate flavor which makes the natural maple
rup a valuable flavoring for certain preparations.

Sucrose can also be produced from the cane of the sorghum plant,
hich is related to the sugar cane but resembles the corn plant.

ORN SYRUP AND SUGAR

Although corn is not a source of sucrose, the value of corn sugar as a
veetener in the food industry makes it worthy of mention. Corn syrup
nainly dextrose but with maltose and other oligosaccharides) is pro-
iced through the hydrolysis of corn starch by either of two processes. In
ie process, corn starch is made into a water slurry of 35–40% solids, to
hich hydrochloric acid is added. The mixture is then heated in steam
ider pressure. When the conversion of the starch to syrup reaches a
rtain point, the acid is neutralized. The acid conversion cannot be
lowed to proceed unchecked, otherwise some of the dextrose recom-
nes to form undesirable high molecular weight products, known as
version products, which could impart a bitter taste to the syrup. Fatty
:ids are removed by flotation and the syrup is then concentrated,
arified and decolorized. The syrup may be passed through ion exchange
sins to remove salts. In the final step, the syrup is evaporated, usually
ider vacuum, to a particular solids content. Crystalline sugar can be
·oduced by spray drying the syrup.

A second hydrolytic process for producing corn syrup from corn starch
volves the use of amylolytic enzymes (alpha and beta amylases). Beta
nylase is specific in its action, attacking starch molecules at their
onreducing ends and causing progressive breaks in the molecule at 12-
rbon intervals, with the resultant release of units of maltose (a 12-
.rbon sugar). Thus, by this conversion, the corn syrup will have a high
altose content. Other enzymes, glucosidases, may also be used to
ipplement either of the two conversion processes. In this case, the syrup
ill contain significant amounts of glucose.

Corn syrups are divided into five commercial classes, depending on
.eir D.E. value (dextrose equivalent value). By this classification
heme, pure dextrose is given a value of 100 D.E.

Type	D.E. Value
Low conversion	28–38
Regular conversion	38–48
Intermediate conversion	48–58
High conversion	58–68
Extra high conversion	68–100

In addition to classifying syrups by their D.E. values, they are also classified by their solids contents.

Corn syrup and corn sugar are used to a large extent in the food industry. They are used to supplement sucrose because they are less expensive, while at the same time nearly as effective as sucrose in sweetening characteristics. In addition, they inhibit crystallization of sucrose, especially when their maltose content is high. They are especially useful in the baking and brewing industries because of the quick and complete fermentability of dextrose. Their use in preserves minimizes oxidative discoloration, and other unique properties make them useful in many other applications.

24

Fats and Oils

Fats and oils are classified as lipids, which comprise an important category of nutrients for man; the other two important categories of nutrients, of course, are proteins and carbohydrates. As the reader may recall from Chapter 9, lipids are esters of glycerol and fatty acids, and fatty acids are mostly long, straight, hydrocarbon chains, with varying degrees of hydrogen saturation of the carbon atoms, having a carboxyl group linked to one of the end carbon atoms. The esterification reaction may be reversed with the addition of alkali, which results in a combination of the fatty acids with the alkali to form soap, a reaction called saponification.

$$
\begin{array}{ccccc}
\text{CH}_2\text{OOCR}^1 & & & \text{CH}_2\text{OH} & \text{N}_\text{A}\text{OOCR}^1 \\
| & & & | & \\
\text{CHOOCR}^2 & + & 3\text{N}_\text{A}\text{OH} & \longrightarrow & \text{CHOH} & + & \text{N}_\text{A}\text{OOCR}^2 \\
| & & & | & \\
\text{CH}_2\text{OOCR}^3 & & & \text{CH}_2\text{OH} & \text{N}_\text{A}\text{OOCR}^3 \\
\text{Lipid} & & \text{Alkali} & & \text{Glycerol} & \text{Soaps}
\end{array}
$$

The specific gravity of lipids is less than that of water, and these compounds are generally insoluble in water; therefore, they float in water. They are soluble in a variety of organic solvents (e.g., ether). They may be suspended in water as a stable emulsion in the presence of emulsifying agents (e.g., bile salts, alkali) which function by lowering the surface tension and by coating the lipid particles preventing their coalescence.

The members of the fats and oils may differ from one another quite distinctly in physical, chemical and dietary properties (melting points,

caloric value, reactivity, mineral content, etc.). Some of the properti
can be used to identify specific fats. For example, fats can abso:
halogens at the point of unsaturation (a point of unsaturation where tv
neighboring carbon atoms in the chain each have only one hydrog
atom attached). Thus the "iodine number" of a fat refers to the amou:
of iodine (iodine being a halogen) in grams that 100 g of the fat c;
absorb. The "saponification value" of a fat refers to the amount
potassium hydroxide in milligrams that is neutralized by saponificatic
of 1 g fat.

The fatty acids occurring in natural fats may contain 4–26 carb
atoms (generally an even number—see Table 24.1), and they may or m;
not be completely saturated with hydrogen. The general form·la f
saturated fatty acids is $C_nH_{2n}O_2$. For unsaturated fatty acids, t'
number of H atoms is less than 2n (e.g., H_{2n-2}, H_{2n-4}, etc). Most fat·
acids are straight chains, although a few may contain a ring of carb·
atoms in the straight chain. An example of this type is chaulmoogr
acid, which has been used in the treatment of leprosy, and is obtain·
from the chaulmoogra tree of east India.

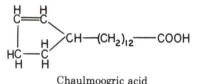

Chaulmoogric acid

The unsaturated oils, such as some vegetable oils that are liquid ;
room temperature and are subject to oxidation (a deteriorative process
may be transformed to a solid at room temperature and at the same tin·
stabilized against oxidative spoilage by "hydrogenation." In hydrogen;
tion, hydrogen is linked to unsaturated carbon atoms, though the proce:
generally is not allowed to go to complete saturation. An example of
hydrogenation process follows.

$$C_3H_5(C_{17}H_{33}COO)_3 + 3H_2 \rightarrow C_3H_5(C_{17}H_{35}COO)_3$$

Liquid lipid (oil)   Solid lipid (fat)

Since the fatty acid part of fat molecules is the largest part, the fat·
acids influence the properties of the fat, and the chemistry of fats and oi
consequently is governed by the chemistry of the fatty acids the·
comprise.

Fats and oils have "smoke-," "fire-," and "flash-points" that defi·
their thermal stability when heated in the presence of air. The "smok·
point" is reached when the temperature is sufficient to drive o·

TABLE 24.1. COMMON NAMES OF SATURATED FATTY ACIDS

Name	No. of C Atoms
Butyric acid	4
Caproic acid	6
Caprylic acid	8
Capric acid	10
Lauric acid	12
Myristic acid	14
Palmitic acid	16
Stearic acid	18
Arachidic acid	20
Behenic acid	22
Lignoceric acid	24
Cerotic acid	26

ecomposition products that can be seen as a smoke. As the temperature is increased to the "flash-point," the decomposition products can be ignited but do not perpetuate fire. When the temperature is increased to the "fire-point," the decomposition products will perpetuate a fire.

Lipids contribute to the diet in many ways. They are a primary source of energy, possessing more than twice the calories occurring in either proteins or carbohydrates. They act as vehicles for the fat-soluble vitamins A, D, E and K and are important in the absorption of calcium, carotene and thiamin. They also provide certain essential fatty acids that are required but cannot be produced by the body. Fatty acids, in the form of phospholipids, are essential to the body. These include phosphoglycerides (e.g., lecithin), phosphoinositides (e.g., diphospho-inositide) and phosphosphingosides (e.g., sphingomyelin).

Essentially, solid fats are of animal origin, and oils or fats, liquid at room temperature, are of vegetable origin. There are some exceptions, however. For instance, coconut oil is solid at ordinary room temperatures, having a melting point of 75°–80°F (23.9°–26.7°C). On the other hand, fish fats or oils are liquid at room temperature. Generally, the reason that some fats are solids and others are liquids at room temperature has to do with the percentage of saturated or unsaturated fatty acid in the fat molecules composing the fats. Stearic acid: CH_3—$CH_2)_{16}$—(COOH), is a saturated fatty acid. Linoleic acid: CH_3—$CH_2)_4$—CH=CH—CH_2—CH=CH—$(CH_3)_7$—COOH, is an unsaturated fatty acid. It should be noted that these two fatty acids have an equal number of carbons but that all the carbons in stearic acid have been saturated with hydrogen or have taken on all the hydrogen that they can accept, while 4 of the carbons in linoleic acid are unsaturated or could each accept one more hydrogen. Oleic acid: CH_3—$(CH_2)_7$— CH

=CH—(CH$_2$)$_7$—COOH, which has only 2 carbons that could accep another hydrogen, has a melting point of 61°–62°F (16.1°–16.7°C). Whe oleic acid is present in fats, even in comparatively large amounts, the fa are generally solid at room temperature if no more unsaturated fatt acids are present.

Fats containing only saturated fatty acids of short chain length (carbons or fewer) are liquid at room temperature, but generally such fa are not found in nature. However, such fats are found as components some natural fats, such as butter made from cattle or goats' milk. should be recognized that unsaturated fatty acids with 4 or more carbon which could accept another hydrogen are more reactive than saturate fatty acids and are especially apt to combine with the oxygen present the atmosphere. This is especially the case when a carbon saturated wit hydrogen is present between two groups of two carbons, each of whic could accept another hydrogen: —CH=CH—CH$_2$—CH=CH— When the unsaturated fatty acids in fats become oxidized, the fa generally becomes rancid or has an off-flavor.

Butterfat, which is discussed in Chapter 16, will not be covered in th chapter. Nonedible lipid products, such as soaps, will also be omitted

LARD

Lard, a solid fat (at room temperature), is obtained from animal (mainly hogs) by rendering (see Fig. 24.1). Rendering involves th melting out of the fat from fatty tissues and removal of the nonfa material by mechanical means. In wet rendering, the fatty tissues ar heated in water with steam under pressure. After heating, the fat laye (on top) is pumped into one tank and the watery portion into another, fo purposes of settling and removal of water in one case and skimming the fat in the other. Dry rendering is accomplished by heating fatt animal tissues under vacuum until all of the water has been evaporate from the mass of material. The fat, while still warm enough to be liquid is then filtered off from the residual tissues. Low-temperature renderin may be used to obtain animal fats. In this case, the tissue is first ground then heated to a temperature only slightly higher than that of th melting point of the fat. The nonfat tissue is then removed from the fa by centrifugation. Most animal fats to be used in food materials ar deodorized by blowing high-temperature steam through the liquid mate rial before packaging. Beef tallow is rendered from the various fatt trimmings from sides or quarters of beef. Comparatively few cattle fat are used in foods today, but beef tallow has various other industrial uses

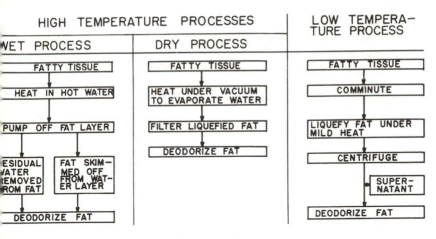

FIG. 24.1. RECOVERY OF ANIMAL FATS BY RENDERING

Prime lard is rendered from fats deposited in body tissues, while ordinary lard is rendered mainly from the back fat portions, but also from the trimmings. It was once the custom to use lard in cooked foods or for cooking foods, and it is still, to some extent. However, the use of lard has diminished with the increasing prevalence of vegetable fats and oils. Both lard and oleostearin may be seeded with crystals of tristearin while held at about 60°F (15.6°C) (lard) or 100°F (37.8°C) (oleostearin). The semi-solid material is then shoveled onto sheets of canvas which are then folded over the top and subjected to pressure in a press. About 50–60% oil can be separated in this manner from lard. Oleostearin oil may be used in candy. Lard oil may be used in baking bread or pies or for greasing baking pans.

Natural fats, including animal fats, tend to have some degree of organization with regard to the pattern in which the fatty acids are distributed on the glycerine molecules. This being the case, fat molecules containing three saturated fatty acids (fatty acid unable to accept more hydrogen) do not occur to any extent unless the content of saturated fatty acids exceeds ⅔ of the total fatty acids. Lard has about 40% saturated fatty acids. Interesterification of fats tends to distribute the fatty acids in a random manner as they are attached to the glycerine molecules. Interesterification usually raises the melting point of fats and oils. While the melting point of such fats as beef oleo and lard are not raised by interesterification, their characteristics and behavior in foods are changed. Interesterification can be brought about by heating the fat to the desired temperature, then adding a catalyst (sodium methoxide is often used as the catalyst). When the reaction is judged to be complete, water or acid (usually phosphoric) is added to destroy the catalyst. Natural lard occurs as clusters of crystals which grow larger as the lard is aged. Interesterification of lard changes the structure in such a manner that the crystals exist as finely dispersed particles. The grain of the lard is, therefore, much finer than that of the natural product. Fine-grain interesterified lard has good creaming qualities and is desirable for the production of cakes and icings. On the other hand, the coarse-grained natural lard has properties which make it better suited for the manufacture of flaky pastries and pie crusts.

OILS

There are many oils, most of which are of vegetable origin. Included among these are coconut oil and palm kernel oil, in both of which the fatty acid composition consists of 45–80% lauric acid, a 12-carbon saturated acid. Other vegetable oils include cottonseed, peanut, olive, sesame, corn and soybean. These oils contain a preponderance of the unsaturated acids—linoleic (18 carbons, with 4 carbons containing 1 less hydrogen than they can accept, or 2 double bonds), or oleic (18 carbons, with 2 carbons containing 1 less hydrogen than they can accept, or 1 double bond). Remember that each carbon in the chain, except for the end carbons, can accept two H atoms ($-CH_2-CH_2-CH_2-CH_2-$). Also among the oils are the marine oils (chiefly whale oil or menhaden oil) which contain some fatty acids with 20, 22, 24 or 26 carbons, and many of the fatty acids have 3, 4, or 5 double bonds (6, 8 or 10 carbons which could accept another hydrogen).

In the extraction of vegetable oils (see Fig. 24.2), some preparation of the material may be required. When corn oil is produced, the germ must first be separated from the corn kernels. The germs are then crushed prior to extraction. Some seeds, such as cottonseed, are prepressed at low pressure to remove part of the oil prior to extraction. Other seeds are crushed or flaked before subjecting to the extraction process. Soybeans are cut into thin flakes and extracted without pressing. Hexane, a 6-carbon saturated hydrocarbon straight chain, which is similar to gasoline ($CH_3-(CH_2)_4-CH_3$), is ordinarily used to extract oils from vegetable sources. The prepared seeds or other oil-containing substances are placed in perforated containers, and the solvent (hexane) is allowed to percolate through the material in the containers. The extraction is started with solvent which has already taken up a high level of oil and finished off with fresh solvent. In order to obtain the extracted oil, the solvent (which

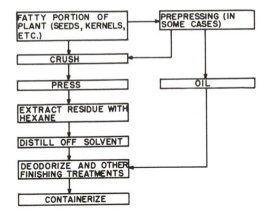

FIG. 24.2. RECOVERY OF PLANT OILS

as a much lower boiling point than the oil) is distilled off and recovered for reuse in the extraction of more oil. Oil from olives and cocoa beans is obtained by pressing the cooked pulp or beans. An expeller, which subjects the cooked material to high pressure, is used to squeeze out the oil. Oil from seeds may also be extracted in this manner. Due to the pressure used in expelling oil this way, the temperature of the press cake may rise above 250°F (121.1°C). Expressed oils are usually darker than solvent-extracted oils.

Extracted or expelled oils contain phospholipids, gums and other materials which are soluble in oil but which settle out when wet with water. The natural oil may, therefore, be degummed by mixing with water and centrifuging to separate out the water and the insoluble material. If oil is to be deodorized, it must first be subjected to several degumming operations, otherwise the oil will become cloudy and difficult to filter.

Oil is usually refined prior to sale as a food product. This is done by mixing a concentrated aqueous solution of sodium hydroxide or sodium carbonate with the hot crude oil, and centrifuging out the separated gums and soaps. The oil is then washed with water to remove traces of soaps, and then dried under vacuum. Oils may be bleached prior to sale for use as food products. Neutral clays, sulfuric acid clays, or charcoal may be used in bleaching. Certain yellow colors or green chlorophyll colors may be removed by filtering oils through clays. The natural yellow color of oils (carotenoids) may be removed with activated charcoal or by heating to high temperature. In bleaching, the oil is mixed with the clay or charcoal and heated under vacuum. After sufficient time has elapsed to provide for contact of the oil and bleaching material, the oil is filtere off from the treatment material. Oils are often winterized to remove th saturated fats, in this case, by cooling and holding at 40°F (4.4°C), the filtering to remove the solid fats.

HYDROGENATION

Large quantities of oil are hydrogenated, usually for purposes c forming fats which are solid at room temperature. During hydrogenatior hydrogen is attached to those carbons which are still deficient in thi element. Also, during hydrogenation, the more unsaturated fatty acids i: the oils are the first to add hydrogen (first those with 3 double bonds o 6 unsaturated carbons, followed by those with 2 double bonds or unsaturated carbons). The result is that in a fully hydrogenated fat, th only fatty acids that may remain unsaturated have only one doubl bond, or two carbons, each of which can accept one hydrogen atom Hydrogenation raises the melting point of the fat, but, in addition to this the fat is changed. For example, oleic acid (one double bond) in the fa has been changed to elaidic acid, which has the same formula as olei acid and still has one double bond, but the spatial distribution of th atoms differs as to the type of symmetry. Oleic acid is the "cis" form c the molecule and elaidic acid is the "trans" form of the same molecule Since the melting point of elaidic acid is much higher than that of olei acid, the melting point of the fat is raised. Oils are usually not full: hydrogenated when solid fats are produced, and in the preparation o margarine, some polyunsaturated fatty acids are left. Soybean oil, whicl may be used for frying, is slightly hydrogenated to eliminate linoleni acid (3 double bonds); otherwise, the oil would develop a fishy flavo: when heated. This is probably due to a primary or intermediate stage o oxidation.

In hydrogenation, the oil is placed in a container and the catalys (usually finely divided metallic nickel) is added. The oil is then heate under vacuum to 200°–400°F (93.3°–204.4°C). When the proper tem perature is attained, the vacuum is discontinued, and hydrogen, under a pressure of 5–50 psi (0.35–3.52 kg/cm²), is forced through the oil. Th mixture is whipped, to expose as much oil surface to the gas as possible Hydrogenation is allowed to proceed to the desired point as determine by periodic tests made on the material. After hydrogenation, the material is cooled to a point where it is still liquid, then filtered througl

eaching clay to remove the nickel catalyst and nickel soaps. Fats are
ﾀdinarily cooled in a heat exchanger after filtering. This is done to retain
homogeneous material, since, if allowed to cool slowly, the harder fats
ould crystallize out at the bottom of the mass, the more liquid fats
ﾀlidifying at the top. After cooling, the fats are pumped through a unit
hich whips them. This is done to prevent large crystals from forming
ﾀd to maintain a smooth creamy texture. The whipped fat (shortening)
 then packaged in containers of various sizes (1–100 lb [0.5–45.4 kg]),
ﾀpending upon whether the product is to be sold at retail or to be used
ﾀ a food product manufacturer. After packaging, the fat must be
ﾀmpered. This is done by holding it, in its containers, at 80°–85°F
ﾀ6.7°–29.4°C) for a period of 24–72 hr. Tempering is done to provide
ﾀitable creaming properties in the shortening. Edible fats and oils are
ﾀed as bakery shortening to make bread, cakes, icing and pastries; as a
ﾀying vehicle used in shallow- or deep-fat frying; for ice cream-like
ﾀods; as oil for salads; for the addition to canned fish products (sardines,
ﾀna, etc.); for the manufacture of margarine; as special coatings for
ﾀeat; for greasing pans used in bakeries; and, in the home, for baking.

MARGARINE

Large quantities of margarine are manufactured. Actually, margarine
ﾀas largely replaced butter, both in the home and in the manufacturing
ﾀ foods. By law, margarine may contain edible fats or oils, whole milk,
ﾀim milk, cream or reconstituted milk solids cultured with bacteria (for
ﾀavor) or combinations of these materials; also, emulsifiers (mono- or
ﾀiglycerides or lecithin), citric acid or citrates to tie up metals which
ﾀccelerate oxidation, salt or benzoates or both (to inhibit bacterial
ﾀpoilage), vitamins A and D, artificial coloring (carotene or annato) and
ﾀrtificial flavoring (distillates from cultured milk or cream). The amount
ﾀ fat present in margarine must be at least 80% of the finished product.
ﾀll the above ingredients are not necessarily included in margarine, but
ﾀme combination will be used. Blending of the materials is carried out
ﾀ two steps. All the fat-soluble ingredients are mixed with the liquid fats
ﾀ one container, the water-soluble materials in another. The two batches
ﾀre then mixed to form a loose emulsion with the aid of a high-speed
ﾀgitator which whips and beats the mixture. The emulsified material is
ﾀhen solidified in a heat exchanger, after which it may be kneaded. In any
ﾀase, the cooled material is extruded into sheets or bars, then the
ﾀxtruded material is cut to length, wrapped and packaged for shipment.

LIPID EMULSIFIERS

The mono- and diglycerides used as emulsifiers are made by heating fats or oil and glycerine, with some sodium hydroxide, under vacuum a approximately 400°F (204.4°C). Under these conditions, some of th fatty acids, attached to the glycerine in the fat, migrate and attach to th free glycerine present. Commercial monoglycerides consist of about 50% monoglycerides (1 fatty acid attached to the glycerine molecule), 40% diglycerides (2 fatty acids attached to the glycerine) and 10% tri glycerides or fats (3 fatty acids attached to the glycerine). Pure mono- o diglycerides can be obtained by distillation.

SALAD DRESSINGS

Salad dressings include mayonnaise, and other products which diffe from mayonnaise mainly in that they do not contain sufficient oil to forr a true emulsion. Mayonnaise is made from vegetable oil (cottonseed o corn), vinegar, sugar, salt, mustard, white pepper and egg yolk. Th proportion of components may vary among different formulations. I popular usage, mayonnaise is not called a salad dressing. Salad dressing other than mayonnaise are made from vegetable oils, vinegar, spices, an in many cases, starch. The oil content in salad dressings—about 40%— is about 1/2 the amount used in mayonnaise.

Suggestions for Further Reading

AMERICAN MEAT INSTITUTE FOUNDATION. 1960. The Science of Meat and Meat Products. W.H. Freeman and Co., San Francisco.

AMERICAN SOCIETY OF HEATING, REFRIGERATING AND AIR-CONDITIONING ENGINEERS. 1974. Guide and Data Book—Applications. Am. Soc. Heating, Refrig., Air-cond. Engrs., New York.

ARBUCKLE, W.S. 1977. Ice Cream, 3rd Edition. AVI Publishing Co., Westport, Conn.

ASSOC. OF FOOD INDUSTRY SANITARIANS. 1952. Sanitation for the Food Preservation Industry. McGraw-Hill Book Co., New York.

BAILEY, A.E. 1951. Industrial Oil and Fat Products, 2nd Edition. Interscience Publishers, New York.

BORGSTROM, G. 1961. Fish as Food, Vol. 1, Biochemistry and Microbiology. Academic Press, New York.

BORGSTROM, G. 1962. Fish as Food, Vol. 2, Nutrition, Sanitation and Utilization. Academic Press, New York.

BORGSTROM, G. 1965. Fish as Food, Vol. 3, Processing, Parts I and II. Academic Press, New York.

BRAVERMAN, J.B.S. 1963. Introduction to the Biochemistry of Foods. Elsevier Publishing Co., New York.

BRODY, J. 1965. Fishery By-products Technology. AVI Publishing Co., Westport, Conn.

BURTON, B.T. 1965. The Heinz Handbook of Nutrition, 2nd Edition. McGraw-Hill Book Co., New York.

CRUESS, W.V. 1958. Commercial Fruit and Vegetable Products, 3rd Edition. McGraw-Hill Book Co., New York.

DANIELS, R. 1974. Breakfast Cereal Technology. Noyes Data Corp., Park Ridge, N.J.

DAVIS, J.G. 1965. Cheese. Elsevier Publishing Co., New York.

DUFFY, M.P. 1963. Federal and state regulation of processed foods. *In* Food Processing Operations, Vol. 2. M.A. Joslyn and J.L. Heid (Editors). AVI Publishing Co., Westport, Conn.

ECKEY, E.W. 1954. Vegetable Fats and Oils. Reinhold Publishing Corp., New York.

FANCE, W.J. 1969. Breadmaking and Flour Confectionery, 3rd Edition. AVI Publishing Co., Westport, Conn.

FURIA, T.E. 1973. Handbook of Food Additives, 2nd Edition. Chemical Rubber Co., Cleveland, Ohio.

GILLIES, M. 1971. Seafood Processing. Noyes Data Corp., Park Ridge, N.J.

GRAHAM-RACK, B. and BINSTED, R. 1964. Hygiene in food manufacturing and handling. Food Trade Rev. *34.*

GUNDERSON, F.L., GUNDERSON, H.W. and FERGUSON, E.R., JR. 196
Food Standards and Definitions in the United States. Academic Press, Ne
York.

GUTCHO, M.H. 1973. Feeds for Livestock, Poultry and Pets. Noyes Data Corp
Park Ridge, N.J.

GUTCHO, S.J. 1974. Microbial Enzyme Production. Noyes Data Corp., Pa
Ridge, N.J.

GUTHRIE, R.K. 1972. Food Sanitation. AVI Publishing Co., Westport, Conn

GUTTERSON, M. 1971. Vegetable Processing. Noyes Data Corp., Park Ridg
N.J.

GUTTERSON, M. 1971. Fruit Processing. Noyes Data Corp., Park Ridge, N.

GUTTERSON, M. 1972. Food Canning Technology. Noyes Data Corp., Pa
Ridge, N.J.

HARPER, W.J. and HALL, C.W. 1976. Dairy Technology and Engineering. A'
Publishing Co., Westport, Conn.

HEID, J.L. and JOSLYN, M.A. 1967. Fundamentals of Food Processing Oper
tions. AVI Publishing Co., Westport, Conn.

HENDERSON, J.L. 1971. The Fluid Milk Industry, 3rd Edition. AVI Publishir
Co., Westport, Conn.

HERSOM, A.C. and HULLAND, E.D. 1964. Canned Foods, an Introduction
Their Microbiology. Chemical Publishing Co., New York.

JENNESS, R. and PATTON, S. 1959. Principles of Dairy Chemistry. Joh
Wiley & Sons, New York.

JOHNSON, J.C. 1975. Antioxidants, Synthesis and Applications. Noyes Da
Corp., Park Ridge, N.J.

JUNK, W.R. and PANCOAST, H.M. 1973. Handbook of Sugars for Processor
Chemists and Technologists. AVI Publishing Co., Westport, Conn.

KARMAS, E. 1972. Sausage Processing. Noyes Data Corp., Park Ridge, N.J.

KELLY, N. 1964. Sugar. In Food Processing Operations, Vol. 3. M.A. Joslyn ar
J.L. Heid (Editors). AVI Publishing Co., Westport, Conn.

KING, C.J. 1971. Freeze Drying of Foods. Chemical Rubber Co., Cleveland.

KRAMLICH, W.E., PEARSON, A.M. and TAUBER, F.W. 1973. Processe
Meats. AVI Publishing Co., Westport, Conn.

LASKIN, A., and LECHEVALIER, H. 1974. Microbial Ecology. Chemic
Rubber Co., Cleveland, Ohio.

LEVIE, A. 1979. The Meat Handbook, 4th Edition. AVI Publishing Cc
Westport, Conn.

LEWIS, K.H. and CASSELL, K., JR. 1964. Botulism. U.S. Dep. of Healt.
Education and Welfare, U.S. Public Health Serv., Publ. 999-FP-1. U.S. Gov
Printing Office, Washington, D.C.

MATZ, S.A. 1969. Cereal Science. AVI Publishing Co., Westport, Conn.

MATZ, S.A. 1970. Cereal Technology. AVI Publishing Co., Westport, Conn.

MATZ, S.A. 1972. Bakery Technology and Engineering, 2nd Edition. A'
Publishing Co., Westport, Conn.

MATZ, S.A. and MATZ, T.D. 1978. Cookie and Cracker Technology, 2r
Edition. AVI Publishing Co., Westport, Conn.

EYER, L.H. 1974. Food Chemistry. AVI Publishing Co., Westport, Conn.

OUNTNEY, G.J. 1976. Poultry Products Technology, 2nd Edition. AVI Publishing Co., Westport, Conn.

ATIONAL CANNERS ASSOC. 1968. Laboratory Manual for Food Canners and Processers, Vol. 1 and 2, 3rd Edition. AVI Publishing Co., Westport, Conn.

ICKERSON, J.T.R. and SINSKEY, A.J. 1974. Microbiology of Foods and Food Processing. Elsevier Publishing Co., New York.

EDERSON, C.S. 1979. Microbiology of Food Fermentations, 2nd Edition. AVI Publishing Co., Westport, Conn.

ENTZER, W.T. 1973. Progress in Refrigeration Science and Technology, Vol. 1, 2, 3, and 4. AVI Publishing Co., Westport, Conn.

INTAURO, N.D. 1974. Food Additives to Extend Shelf Life. Noyes Data Corp., Park Ridge, N.J.

OMERANZ, Y. and SHELLENBERGER, J.A. 1971. Bread Science and Technology. AVI Publishing Co., Westport, Conn.

YLER, E.J. 1952. Baking Science and Technology. Siebel Publishing Co., Chicago.

ECHIGT, M., JR. 1973. Man, Food and Nutrition. Chemical Rubber Co., Cleveland.

EED, G. and UNDERKOFLER, L.A. 1966. Enzymes in Food Processing. Academic Press, New York.

EES, G.H. 1963. Edible crabs of the United States. U.S. Dep. of the Interior, Fishery Leaflet 550. U.S. Govt. Printing Office, Washington, D.C.

IEMANN, H. 1969. Food-borne Infections and Intoxications. Academic Press, New York.

CHULTZ, H.W. 1960. Food Enzymes. AVI Publishing Co., Westport, Conn.

CHULTZ, H.W. 1962. Lipids and Their Oxidation. AVI Publishing Co., Westport, Conn.

CHULTZ, H.W. and ANGLEMIER, A.F. 1964. Proteins and Their Reactions. AVI Publishing Co., Westport, Conn.

EBRELL, W.H., JR. and HARRIS, R.S. 1954. The Vitamins. Academic Press, New York.

HALLENBERGER, R.S. and BIRCH, G.G. 1975. Sugar Chemistry. AVI Publishing Co., Westport, Conn.

HAPIRO, S. 1971. Our Changing Fisheries. U.S. Govt. Printing Office, Washington, D.C.

HOEMAKER, J.S. 1978. Small Fruit Culture, 5th Edition. AVI Publishing Co., Westport, Conn.

TADELMAN, W.J. and COTTERILL, O.J. 1977. Egg Science and Technology, 2nd Edition. AVI Publishing Co., Westport, Conn.

TANSBY, M.E. 1976. Industrial Fisheries Technology, 2nd Edition. R.E. Krieger Publishing Co., Huntington, N.Y.

WERN, D. 1964. Bailey's Industrial Oil and Fat Products, 3rd Edition. John Wiley & Sons, New York.

TALBURT, W.F. and SMITH, O. 1975. Potato Processing, 3rd Edition. A*
Publishing Co., Westport, Conn.

TESKEY, B.J.E. and SHOEMAKER, J.S. 1978. Tree Fruit Production, 3
Edition. AVI Publishing Co., Westport, Conn.

TORREY, M. 1974. Dehydration of Fruits and Vegetables. Noyes Data Cor
Park Ridge, N.J.

TRESSLER, D.K., VAN ARSDEL, W.B. and COPLEY, M.J. 1968. T
Freezing Preservation of Foods, Vol. 1, 2, 3, and 4. AVI Publishing C
Westport, Conn.

U.S. DEP. AGRIC. 1958. Regulations governing inspection and certification
processed fruits and vegetables and related products. USDA Agr. Marketi
Serv., SRA-AMS 155. U.S. Govt. Printing Office, Washington, D.C.

U.S. DEP. AGRIC. 1960. Regulations governing the meat inspection of the U.
Department of Agriculture. USDA Agr. Res. Serv., U.S. Govt. Printing Offic
Washington, D.C.

U.S. DEPT. AGRIC. 1972. Farm and poultry management. Farmer's Bull. 219
U.S. Govt. Printing Office, Washington, D.C.

U.S. DEP. COMMER. 1972. Regulations governing processed fishery product
Code of Federal Regulations Title 50. U.S. Govt. Printing Office, Washingto
D.C.

VAN ARSDEL, W.B., COPLEY, M.J. and MORGAN, A.I., JR. 1973. Foo
Dehydration, Vol. 1 and 2. AVI Publishing Co., Westport, Conn.

WEBB, B.H., JOHNSON, A.H. and ALFORD, J.A. 1974. Fundamentals
Dairy Chemistry, 2nd Edition. AVI Publishing Co., Westport, Conn.

WEISS, G.H. 1971. Poultry Processing. Noyes Data Corp., Park Ridge, N.J.

WEISS, T.J. 1963. Fats and oils. In Food Processing Operations, Vol. 2. M.
Joslyn and J.L. Heid (Editors). AVI Publishing Co., Westport, Conn.

WEISS, T.J. 1970. Food Oils and Their Uses. AVI Publishing Co., Westpor
Conn.

WILCOX, G. 1971. Milk, Cream and Butter Technology. Noyes Data Cor
Park Ridge, N.J.

WOODROOF, J.G. 1963. Production, harvesting and delivery of vegetable crop
In Food Processing Operations, Vol. 1. M.A. Joslyn and J.L. Heid (Editors
AVI Publishing Co., Westport, Conn.

WOODROOF, J.G. and LUH, B.S. 1975. Commercial Fruit Processing. A*
Publishing Co., Westport, Conn.

Index

Appendix

TEMPERATURE CONVERSION

The numbers in boldface type in the center column refer to the temperature, either in degree Celsius or Fahrenheit, wh... is to be converted to the other scale. If converting Fahrenheit to degree Celsius, the equivalent temperature will be fo... in the left column. If converting degree Celsius to Fahrenheit, the equivalent temperature will be found in the column... the right.

Celsius	°C or F	Fahr	Celsius	°C or F	Fahr	Celsius	°C or F	Fahr	Celsius	°C or F	F...
-40.0	-40	-40.0	+1.7	+35	+95.0	+43.3	+110	+230.0	+85.0	+185	+36
-39.4	-39	-38.2	+2.2	+36	+96.8	+43.9	+111	+231.8	+85.6	+186	+36
-38.9	-38	-36.4	+2.8	+37	+98.6	+44.4	+112	+233.6	+86.1	+187	+36
-38.3	-37	-34.6	+3.3	+38	+100.4	+45.0	+113	+235.4	+86.7	+188	+37
-37.8	-36	-32.8	+3.9	+39	+102.2	+45.6	+114	+237.2	+87.2	+189	+37
-37.2	-35	-31.0	+4.4	+40	+104.0	+46.1	+115	+239.0	+87.8	+190	+37
-36.7	-34	-29.2	+5.0	+41	+105.8	+46.7	+116	+240.8	+88.3	+191	+37
-36.1	-33	-27.4	+5.5	+42	+107.6	+47.2	+117	+242.6	+88.9	+192	+37
-35.6	-32	-25.6	+6.1	+43	+109.4	+47.8	+118	+244.4	+89.4	+193	+37
-35.0	-31	-23.8	+6.7	+44	+111.2	+48.3	+119	+246.2	+90.0	+194	+38
-34.4	-30	-22.0	+7.2	+45	+113.0	+48.9	+120	+248.0	+90.6	+195	+38
-33.9	-29	-20.2	+7.8	+46	+114.8	+49.4	+121	+249.8	+91.1	+196	+38
-33.3	-28	-18.4	+8.3	+47	+116.6	+50.0	+122	+251.6	+91.7	+197	+38
-32.8	-27	-16.6	+8.9	+48	+118.4	+50.6	+123	+253.4	+92.2	+198	+38
-32.2	-26	-14.8	+9.4	+49	+120.2	+51.1	+124	+255.2	+92.8	+199	+39
-31.7	-25	-13.0	+10.0	+50	+122.0	+51.7	+125	+257.0	+93.3	+200	+39
-31.1	-24	-11.2	+10.6	+51	+123.8	+52.2	+126	+258.8	+93.9	+201	+39
-30.6	-23	-9.4	+11.1	+52	+125.6	+52.8	+127	+260.6	+94.4	+202	+39
-30.0	-22	-7.6	+11.7	+53	+127.4	+53.3	+128	+262.4	+95.0	+203	+39
-29.4	-21	-5.8	+12.2	+54	+129.2	+53.9	+129	+264.2	+95.6	+204	+39
-28.9	-20	-4.0	+12.8	+55	+131.0	+54.4	+130	+266.0	+96.1	+205	+40
-28.3	-19	-2.2	+13.3	+56	+132.8	+55.0	+131	+267.8	+96.7	+206	+40
-27.8	-18	-0.4	+13.9	+57	+134.6	+55.6	+132	+269.6	+97.2	+207	+40
-27.2	-17	+1.4	+14.4	+58	+136.4	+56.1	+133	+271.4	+97.8	+208	+40
-26.7	-16	+3.2	+15.0	+59	+138.2	+56.7	+134	+273.2	+98.3	+209	+40
-26.1	-15	+5.0	+15.6	+60	+140.0	+57.2	+135	+275.0	+98.9	+210	+41
-25.6	-14	+6.8	+16.1	+61	+141.8	+57.8	+136	+276.8	+99.4	+211	+41
-25.0	-13	+8.6	+16.7	+62	+143.6	+58.3	+137	+278.6	+100.0	+212	+41
-24.4	-12	+10.4	+17.2	+63	+145.4	+58.9	+138	+280.4	+100.6	+213	+41
-23.9	-11	+12.2	+17.8	+64	+147.2	+59.4	+139	+282.2	+101.1	+214	+41
-23.3	-10	+14.0	+18.3	+65	+149.0	+60.0	+140	+284.0	+101.7	+215	+41
-22.8	-9	+15.8	+18.9	+66	+150.8	+60.6	+141	+285.8	+102.2	+216	+42
-22.2	-8	+17.6	+19.4	+67	+152.6	+61.1	+142	+287.6	+102.8	+217	+42
-21.7	-7	+19.4	+20.0	+68	+154.4	+61.7	+143	+289.4	+103.3	+218	+42
-21.1	-6	+21.2	+20.6	+69	+156.2	+62.2	+144	+291.2	+103.9	+219	+42
-20.6	-5	+23.0	+21.1	+70	+158.0	+62.8	+145	+293.0	+104.4	+220	+42
-20.0	-4	+24.8	+21.7	+71	+159.8	+63.3	+146	+294.8	+105.6	+222	+43
-19.4	-3	+26.6	+22.2	+72	+161.6	+63.9	+147	+296.6	+106.7	+224	+43
-18.9	-2	+28.4	+22.8	+73	+163.4	+64.4	+148	+298.4	+107.8	+226	+43
-18.3	-1	+30.2	+23.3	+74	+165.2	+65.0	+149	+300.2	+108.9	+228	+44
-17.8	0	+32.0	+23.9	+75	+167.0	+65.6	+150	+302.0	+110.0	+230	+44
-17.2	+1	+33.8	+24.4	+76	+168.8	+66.1	+151	+303.8	+111.1	+232	+44
-16.7	+2	+35.6	+25.0	+77	+170.6	+66.7	+152	+305.6	+112.2	+234	+45
-16.1	+3	+37.4	+25.6	+78	+172.4	+67.2	+153	+307.4	+113.3	+236	+45
-15.6	+4	+39.2	+26.1	+79	+174.2	+67.8	+154	+309.2	+114.4	+238	+46
-15.0	+5	+41.0	+26.7	+80	+176.0	+68.3	+155	+311.0	+115.6	+240	+46
-14.4	+6	+42.8	+27.2	+81	+177.8	+68.9	+156	+312.8	+116.7	+242	+46
-13.9	+7	+44.6	+27.8	+82	+179.6	+69.4	+157	+314.6	+117.8	+244	+47
-13.3	+8	+46.4	+28.3	+83	+181.4	+70.0	+158	+316.4	+118.9	+246	+47
-12.8	+9	+48.2	+28.9	+84	+183.2	+70.6	+159	+318.2	+120.0	+248	+47
-12.2	+10	+50.0	+29.4	+85	+185.0	+71.1	+160	+320.0	+121.1	+250	+48
-11.7	+11	+51.8	+30.0	+86	+186.8	+71.7	+161	+321.8	+122.4	+252	+48
-11.1	+12	+53.6	+30.6	+87	+188.6	+72.2	+162	+323.6	+123.3	+254	+48
-10.6	+13	+55.4	+31.1	+88	+190.4	+72.8	+163	+325.4	+124.4	+256	+49
-10.0	+14	+57.2	+31.7	+89	+192.2	+73.3	+164	+327.2	+125.5	+258	+49
-9.4	+15	+59.0	+32.2	+90	+194.0	+73.9	+165	+329.0	+126.7	+260	+50
-8.9	+16	+60.8	+32.8	+91	+195.8	+74.4	+166	+330.8	+127.8	+262	+50
-8.3	+17	+62.6	+33.3	+92	+197.6	+75.0	+167	+332.6	+128.9	+264	+50
-7.8	+18	+64.4	+33.9	+93	+199.4	+75.6	+168	+334.4	+130.0	+266	+51
-7.2	+19	+66.2	+34.4	+94	+201.2	+76.1	+169	+336.2	+131.3	+268	+51
-6.7	+20	+68.0	+35.0	+95	+203.0	+76.7	+170	+338.0	+132.2	+270	+51
-6.1	+21	+69.8	+35.6	+96	+204.8	+77.2	+171	+339.8	+133.3	+272	+52
-5.5	+22	+71.6	+36.1	+97	+206.6	+77.8	+172	+341.6	+134.4	+274	+52
-5.0	+23	+73.4	+36.7	+98	+208.4	+78.3	+173	+343.4	+135.6	+276	+52
-4.4	+24	+75.2	+37.2	+99	+210.2	+78.9	+174	+345.2	+136.7	+278	+53
-3.9	+25	+77.0	+37.8	+100	+212.0	+79.4	+175	+347.0	+137.8	+280	+53
-3.3	+26	+78.8	+38.3	+101	+213.8	+80.0	+176	+348.8	+138.9	+282	+53
-2.8	+27	+80.6	+38.9	+102	+215.6	+80.6	+177	+350.6	+140.0	+284	+54
-2.2	+28	+82.4	+39.4	+103	+217.4	+81.1	+178	+352.4	+141.1	+286	+54
-1.7	+29	+84.2	+40.0	+104	+219.2	+81.7	+179	+354.2	+142.2	+288	+55
-1.1	+30	+86.0	+40.6	+105	+221.0	+82.2	+180	+356.0	+143.3	+290	+55
-0.6	+31	+87.8	+41.1	+106	+222.8	+82.8	+181	+357.8	+144.4	+292	+55
.0	+32	+89.6	+41.7	+107	+224.6	+83.3	+182	+359.6	+145.6	+294	+56
+0.6	+33	+91.4	+42.2	+108	+226.4	+83.9	+183	+361.4	+146.7	+296	+56
+1.1	+34	+93.2	+42.8	+109	+228.2	+84.4	+184	+363.2	+147.8	+298	+56

COMPARISON OF AVOIRDUPOIS AND METRIC UNITS OF WEIGHT

1 oz = 0.06 lb = 28.35 g	1 lb = 0.454 kg	1 g = 0.035 oz	1 kg = 2.205 lb
2 oz = 0.12 lb = 56.70 g	2 lb = 0.91 kg	2 g = 0.07 oz	2 kg = 4.41 lb
3 oz = 0.19 lb = 85.05 g	3 lb = 1.36 kg	3 g = 0.11 oz	3 kg = 6.61 lb
4 oz = 0.25 lb = 113.40 g	4 lb = 1.81 kg	4 g = 0.14 oz	4 kg = 8.82 lb
5 oz = 0.31 lb = 141.75 g	5 lb = 2.27 kg	5 g = 0.18 oz	5 kg = 11.02 lb
6 oz = 0.38 lb = 170.10 g	6 lb = 2.72 kg	6 g = 0.21 oz	6 kg = 13.23 lb
7 oz = 0.44 lb = 198.45 g	7 lb = 3.18 kg	7 g = 0.25 oz	7 kg = 15.43 lb
8 oz = 0.50 lb = 226.80 g	8 lb = 3.63 kg	8 g = 0.28 oz	8 kg = 17.64 lb
9 oz = 0.56 lb = 255.15 g	9 lb = 4.08 kg	9 g = 0.32 oz	9 kg = 19.84 lb
10 oz = 0.62 lb = 283.50 g	10 lb = 4.54 kg	10 g = 0.35 oz	10 kg = 22.05 lb
11 oz = 0.69 lb = 311.85 g	11 lb = 4.99 kg	11 g = 0.39 oz	11 kg = 24.26 lb
12 oz = 0.75 lb = 340.20 g	12 lb = 5.44 kg	12 g = 0.42 oz	12 kg = 26.46 lb
13 oz = 0.81 lb = 368.55 g	13 lb = 5.90 kg	13 g = 0.46 oz	13 kg = 28.67 lb
14 oz = 0.88 lb = 396.90 g	14 lb = 6.35 kg	14 g = 0.49 oz	14 kg = 30.87 lb
15 oz = 0.94 lb = 425.25 g	15 lb = 6.81 kg	15 g = 0.53 oz	15 kg = 33.08 lb
16 oz = 1.00 lb = 453.59 g	16 lb = 7.26 kg	16 g = 0.56 oz	16 kg = 35.28 lb

COMPARISON OF U.S. AND METRIC UNITS OF LIQUID MEASURE

1 fl oz = 29.573 ml	1 qt = 0.946 liter	1 gal. = 3.785 liters
2 fl oz = 59.15 ml	2 qt = 1.89 liters	2 gal. = 7.57 liters
3 fl oz = 88.72 ml	3 qt = 2.84 liters	3 gal. = 11.36 liters
4 fl oz = 118.30 ml	4 qt = 3.79 liters	4 gal. = 15.14 liters
5 fl oz = 147.87 ml	5 qt = 4.73 liters	5 gal. = 18.93 liters
6 fl oz = 177.44 ml	6 qt = 5.68 liters	6 gal. = 22.71 liters
7 fl oz = 207.02 ml	7 qt = 6.62 liters	7 gal. = 26.50 liters
8 fl oz = 236.59 ml	8 qt = 7.57 liters	8 gal. = 30.28 liters
9 fl oz = 266.16 ml	9 qt = 8.52 liters	9 gal. = 34.07 liters
10 fl oz = 295.73 ml	10 qt = 9.46 liters	10 gal. = 37.85 liters

1 ml = 0.034 fl oz	1 liter = 1.057 qt	1 liter = 0.264 gal.
2 ml = 0.07 fl oz	2 liters = 2.11 qt	2 liters = 0.53 gal.
3 ml = 0.10 fl oz	3 liters = 3.17 qt	3 liters = 0.79 gal.
4 ml = 0.14 fl oz	4 liters = 4.23 qt	4 liters = 1.06 gal.
5 ml = 0.17 fl oz	5 liters = 5.28 qt	5 liters = 1.32 gal.
6 ml = 0.20 fl oz	6 liters = 6.34 qt	6 liters = 1.59 gal.
7 ml = 0.24 fl oz	7 liters = 7.40 qt	7 liters = 1.85 gal.
8 ml = 0.27 fl oz	8 liters = 8.45 qt	8 liters = 2.11 gal.
9 ml = 0.30 fl oz	9 liters = 9.51 qt	9 liters = 2.38 gal.
10 ml = 0.34 fl oz	10 liters = 10.57 qt	10 liters = 2.64 gal.

CONVERSION OF OVEN TEMPERATURES

Conventional (Fahrenheit)		Metric (Celsius)
200 F		93 C
225 F		107 C
250 F	Very low	121 C
300 F	Low	149 C
325 F		163 C
350 F	Moderate	177 C
400 F	Hot	204 C
450 F	Very high	232 C
500 F	Extremely high	260 C

VOLUME CONVERSION DIFFERENCES CONVENTIONAL VS. METRIC MEASUREMENTS

Utensil	Capacity (ml)	Tolerance (ml)
1 cup	236.6	11.8
½ cup	118.3	5.9
⅓ cup	78.9	3.9
¼ cup	59.2	3.0
1 tablespoon	14.79	0.73
1 teaspoon	4.93	0.24
½ teaspoon	2.46	0.12
¼ teaspoon	1.23	0.06

Related AVI Books

BASIC FOOD MICROBIOLOGY
  Banwart
COMMERCIAL FRUIT PROCESSING
  Woodroof, Luh
COMMERCIAL VEGETABLE PROCESSING
  Luh, Woodroof
COMMERCIAL WINEMAKING
  Vine
DRYING CEREAL GRAINS
  Brooker, Bakker-Arkema, Hall
EGG SCIENCE AND TECHNOLOGY
  2nd Edition   *Stadelman, Cotterill*
ENCYCLOPEDIA OF FOOD SCIENCE
  Peterson, Johnson
EXPERIMENTAL FOOD CHEMISTRY
  Mondy
FOOD AND BEVERAGE MYCOLOGY
  Beuchat
FOOD SANITATION
  2nd Edition   *Guthrie*
FOOD SCIENCE
  3rd Edition   *Potter*
FUNDAMENTALS OF FOOD CANNING TECHNOLOGY
  Jackson, Shinn
FUNDAMENTALS OF FOOD PROCESS ENGINEERING
  Toledo
LABORATORY MANUAL IN FOOD PRESERVATION
  Fields
LABORATORY MANUAL FOR FOOD PROCESSORS
  3rd Edition   *NFPA*
MEAT HANDBOOK
  4th Edition   *Levie*
NUTRITIONAL EVALUATION OF FOOD PROCESSING
  2nd Edition   *Harris, Karmas*
PRESCOTT & DUNN'S INDUSTRIAL MICROBIOLOGY
  4th Edition   *Reed*
PROCESSED MEATS
  Kramlich, Pearson, Tauber
TECHNOLOGY OF WINE MAKING
  4th Edition   *Amerine et al.*